Mathematical Tools for Theoretical Physics

A Comprehensive Reference Guide

Chokri Ben Dhaou
chokri@bendhaou.net

From Foundations to Advanced Methods

$$\nabla \times \mathbf{E} = -\frac{\partial \mathbf{B}}{\partial t}$$

$\int_\Sigma \omega$ $[\hat{x}, \hat{p}] = i\hbar$

$R_{\mu\nu} - \frac{1}{2} g_{\mu\nu} R = 8\pi G T_{\mu\nu}$

First Edition

2025

Mathematical Tools for Theoretical Physics
First Edition

Typeset in LaTeX.

Mathematical Tools for Theoretical Physics

Chokri Ben Dhaou
chokri@bendhaou.net

Edited by NAWVA
contact@nawva.com

ISBN: 978-2-488616-12-6

to NEYL, and ARMEEN

Preface

The relationship between mathematics and physics is one of the most profound and productive partnerships in intellectual history. From Newton's invention of calculus to describe planetary motion, to the geometric language of general relativity, to the operator algebras of quantum mechanics, physics has both drawn from and inspired the development of mathematics.

This book is designed as a practical reference—a mathematical companion for anyone working in theoretical physics. It is not meant to replace rigorous mathematical textbooks, nor is it a physics course in disguise. Rather, it occupies the essential middle ground: providing the mathematical tools that physicists need, with the level of rigor that ensures correct application, while maintaining focus on utility and clarity.

Philosophy and Approach. Several principles have guided the writing of this book:

Completeness over depth. We cover a broad range of topics rather than exploring any single area exhaustively. The goal is to ensure that readers have at least an introduction to every mathematical tool they are likely to encounter in graduate-level theoretical physics.

Definitions and theorems. We state precise definitions and key theorems, because clarity in fundamentals prevents errors in applications. However, we often omit proofs or provide only sketches, directing interested readers to specialized texts.

Connection to physics. Throughout the book, we indicate where each mathematical structure appears in physics. This serves both to motivate the material and to help readers recognize mathematical structures in physical contexts.

Self-contained sections. Each chapter is designed to be relatively self-contained, allowing the book to function as a reference that can be consulted as needed rather than read linearly.

Prerequisites. We assume familiarity with undergraduate mathematics: single-variable calculus, basic linear algebra, and elementary differential equations. Some exposure to multivariable calculus is helpful but not strictly required, as we develop these tools systematically.

How to Use This Book. For students beginning graduate studies, we recommend reading Part I carefully to ensure solid foundations. The remaining parts can then be consulted as needed based on your coursework and research interests.

For researchers, this book serves as a quick reference. The detailed table of contents and index should help locate specific topics rapidly.

Acknowledgments. Mathematics is a cumulative enterprise, and this book draws on centuries of work by mathematicians and physicists. While we cannot acknowledge all sources, we are particularly indebted to the classic texts that have shaped how these subjects are taught.

The Author

Contents

Introduction

The Language of Physics

Physics seeks to describe the natural world through mathematical models. This description operates at multiple levels: we use numbers to quantify measurements, functions to describe how quantities vary, equations to express relationships, and abstract structures to capture symmetries and invariances.

The mathematics used in physics has evolved dramatically over centuries. Classical mechanics can be formulated with calculus and ordinary differential equations. Electromagnetism requires vector calculus and partial differential equations. Quantum mechanics demands linear algebra in infinite dimensions—Hilbert space theory. General relativity is built on differential geometry. Quantum field theory combines all of these with group theory, functional analysis, and distribution theory.

Structure of This Book

This book is organized into seven parts, progressing from foundational to advanced material:

Part I: Foundations covers the basic mathematical structures—sets, logic, linear algebra, and analysis—that underlie everything that follows.

Part II: Calculus on Manifolds develops the geometry needed for classical mechanics, electromagnetism, and general relativity: multivariable calculus, differential forms, and Riemannian geometry.

Part III: Algebra and Symmetry introduces the algebraic structures central to modern physics: group theory, Lie algebras, representation theory, and spinors.

Part IV: Functional Analysis provides the rigorous framework for quantum mechanics: Hilbert spaces, operators, distributions, and spectral theory.

Part V: Differential Equations covers both ordinary and partial differential equations, including the special functions that pervade physics.

Part VI: Probability and Statistics introduces probability theory and stochastic processes, essential for statistical mechanics and quantum measurement.

Part VII: Advanced Tools collects specialized techniques: integral transforms, asymptotic methods, and the mathematics of gauge theories.

Notation and Conventions

We generally follow standard mathematical and physical conventions:

The natural numbers $\mathbb{N}$ include zero: $\mathbb{N} = \{0, 1, 2, \ldots\}$. The integers are $\mathbb{Z}$, rationals $\mathbb{Q}$, reals $\mathbb{R}$, and complex numbers $\mathbb{C}$.

We use $:=$ to indicate a definition (the left side is being defined).

Vectors may be written in bold ($\mathbf{v}$) or with indices (v^i or v_i). Context determines the choice.

The Einstein summation convention is used where indicated: repeated indices are summed over.

We generally use "natural units" in physical examples, setting $c = \hbar = 1$ unless otherwise stated.

A Note on Rigor

This is a mathematics book written for physicists. We strive for precision in definitions and careful statements of theorems. However, we do not always provide complete proofs, and we sometimes prioritize intuition over full generality.

The working physicist needs to know *what* is true and *how* to use it correctly. Understanding *why* it is true is valuable—it builds intuition and helps avoid errors—but the full machinery of proof is often not necessary for applications.

Where we omit proofs, we try to give at least a sense of why a result holds. And we always aim for correct statements: a theorem may be stated in less than full generality, but it should never be wrong.

Part I

Foundations

Chapter 1

Sets, Logic, and Proofs

Mathematics is built on a foundation of sets and logic. While physicists rarely need to engage deeply with foundational issues, a clear understanding of basic set theory and logical reasoning is essential for precise thinking. This chapter establishes the language and methods that will be used throughout this book.

1.1 Naive Set Theory

1.1.1 Sets and Elements

A **set** is a collection of objects, called **elements** or **members**. We write $x \in A$ to indicate that x is an element of the set A, and $x \notin A$ to indicate that it is not.

Definition 1.1.1 (Set equality)**.** Two sets A and B are **equal**, written $A = B$, if and only if they have exactly the same elements:

$$A = B \quad \Leftrightarrow \quad \forall x : (x \in A \Leftrightarrow x \in B).$$

Sets can be specified in several ways. *Enumeration* lists the elements explicitly:

$$A = \{1, 2, 3\}, \qquad B = \{a, b, c, d\}.$$

Set-builder notation describes elements by a property:

$$C = \{x \in \mathbb{R} : x^2 < 4\} = \{x \in \mathbb{R} : -2 < x < 2\}.$$

The **empty set**, denoted $\varnothing$ or $\{\}$, contains no elements. It is unique: there is only one empty set.

Remark 1.1.1. A set is determined entirely by its elements, not by how they are described. Thus $\{1, 2, 3\} = \{3, 2, 1\} = \{1, 1, 2, 3\}$—order and repetition do not matter. When order matters, we use *sequences* or *tuples*; when repetition matters, we use *multisets*.

1.1.2 Subsets

Definition 1.1.2 (Subset). A set A is a **subset** of a set B, written $A \subseteq B$, if every element of A is also an element of B:

$$A \subseteq B \quad \Leftrightarrow \quad \forall x : (x \in A \Rightarrow x \in B).$$

If $A \subseteq B$ and $A \neq B$, we say A is a **proper subset** of B, written $A \subsetneq B$ or $A \subset B$.

Proposition 1.1.1. *For any sets A, B, C:*

(i) $A \subseteq A$ *(reflexivity)*

(ii) $A \subseteq B$ *and* $B \subseteq A$ *implies* $A = B$ *(antisymmetry)*

(iii) $A \subseteq B$ *and* $B \subseteq C$ *implies* $A \subseteq C$ *(transitivity)*

(iv) $\varnothing \subseteq A$ *(empty set is subset of everything)*

Property (ii) is particularly useful: to prove two sets are equal, we often prove mutual inclusion—that each is a subset of the other.

1.1.3 Set Operations

Definition 1.1.3 (Set operations). Let A and B be sets.

- The **union** of A and B is

$$A \cup B := \{x : x \in A \text{ or } x \in B\}.$$

- The **intersection** of A and B is

$$A \cap B := \{x : x \in A \text{ and } x \in B\}.$$

- The **set difference** (or **relative complement**) is

$$A \setminus B := \{x : x \in A \text{ and } x \notin B\}.$$

- The **symmetric difference** is

$$A \triangle B := (A \setminus B) \cup (B \setminus A) = (A \cup B) \setminus (A \cap B).$$

When working within a fixed **universal set** U, the **complement** of A is $A^c := U \setminus A$.

Theorem 1.1.2 (De Morgan's laws). *For any sets A, B within a universal set U:*

$$(A \cup B)^c = A^c \cap B^c, \qquad (A \cap B)^c = A^c \cup B^c.$$

Proof. We prove the first identity; the second is similar. Let x be arbitrary.

$$\begin{aligned} x \in (A \cup B)^c &\Leftrightarrow x \notin A \cup B \\ &\Leftrightarrow \neg(x \in A \text{ or } x \in B) \\ &\Leftrightarrow x \notin A \text{ and } x \notin B \\ &\Leftrightarrow x \in A^c \text{ and } x \in B^c \\ &\Leftrightarrow x \in A^c \cap B^c. \end{aligned}$$

□

These operations satisfy many algebraic properties:

Proposition 1.1.3 (Algebra of sets). *For any sets A, B, C:*

$$\begin{array}{lll} A \cup B = B \cup A & A \cap B = B \cap A & \textit{(commutativity)} \\ A \cup (B \cup C) = (A \cup B) \cup C & A \cap (B \cap C) = (A \cap B) \cap C & \textit{(associativity)} \\ A \cup (B \cap C) = (A \cup B) \cap (A \cup C) & A \cap (B \cup C) = (A \cap B) \cup (A \cap C) & \textit{(distributivity)} \\ A \cup A = A & A \cap A = A & \textit{(idempotence)} \\ A \cup \varnothing = A & A \cap \varnothing = \varnothing & \textit{(identity)} \end{array}$$

1.1.4 Power Sets and Cartesian Products

Definition 1.1.4 (Power set). The **power set** of A, denoted $\mathcal{P}(A)$ or 2^A, is the set of all subsets of A:

$$\mathcal{P}(A) := \{B : B \subseteq A\}.$$

Example 1.1.1. If $A = \{1, 2\}$, then $\mathcal{P}(A) = \{\varnothing, \{1\}, \{2\}, \{1, 2\}\}$. Note that $|\mathcal{P}(A)| = 4 = 2^2 = 2^{|A|}$.

Proposition 1.1.4. *If A is a finite set with $|A| = n$ elements, then $|\mathcal{P}(A)| = 2^n$.*

Definition 1.1.5 (Cartesian product). The **Cartesian product** of sets A and B is

$$A \times B := \{(a, b) : a \in A \text{ and } b \in B\},$$

where (a, b) denotes an **ordered pair**.

Ordered pairs have the property that $(a, b) = (c, d)$ if and only if $a = c$ and $b = d$. This distinguishes them from sets, where $\{a, b\} = \{b, a\}$.

The Cartesian product generalizes to any finite number of sets:

$$A_1 \times A_2 \times \cdots \times A_n = \{(a_1, a_2, \ldots, a_n) : a_i \in A_i \text{ for each } i\}.$$

We write $A^n := A \times A \times \cdots \times A$ (n times).

Example 1.1.2. The Euclidean plane is $\mathbb{R}^2 = \mathbb{R} \times \mathbb{R} = \{(x, y) : x, y \in \mathbb{R}\}$. More generally, $\mathbb{R}^n$ is n-dimensional Euclidean space.

1.1.5 Indexed Families and Generalized Operations

Often we need unions or intersections over more than two sets. Let I be an **index set**, and suppose that for each $i \in I$, we have a set A_i. The collection $\{A_i\}_{i\in I}$ is an **indexed family** of sets.

Definition 1.1.6 (Generalized union and intersection).

$$\bigcup_{i\in I} A_i := \{x : x \in A_i \text{ for some } i \in I\}, \qquad \bigcap_{i\in I} A_i := \{x : x \in A_i \text{ for all } i \in I\}.$$

Example 1.1.3. Let $A_n = \{x \in \mathbb{R} : -1/n < x < 1/n\}$ for $n \in \mathbb{N}^* = \{1, 2, 3, \ldots\}$. Then

$$\bigcap_{n=1}^{\infty} A_n = \{0\}, \qquad \bigcup_{n=1}^{\infty} A_n = (-1, 1).$$

De Morgan's laws generalize:

$$\left(\bigcup_{i\in I} A_i\right)^c = \bigcap_{i\in I} A_i^c, \qquad \left(\bigcap_{i\in I} A_i\right)^c = \bigcup_{i\in I} A_i^c.$$

1.2 Relations

1.2.1 Binary Relations

Definition 1.2.1 (Binary relation). A **binary relation** R from a set A to a set B is a subset $R \subseteq A \times B$. If $(a, b) \in R$, we say a is **related to** b and write $a \mathrel{R} b$ or aRb.

A relation on a set A is a relation from A to itself, i.e., a subset of $A \times A$.

Example 1.2.1. The relation $\leq$ on $\mathbb{R}$ is the set $\{(x, y) \in \mathbb{R}^2 : x \leq y\}$. The relation "divides" on $\mathbb{Z}$ is $\{(a, b) \in \mathbb{Z}^2 : a \mid b\}$.

1.2.2 Properties of Relations

Let R be a relation on a set A.

Definition 1.2.2 (Properties of relations). R is:

- **Reflexive** if aRa for all $a \in A$.

- **Symmetric** if aRb implies bRa for all $a, b \in A$.
- **Antisymmetric** if aRb and bRa implies $a = b$ for all $a, b \in A$.
- **Transitive** if aRb and bRc implies aRc for all $a, b, c \in A$.

Example 1.2.2. On $\mathbb{R}$: the relation $\leq$ is reflexive, antisymmetric, and transitive, but not symmetric. The relation $<$ is transitive and antisymmetric, but not reflexive or symmetric. The relation $=$ is reflexive, symmetric, antisymmetric, and transitive.

1.2.3 Equivalence Relations

Definition 1.2.3 (Equivalence relation). An **equivalence relation** on A is a relation that is reflexive, symmetric, and transitive.

We typically denote equivalence relations by $\sim$ or $\equiv$.

Example 1.2.3. (a) Equality $=$ is an equivalence relation on any set.

(b) For fixed $n \in \mathbb{N}^*$, congruence modulo n is an equivalence relation on $\mathbb{Z}$:

$$a \equiv b \pmod{n} \quad \Leftrightarrow \quad n \mid (a - b).$$

(c) On the set of lines in the plane, parallelism (including self-parallelism) is an equivalence relation.

(d) On a set of people, "has the same birthday as" is an equivalence relation.

Definition 1.2.4 (Equivalence class). Let $\sim$ be an equivalence relation on A. The **equivalence class** of $a \in A$ is

$$[a] := \{x \in A : x \sim a\}.$$

The set of all equivalence classes is the **quotient set**:

$$A/\sim := \{[a] : a \in A\}.$$

Theorem 1.2.1 (Fundamental theorem on equivalence relations). *Let $\sim$ be an equivalence relation on A. Then:*

(i) Every element belongs to some equivalence class: $a \in [a]$ for all $a \in A$.

(ii) Two equivalence classes are either equal or disjoint: $[a] = [b]$ or $[a] \cap [b] = \varnothing$.

(iii) A is the disjoint union of its equivalence classes.

Thus equivalence classes form a **partition** *of A.*

Proof. (i) Reflexivity gives $a \sim a$, so $a \in [a]$.

(ii) Suppose $[a] \cap [b] \neq \varnothing$, and let $c \in [a] \cap [b]$. Then $c \sim a$ and $c \sim b$. For any $x \in [a]$, we have $x \sim a \sim c \sim b$ by symmetry and transitivity, so $x \in [b]$. Thus $[a] \subseteq [b]$. By symmetry, $[b] \subseteq [a]$, so $[a] = [b]$.

(iii) By (i), $A = \bigcup_{a\in A}[a]$. By (ii), distinct classes are disjoint. □

Example 1.2.4. The equivalence classes of $\mathbb{Z}$ under congruence modulo 3 are

$$[0] = \{\dots,-6,-3,0,3,6,\dots\}, \quad [1] = \{\dots,-5,-2,1,4,7,\dots\}, \quad [2] = \{\dots,-4,-1,2,5,$$

The quotient set is $\mathbb{Z}/3\mathbb{Z} = \{[0],[1],[2]\}$, often written as $\mathbb{Z}_3 = \{0,1,2\}$ with arithmetic modulo 3.

Conversely, any partition of a set defines an equivalence relation:

Proposition 1.2.2. *Let $\{A_i\}_{i\in I}$ be a partition of A (i.e., the A_i are pairwise disjoint and their union is A). Define $a \sim b$ if and only if a and b belong to the same A_i. Then $\sim$ is an equivalence relation, and its equivalence classes are exactly the A_i.*

1.2.4 Order Relations

Definition 1.2.5 (Partial order)**.** A **partial order** on A is a relation that is reflexive, antisymmetric, and transitive. A set with a partial order is called a **partially ordered set** or **poset**.

Definition 1.2.6 (Total order)**.** A partial order $\leq$ on A is a **total order** (or **linear order**) if for all $a, b \in A$, either $a \leq b$ or $b \leq a$. We say any two elements are **comparable**.

Example 1.2.5. (a) The usual $\leq$ on $\mathbb{R}$ is a total order.

(b) Subset inclusion $\subseteq$ on $\mathcal{P}(A)$ is a partial order but generally not a total order.

(c) Divisibility on $\mathbb{N}^*$: $a \leq b$ iff $a \mid b$. This is a partial order but not total (e.g., 2 and 3 are not comparable).

Definition 1.2.7 (Bounds and extrema)**.** Let $(A, \leq)$ be a poset and $S \subseteq A$.

- An **upper bound** of S is an element $u \in A$ such that $s \leq u$ for all $s \in S$.

- The **least upper bound** or **supremum**, if it exists, is an upper bound u such that $u \leq v$ for every upper bound v. Notation: $\sup S$ or $\bigvee S$.
- A **maximum** of S is an element $m \in S$ such that $s \leq m$ for all $s \in S$.

Lower bound, greatest lower bound (infimum, $\inf S$ or $\bigwedge S$), and minimum are defined dually.

Remark 1.2.1. A maximum is always a supremum, but a supremum need not be a maximum (it may not belong to S). For example, $\sup\{x \in \mathbb{R} : x < 1\} = 1$, but this set has no maximum.

Definition 1.2.8 (Well-ordering)**.** A total order $\leq$ on A is a **well-ordering** if every nonempty subset of A has a minimum element.

Example 1.2.6. The usual order on $\mathbb{N}$ is a well-ordering. The usual order on $\mathbb{Z}$ is not (no minimum), nor is the usual order on $\mathbb{R}^+$ (the set $(0, 1)$ has no minimum).

The **well-ordering principle**—that $\mathbb{N}$ is well-ordered—is equivalent to the principle of mathematical induction.

1.3 Functions (Mappings)

1.3.1 Definition and Basic Concepts

Definition 1.3.1 (Function)**.** A **function** (or **mapping**) f from a set A to a set B, written $f : A \to B$, is a rule that assigns to each element $a \in A$ exactly one element $f(a) \in B$.

The set A is the **domain** of f, denoted $\mathrm{dom}(f)$. The set B is the **codomain**. The **image** (or **range**) of f is

$$\mathrm{im}(f) := \{f(a) : a \in A\} = \{b \in B : b = f(a) \text{ for some } a \in A\}.$$

We write $a \mapsto f(a)$ to indicate that a is mapped to $f(a)$. For example, we might define $f : \mathbb{R} \to \mathbb{R}$ by $x \mapsto x^2$.

Definition 1.3.2 (Equality of functions)**.** Two functions $f : A \to B$ and $g : A \to B$ are **equal** if $f(a) = g(a)$ for all $a \in A$. Note that equality requires the same domain and codomain.

Definition 1.3.3 (Image and preimage of sets)**.** Let $f : A \to B$, $S \subseteq A$, and $T \subseteq B$.

- The **image of** S under f is $f(S) := \{f(a) : a \in S\}$.

- The **preimage of** T under f is $f^{-1}(T) := \{a \in A : f(a) \in T\}$.

Remark 1.3.1. The notation $f^{-1}(T)$ for preimage does not require f to have an inverse function. It is simply notation for the set of elements mapping into T.

1.3.2 Injections, Surjections, Bijections

Definition 1.3.4 (Types of functions). Let $f : A \to B$.

- f is **injective** (or **one-to-one**) if $f(a_1) = f(a_2)$ implies $a_1 = a_2$.
- f is **surjective** (or **onto**) if $\text{im}(f) = B$, i.e., every $b \in B$ is $f(a)$ for some $a \in A$.
- f is **bijective** if it is both injective and surjective.

Example 1.3.1. Consider functions $\mathbb{R} \to \mathbb{R}$:

- $f(x) = x^2$ is neither injective (since $f(1) = f(-1)$) nor surjective (since $-1 \notin \text{im}(f)$).
- $f(x) = x^3$ is bijective.
- $f(x) = e^x$ is injective but not surjective (range is $(0, \infty)$).
- $f(x) = \sin x$ is neither injective nor surjective.

Proposition 1.3.1 (Characterization by preimages). *$f : A \to B$ is injective if and only if $|f^{-1}(\{b\})| \leq 1$ for all $b \in B$. It is surjective if and only if $|f^{-1}(\{b\})| \geq 1$ for all $b \in B$. It is bijective if and only if $|f^{-1}(\{b\})| = 1$ for all $b \in B$.*

1.3.3 Composition and Inverses

Definition 1.3.5 (Composition). If $f : A \to B$ and $g : B \to C$, the **composition** $g \circ f : A \to C$ is defined by

$$(g \circ f)(a) := g(f(a)).$$

Composition is associative: $(h \circ g) \circ f = h \circ (g \circ f)$.

Definition 1.3.6 (Identity function). The **identity function** on A is $\text{id}_A : A \to A$ defined by $\text{id}_A(a) = a$.

Proposition 1.3.2. *Let $f : A \to B$ and $g : B \to C$.*

(i) If f and g are injective, so is $g \circ f$.

(ii) *If f and g are surjective, so is $g \circ f$.*

(iii) *If f and g are bijective, so is $g \circ f$.*

Definition 1.3.7 (Inverse function)**.** If $f : A \to B$ is bijective, there exists a unique function $f^{-1} : B \to A$ such that

$$f^{-1} \circ f = \mathrm{id}_A \quad \text{and} \quad f \circ f^{-1} = \mathrm{id}_B\,.$$

This f^{-1} is the **inverse** of f.

Proposition 1.3.3. *$f : A \to B$ has an inverse if and only if f is bijective. When the inverse exists:*

(i) *$(f^{-1})^{-1} = f$.*

(ii) *If $g : B \to C$ is also bijective, then $(g \circ f)^{-1} = f^{-1} \circ g^{-1}$.*

Definition 1.3.8 (Left and right inverses)**.** $g : B \to A$ is a **left inverse** of $f : A \to B$ if $g \circ f = \mathrm{id}_A$. It is a **right inverse** if $f \circ g = \mathrm{id}_B$.

Proposition 1.3.4. (i) *$f : A \to B$ has a left inverse if and only if f is injective.*

(ii) *$f : A \to B$ has a right inverse if and only if f is surjective.*

(iii) *If f has both a left inverse g and a right inverse h, then $g = h = f^{-1}$.*

1.3.4 Cardinality

Definition 1.3.9 (Cardinality for finite sets)**.** Two sets A and B have the same **cardinality**, written $|A| = |B|$, if there exists a bijection $f : A \to B$.

For finite sets, this matches the intuitive notion of "having the same number of elements." But cardinality extends to infinite sets:

Definition 1.3.10 (Countable and uncountable)**.** A set A is **countably infinite** if $|A| = |\mathbb{N}|$. It is **countable** if it is finite or countably infinite. It is **uncountable** if it is not countable.

Theorem 1.3.5 (Cantor)**.** *For any set A, there is no surjection $A \to \mathcal{P}(A)$. Hence $|A| < |\mathcal{P}(A)|$.*

Proof. Suppose $f : A \to \mathcal{P}(A)$ is surjective. Consider $D := \{a \in A : a \notin f(a)\}$. Since f is surjective, $D = f(d)$ for some $d \in A$. Is $d \in D$?

If $d \in D$, then by definition of D, $d \notin f(d) = D$. Contradiction.
If $d \notin D$, then $d \notin f(d)$, so by definition of D, $d \in D$. Contradiction.
Either way we have a contradiction, so no such surjection exists. □

Corollary 1.3.6. *$\mathbb{R}$ is uncountable.*

This follows from the fact that $|\mathbb{R}| = |\mathcal{P}(\mathbb{N})|$ (which requires proof).

1.4 Propositional Logic

1.4.1 Propositions and Logical Connectives

Definition 1.4.1 (Proposition). A **proposition** is a statement that is either true or false (but not both). We denote the truth values as T (true) and F (false).

Example 1.4.1. "$2 + 2 = 4$" is a proposition (true). "π is rational" is a proposition (false). "Is this a question?" is not a proposition. "$x > 0$" is not a proposition until x is specified.

Definition 1.4.2 (Logical connectives). Given propositions P and Q:

- **Negation**: $\neg P$ ("not P") is true iff P is false.
- **Conjunction**: $P \wedge Q$ ("P and Q") is true iff both P and Q are true.
- **Disjunction**: $P \vee Q$ ("P or Q") is true iff at least one of P, Q is true.
- **Implication**: $P \Rightarrow Q$ ("if P then Q") is false only when P is true and Q is false.
- **Biconditional**: $P \Leftrightarrow Q$ ("P if and only if Q") is true iff P and Q have the same truth value.

The truth tables are:

P	Q	$\neg P$	$P \wedge Q$	$P \vee Q$	$P \Rightarrow Q$	$P \Leftrightarrow Q$
T	T	F	T	T	T	T
T	F	F	F	T	F	F
F	T	T	F	T	T	F
F	F	T	F	F	T	T

Remark 1.4.1. The implication $P \Rightarrow Q$ is called **material implication**. It is "vacuously true" when P is false. This convention, while sometimes counterintuitive, is essential for mathematical logic.

1.4.2 Logical Equivalence and Laws

Definition 1.4.3 (Logical equivalence). Two compound propositions are **logically equivalent** if they have the same truth value for all possible truth values of their component propositions. We write $P \equiv Q$.

Proposition 1.4.1 (Fundamental logical equivalences).

$$
\begin{aligned}
\neg(\neg P) &\equiv P && \textit{(double negation)}\\
P \vee Q &\equiv Q \vee P, \quad P \wedge Q \equiv Q \wedge P && \textit{(commutativity)}\\
(P \vee Q) \vee R &\equiv P \vee (Q \vee R) && \textit{(associativity)}\\
P \wedge (Q \vee R) &\equiv (P \wedge Q) \vee (P \wedge R) && \textit{(distributivity)}\\
\neg(P \wedge Q) &\equiv \neg P \vee \neg Q && \textit{(De Morgan)}\\
\neg(P \vee Q) &\equiv \neg P \wedge \neg Q && \textit{(De Morgan)}\\
P \Rightarrow Q &\equiv \neg P \vee Q && \textit{(implication)}\\
P \Rightarrow Q &\equiv \neg Q \Rightarrow \neg P && \textit{(contrapositive)}\\
P \Leftrightarrow Q &\equiv (P \Rightarrow Q) \wedge (Q \Rightarrow P) && \textit{(biconditional)}
\end{aligned}
$$

The equivalence $P \Rightarrow Q \equiv \neg Q \Rightarrow \neg P$ is particularly important: the **contrapositive** is logically equivalent to the original implication.

Remark 1.4.2. The **converse** of $P \Rightarrow Q$ is $Q \Rightarrow P$. The **inverse** is $\neg P \Rightarrow \neg Q$. Neither is equivalent to the original. However, the converse and inverse are equivalent to each other (they are contrapositives).

1.5 Predicate Logic and Quantifiers

1.5.1 Predicates

Definition 1.5.1 (Predicate). A **predicate** is a statement containing one or more variables that becomes a proposition when specific values are substituted for the variables. If P is a predicate with variable x, we write $P(x)$.

Example 1.5.1. $P(x)$: "$x > 0$" is a predicate. $P(3)$ is true; $P(-2)$ is false. The **domain of discourse** specifies which values x may take.

1.5.2 Quantifiers

Definition 1.5.2 (Universal and existential quantifiers). Let $P(x)$ be a predicate with domain D.

- $\forall x \in D : P(x)$ means "$P(x)$ is true for all x in D."
- $\exists x \in D : P(x)$ means "there exists at least one x in D such that $P(x)$ is true."

When the domain is clear from context, we may write simply $\forall x : P(x)$.

Example 1.5.2. Let the domain be $\mathbb{R}$.

- $\forall x : x^2 \geq 0$ is true.
- $\exists x : x^2 = 2$ is true (e.g., $x = \sqrt{2}$).
- $\forall x : x^2 = x$ is false (fails for $x = 2$).
- $\exists x : x^2 < 0$ is false.

Definition 1.5.3 (Unique existence). $\exists! x : P(x)$ means "there exists exactly one x such that $P(x)$." Formally:

$$\exists! x : P(x) \quad \Leftrightarrow \quad \exists x : P(x) \wedge \forall y \forall z : (P(y) \wedge P(z) \Rightarrow y = z).$$

1.5.3 Negation of Quantified Statements

Proposition 1.5.1 (Negation rules for quantifiers).

$$\neg(\forall x : P(x)) \equiv \exists x : \neg P(x)$$
$$\neg(\exists x : P(x)) \equiv \forall x : \neg P(x)$$

Example 1.5.3. The negation of "all swans are white" is "there exists a swan that is not white," not "all swans are not white."

For nested quantifiers:

$$\neg(\forall x \exists y : P(x, y)) \equiv \exists x \forall y : \neg P(x, y)$$
$$\neg(\exists x \forall y : P(x, y)) \equiv \forall x \exists y : \neg P(x, y)$$

1.5.4 Order of Quantifiers

The order of different quantifiers matters:

Example 1.5.4. Consider predicates over $\mathbb{R}$:

- $\forall x \exists y : x + y = 0$ is true (for any x, take $y = -x$).
- $\exists y \forall x : x + y = 0$ is false (no single y works for all x).

Proposition 1.5.2. *Same quantifiers can be interchanged:*

$$\forall x \forall y : P(x, y) \equiv \forall y \forall x : P(x, y), \qquad \exists x \exists y : P(x, y) \equiv \exists y \exists x : P(x, y).$$

Mixed quantifiers cannot generally be interchanged:

$$\forall x \exists y : P(x, y) \quad \not\equiv \quad \exists y \forall x : P(x, y)$$

However: $\exists y \forall x : P(x, y) \Rightarrow \forall x \exists y : P(x, y)$ *(but not conversely).*

1.6 Methods of Proof

Mathematics advances by proving theorems. This section surveys the main proof techniques.

1.6.1 Direct Proof

A **direct proof** of $P \Rightarrow Q$ assumes P is true and deduces Q through a sequence of logical steps.

Example 1.6.1. Claim: If n is even, then n^2 is even.

Proof: Assume n is even. Then $n = 2k$ for some integer k. Thus $n^2 = (2k)^2 = 4k^2 = 2(2k^2)$. Since $2k^2$ is an integer, n^2 is even. □

1.6.2 Proof by Contrapositive

To prove $P \Rightarrow Q$, we may instead prove the logically equivalent **contrapositive** $\neg Q \Rightarrow \neg P$.

Example 1.6.2. Claim: If n^2 is odd, then n is odd.

Proof: We prove the contrapositive: if n is even, then n^2 is even. This was shown above. □

1.6.3 Proof by Contradiction

A **proof by contradiction** (or *reductio ad absurdum*) assumes the negation of what we want to prove and derives a contradiction.

Example 1.6.3. Claim: $\sqrt{2}$ is irrational.

Proof: Suppose, for contradiction, that $\sqrt{2}$ is rational. Then $\sqrt{2} = p/q$ for integers p, q with $q \neq 0$ and $\gcd(p, q) = 1$ (lowest terms).

Squaring: $2 = p^2/q^2$, so $p^2 = 2q^2$. Thus p^2 is even, so p is even (by the previous example's contrapositive). Write $p = 2m$.

Then $(2m)^2 = 2q^2$, so $4m^2 = 2q^2$, giving $q^2 = 2m^2$. Thus q^2 is even, so q is even.

But if both p and q are even, then $\gcd(p, q) \geq 2$, contradicting $\gcd(p, q) = 1$. This contradiction shows our assumption was false: $\sqrt{2}$ is irrational. □

1.6.4 Proof by Cases

When a statement naturally divides into cases, we can prove each case separately.

Example 1.6.4. Claim: For any integer n, $n^2 + n$ is even.

Proof: Case 1: n is even. Then $n = 2k$, so $n^2+n = 4k^2+2k = 2(2k^2+k)$, which is even.

Case 2: n is odd. Then $n = 2k + 1$, so $n^2 + n = (2k + 1)^2 + (2k + 1) = 4k^2 + 4k + 1 + 2k + 1 = 4k^2 + 6k + 2 = 2(2k^2 + 3k + 1)$, which is even.

In both cases, $n^2 + n$ is even. □

1.6.5 Existence and Uniqueness Proofs

Existence proofs show that something exists, either constructively (by exhibiting an example) or non-constructively.

Example 1.6.5 (Constructive). **Claim:** There exists an irrational number r such that r^r is rational.

Proof: Consider $r = \sqrt{2}^{\sqrt{2}}$. If this is rational, we're done. If it's irrational, let $r' = r^{\sqrt{2}} = (\sqrt{2}^{\sqrt{2}})^{\sqrt{2}} = \sqrt{2}^2 = 2$, which is rational. Either way, such an r exists. □

Uniqueness proofs show that if something exists, it is unique. A common approach: assume two objects satisfy the condition, then show they are equal.

Example 1.6.6. Claim: The empty set is unique.

Proof: Suppose $\varnothing_1$ and $\varnothing_2$ are both empty sets. Since $\varnothing_1$ has no elements, $\varnothing_1 \subseteq \varnothing_2$ vacuously. Similarly, $\varnothing_2 \subseteq \varnothing_1$. By antisymmetry of $\subseteq$, $\varnothing_1 = \varnothing_2$. □

1.6.6 Mathematical Induction

Mathematical induction proves statements about natural numbers (or more general well-ordered sets).

Theorem 1.6.1 (Principle of Mathematical Induction). *Let $P(n)$ be a predicate on $\mathbb{N}$. If:*

(i) $P(0)$ is true *(base case)*

(ii) $\forall k \in \mathbb{N} : P(k) \Rightarrow P(k+1)$ *(inductive step)*

Then $P(n)$ is true for all $n \in \mathbb{N}$.

Example 1.6.7. Claim: For all $n \geq 1$, $\displaystyle\sum_{i=1}^{n} i = \frac{n(n+1)}{2}$.

Proof: Base case ($n = 1$): $\sum_{i=1}^{1} i = 1 = \frac{1 \cdot 2}{2}$. ✓

Inductive step: Assume the formula holds for some $k \geq 1$. Then:

$$\sum_{i=1}^{k+1} i = \left(\sum_{i=1}^{k} i\right) + (k+1) = \frac{k(k+1)}{2} + (k+1) = \frac{k(k+1) + 2(k+1)}{2} = \frac{(k+1)(k+2)}{2}.$$

This is the formula with $n = k + 1$. ✓

By induction, the formula holds for all $n \geq 1$. □

Theorem 1.6.2 (Strong Induction). *Let $P(n)$ be a predicate on $\mathbb{N}$. If:*

(i) $P(0)$ *is true*

(ii) $\forall k \in \mathbb{N} : [P(0) \wedge P(1) \wedge \cdots \wedge P(k)] \Rightarrow P(k+1)$

Then $P(n)$ is true for all $n \in \mathbb{N}$.

Strong induction allows the inductive hypothesis to assume all previous cases, not just the immediately preceding one. This is useful when $P(k+1)$ depends on $P(j)$ for some $j < k$.

Example 1.6.8. Claim: Every integer $n \geq 2$ can be written as a product of primes.

Proof: Base case ($n = 2$): 2 is prime, so it is (trivially) a product of primes.

Inductive step: Assume every integer m with $2 \leq m \leq k$ is a product of primes. Consider $k + 1$.

If $k + 1$ is prime, it is a product of primes (itself).

If $k + 1$ is composite, then $k + 1 = ab$ for some $2 \leq a, b \leq k$. By the inductive hypothesis, a and b are each products of primes. Therefore $k + 1 = ab$ is also a product of primes.

By strong induction, every $n \geq 2$ is a product of primes. □

1.6.7 The Well-Ordering Principle

Theorem 1.6.3 (Well-Ordering Principle). *Every nonempty subset of $\mathbb{N}$ has a least element.*

This principle is equivalent to mathematical induction. It provides an alternative proof technique: to show $P(n)$ holds for all $n \in \mathbb{N}$, assume the set of counterexamples is nonempty and derive a contradiction using the minimal counterexample.

Example 1.6.9. Claim: For all $n \in \mathbb{N}$, $n < 2^n$.

Proof: Suppose not. Let $S = \{n \in \mathbb{N} : n \geq 2^n\}$. By assumption, S is nonempty. By the well-ordering principle, S has a least element m.

Note $m \neq 0$ since $0 < 1 = 2^0$. So $m \geq 1$, meaning $m - 1 \in \mathbb{N}$ and $m - 1 < m$.

Since m is minimal in S, we have $m - 1 \notin S$, so $m - 1 < 2^{m-1}$.

Then $m = (m-1) + 1 < 2^{m-1} + 1 \leq 2^{m-1} + 2^{m-1} = 2^m$.

But this contradicts $m \in S$ (which requires $m \geq 2^m$). Hence $S = \varnothing$, and $n < 2^n$ for all n. □

1.7 Summary and Physical Applications

The concepts introduced in this chapter—sets, relations, functions, logic, and proof—are the foundation upon which all subsequent mathematics is built. Their applications in physics include:

State spaces: The set of all possible states of a physical system. In classical mechanics, this is phase space $\mathbb{R}^{2n}$; in quantum mechanics, it is a Hilbert space.

Equivalence relations: Gauge equivalence identifies physically equivalent field configurations. Topologically equivalent spaces are related by homeomorphism.

Functions and mappings: Physical observables are functions from state space to $\mathbb{R}$. Symmetry transformations are bijections on state space. Time evolution is a family of mappings.

Cardinality: The countability of quantum states for systems in a finite box (discrete spectrum) versus uncountability for free particles (continuous spectrum) reflects fundamental physics.

Logical structure: The axiomatic method—starting from basic postulates and deriving consequences—is the template for theoretical physics, from Newton's laws to the postulates of quantum mechanics.

Exercises

Exercise 1.7.1. Prove that for any sets A, B, C: $A \cap (B \cup C) = (A \cap B) \cup (A \cap C)$.

Exercise 1.7.2. Let $f : A \to B$ and $g : B \to C$. Prove that:

(a) If $g \circ f$ is injective, then f is injective.

(b) If $g \circ f$ is surjective, then g is surjective.

Give counterexamples to the converses.

Exercise 1.7.3. Prove that $\mathbb{Q}$ is countable. (Hint: arrange positive rationals in a grid and count diagonally.)

Exercise 1.7.4. Write the negation of each statement in simplest form:

(a) $\forall \epsilon > 0 \, \exists \delta > 0 \, \forall x : |x - a| < \delta \Rightarrow |f(x) - L| < \epsilon$

(b) $\exists M > 0 \, \forall x : |f(x)| \leq M$

Exercise 1.7.5. Prove by induction: $\sum_{i=1}^{n} i^2 = \frac{n(n+1)(2n+1)}{6}$ for all $n \geq 1$.

Exercise 1.7.6. Use the well-ordering principle to prove that $\sqrt{3}$ is irrational.

Exercise 1.7.7. Define a relation $\sim$ on $\mathbb{Z} \times (\mathbb{Z} \setminus \{0\})$ by $(a, b) \sim (c, d)$ iff $ad = bc$. Prove that $\sim$ is an equivalence relation. (This construction defines the rational numbers.)

Exercise 1.7.8. Prove that every equivalence relation arises from a partition, and every partition defines an equivalence relation (as stated in Section 1.2).

Chapter 2

Linear Algebra

Linear algebra is arguably the most essential mathematical tool for physics. It provides the language for quantum mechanics (state vectors, operators), classical mechanics (phase space, symplectic structure), relativity (tensor calculus), and virtually every other area of theoretical physics. This chapter develops the theory systematically, emphasizing concepts that recur throughout physics.

2.1 Vector Spaces

2.1.1 Definition and Examples

Definition 2.1.1 (Field). A **field** is a set $\mathbb{F}$ equipped with two operations, addition and multiplication, satisfying the usual arithmetic properties: associativity, commutativity, existence of identity elements (0 and 1), existence of additive inverses (negatives), existence of multiplicative inverses for nonzero elements, and distributivity.

The most important fields for physics are $\mathbb{R}$ (real numbers) and $\mathbb{C}$ (complex numbers). We will work over a general field $\mathbb{F}$ when possible, specializing to $\mathbb{R}$ or $\mathbb{C}$ when necessary.

Definition 2.1.2 (Vector space). A **vector space** over a field $\mathbb{F}$ is a set V together with two operations:

- **Vector addition**: $+ : V \times V \to V$
- **Scalar multiplication**: $\cdot : \mathbb{F} \times V \to V$

satisfying, for all $\mathbf{u}, \mathbf{v}, \mathbf{w} \in V$ and $a, b \in \mathbb{F}$:

(V1) $\mathbf{u} + \mathbf{v} = \mathbf{v} + \mathbf{u}$ (commutativity)

(V2) $(\mathbf{u}+\mathbf{v})+\mathbf{w}=\mathbf{u}+(\mathbf{v}+\mathbf{w})$ (associativity)

(V3) There exists $\mathbf{0} \in V$ such that $\mathbf{v}+\mathbf{0}=\mathbf{v}$ (zero vector)

(V4) For each $\mathbf{v}$, there exists $-\mathbf{v}$ such that $\mathbf{v}+(-\mathbf{v})=\mathbf{0}$ (additive inverse)

(V5) $a(b\mathbf{v})=(ab)\mathbf{v}$ (compatibility)

(V6) $1 \cdot \mathbf{v}=\mathbf{v}$ (identity)

(V7) $a(\mathbf{u}+\mathbf{v})=a\mathbf{u}+a\mathbf{v}$ (distributivity)

(V8) $(a+b)\mathbf{v}=a\mathbf{v}+b\mathbf{v}$ (distributivity)

Elements of V are called **vectors**; elements of $\mathbb{F}$ are called **scalars**.

Example 2.1.1 (Fundamental examples). (a) $\mathbb{F}^n = \{(x_1, \ldots, x_n) : x_i \in \mathbb{F}\}$ with componentwise operations is the **coordinate space**. For $\mathbb{F} = \mathbb{R}$, this is Euclidean n-space.

(b) The set of $m \times n$ matrices $M_{m\times n}(\mathbb{F})$ with entry-wise addition and scalar multiplication.

(c) The set of polynomials $\mathbb{F}[x]$ with coefficients in $\mathbb{F}$, or the subset $\mathbb{F}_n[x]$ of polynomials of degree at most n.

(d) The set $C([a,b])$ of continuous real-valued functions on $[a,b]$, with pointwise operations: $(f+g)(x)=f(x)+g(x)$ and $(af)(x)=a\cdot f(x)$.

(e) The set of solutions to a homogeneous linear differential equation, such as $y''+y=0$.

(f) The trivial vector space $\{\mathbf{0}\}$ containing only the zero vector.

Proposition 2.1.1 (Basic properties). *In any vector space:*

(i) The zero vector is unique.

(ii) Each vector has a unique additive inverse.

(iii) $0 \cdot \mathbf{v} = \mathbf{0}$ *for all* $\mathbf{v}$.

(iv) $a \cdot \mathbf{0} = \mathbf{0}$ *for all* a.

(v) $(-1)\cdot \mathbf{v} = -\mathbf{v}$.

(vi) If $a\mathbf{v}=\mathbf{0}$*, then* $a=0$ *or* $\mathbf{v}=\mathbf{0}$.

2.1.2 Subspaces

Definition 2.1.3 (Subspace). A **subspace** of a vector space V is a nonempty subset $W \subseteq V$ that is itself a vector space under the same operations. Equivalently, W is a subspace if and only if:

(i) $\mathbf{0} \in W$

(ii) $\mathbf{u}, \mathbf{v} \in W \Rightarrow \mathbf{u} + \mathbf{v} \in W$ (closed under addition)

(iii) $\mathbf{v} \in W, a \in \mathbb{F} \Rightarrow a\mathbf{v} \in W$ (closed under scalar multiplication)

Conditions (ii) and (iii) can be combined: W is a subspace iff $W \neq \varnothing$ and $a\mathbf{u} + b\mathbf{v} \in W$ for all $\mathbf{u}, \mathbf{v} \in W$ and $a, b \in \mathbb{F}$.

Example 2.1.2. (a) In $\mathbb{R}^3$: any line through the origin is a 1-dimensional subspace; any plane through the origin is a 2-dimensional subspace.

(b) In $C(\mathbb{R})$: the set of differentiable functions, or the set of solutions to $y' = y$, forms a subspace.

(c) $\{(x, y) \in \mathbb{R}^2 : x + y = 1\}$ is *not* a subspace (doesn't contain $\mathbf{0}$).

Proposition 2.1.2 (Intersection of subspaces). *If $\{W_i\}_{i \in I}$ is any collection of subspaces of V, then $\bigcap_{i \in I} W_i$ is also a subspace.*

Definition 2.1.4 (Sum of subspaces). If W_1, W_2 are subspaces of V, their **sum** is

$$W_1 + W_2 := \{\mathbf{w}_1 + \mathbf{w}_2 : \mathbf{w}_1 \in W_1, \mathbf{w}_2 \in W_2\}.$$

This is the smallest subspace containing both W_1 and W_2.

Definition 2.1.5 (Direct sum). The sum $W_1 + W_2$ is a **direct sum**, written $W_1 \oplus W_2$, if $W_1 \cap W_2 = \{\mathbf{0}\}$. Equivalently, every vector in $W_1 + W_2$ can be written uniquely as $\mathbf{w}_1 + \mathbf{w}_2$ with $\mathbf{w}_i \in W_i$.

2.1.3 Linear Combinations and Span

Definition 2.1.6 (Linear combination). A **linear combination** of vectors $\mathbf{v}_1, \ldots, \mathbf{v}_n \in V$ is any vector of the form

$$a_1\mathbf{v}_1 + a_2\mathbf{v}_2 + \cdots + a_n\mathbf{v}_n = \sum_{i=1}^{n} a_i\mathbf{v}_i$$

where $a_1, \ldots, a_n \in \mathbb{F}$.

Definition 2.1.7 (Span). The **span** of a set of vectors $S = \{\mathbf{v}_1, \dots, \mathbf{v}_n\}$ is the set of all linear combinations:

$$\operatorname{span}(S) := \left\{ \sum_{i=1}^{n} a_i \mathbf{v}_i : a_i \in \mathbb{F} \right\}.$$

For an infinite set S, $\operatorname{span}(S)$ consists of all *finite* linear combinations of elements of S. By convention, $\operatorname{span}(\varnothing) = \{\mathbf{0}\}$.

Proposition 2.1.3. $\operatorname{span}(S)$ *is the smallest subspace containing* S. *If* $S \subseteq W$ *for some subspace* W, *then* $\operatorname{span}(S) \subseteq W$.

Definition 2.1.8 (Spanning set). A set S **spans** V (or is a **spanning set** for V) if $\operatorname{span}(S) = V$.

2.1.4 Linear Independence

Definition 2.1.9 (Linear independence). Vectors $\mathbf{v}_1, \dots, \mathbf{v}_n$ are **linearly independent** if

$$a_1\mathbf{v}_1 + \cdots + a_n\mathbf{v}_n = \mathbf{0} \quad \Rightarrow \quad a_1 = \cdots = a_n = 0.$$

Otherwise, they are **linearly dependent**.

Proposition 2.1.4 (Characterizations of dependence). *The following are equivalent:*

(i) $\mathbf{v}_1, \dots, \mathbf{v}_n$ *are linearly dependent.*

(ii) There exist scalars $a_1, \dots, a_n$, *not all zero, such that* $\sum a_i \mathbf{v}_i = \mathbf{0}$.

(iii) Some $\mathbf{v}_j$ *is a linear combination of the others.*

(iv) $\operatorname{span}\{\mathbf{v}_1, \dots, \mathbf{v}_n\} = \operatorname{span}\{\mathbf{v}_1, \dots, \hat{\mathbf{v}}_j, \dots, \mathbf{v}_n\}$ *for some* j.

Example 2.1.3.

(a) In $\mathbb{R}^3$: $\mathbf{e}_1 = (1, 0, 0)$, $\mathbf{e}_2 = (0, 1, 0)$, $\mathbf{e}_3 = (0, 0, 1)$ are linearly independent.

(b) $(1, 2)$, $(3, 6)$ are linearly dependent in $\mathbb{R}^2$ since $(3, 6) = 3(1, 2)$.

(c) $1, x, x^2$ are linearly independent in $\mathbb{R}[x]$.

(d) $\sin x$, $\cos x$ are linearly independent in $C(\mathbb{R})$ (no constant a satisfies $\sin x = a \cos x$ for all x).

Lemma 2.1.5 (Steinitz exchange lemma). *Let* $\{\mathbf{v}_1, \dots, \mathbf{v}_m\}$ *be linearly independent and* $\{\mathbf{w}_1, \dots, \mathbf{w}_n\}$ *span* V. *Then* $m \leq n$.

This lemma has the crucial consequence that all bases have the same size.

2.1.5 Bases and Dimension

Definition 2.1.10 (Basis). A **basis** for a vector space V is a set $\mathcal{B}$ that is both linearly independent and spans V.

Theorem 2.1.6 (Basis characterizations). *For a set $\mathcal{B} = \{\mathbf{v}_1, \ldots, \mathbf{v}_n\}$ in V, the following are equivalent:*

(i) $\mathcal{B}$ is a basis.

(ii) $\mathcal{B}$ is a maximal linearly independent set.

(iii) $\mathcal{B}$ is a minimal spanning set.

(iv) Every $\mathbf{v} \in V$ can be written uniquely as $\mathbf{v} = \sum_{i=1}^{n} a_i \mathbf{v}_i$.

Proof of (i) $\Leftrightarrow$ (iv). Suppose $\mathcal{B}$ is a basis. Since $\mathcal{B}$ spans V, every $\mathbf{v}$ is a linear combination of the $\mathbf{v}_i$. For uniqueness, suppose $\mathbf{v} = \sum a_i \mathbf{v}_i = \sum b_i \mathbf{v}_i$. Then $\sum (a_i - b_i)\mathbf{v}_i = \mathbf{0}$. By linear independence, $a_i - b_i = 0$ for all i, so $a_i = b_i$.

Conversely, suppose (iv) holds. Then $\mathcal{B}$ spans V by definition. For independence, if $\sum a_i \mathbf{v}_i = \mathbf{0}$, then comparing with $\mathbf{0} = \sum 0 \cdot \mathbf{v}_i$, uniqueness gives $a_i = 0$ for all i. □

Theorem 2.1.7 (Existence of bases). *Every vector space has a basis. More precisely:*

(i) Any linearly independent set can be extended to a basis.

(ii) Any spanning set contains a basis.

For finite-dimensional spaces, this can be proved constructively. For infinite-dimensional spaces, it requires the Axiom of Choice (via Zorn's Lemma).

Theorem 2.1.8 (Invariance of dimension). *Any two bases of a vector space V have the same cardinality.*

Definition 2.1.11 (Dimension). The **dimension** of V, denoted $\dim V$, is the cardinality of any basis. If V has a finite basis, we say V is **finite-dimensional**; otherwise, V is **infinite-dimensional**.

Example 2.1.4.

(a) $\dim \mathbb{F}^n = n$. The **standard basis** is $\{\mathbf{e}_1, \ldots, \mathbf{e}_n\}$ where $\mathbf{e}_i$ has 1 in position i and 0 elsewhere.

(b) $\dim M_{m \times n}(\mathbb{F}) = mn$.

(c) $\dim \mathbb{F}_n[x] = n + 1$ (basis: $1, x, x^2, \ldots, x^n$).

(d) $\dim \mathbb{F}[x] = \infty$ (countably infinite).

(e) $\dim C([0,1]) = \infty$ (uncountably infinite).

Theorem 2.1.9 (Dimension formulas). *Let V be finite-dimensional with subspaces W_1, W_2.*

(i) $\dim(W_1 + W_2) = \dim W_1 + \dim W_2 - \dim(W_1 \cap W_2)$.

(ii) If $V = W_1 \oplus W_2$, then $\dim V = \dim W_1 + \dim W_2$.

(iii) If $W \subseteq V$, then $\dim W \leq \dim V$, *with equality iff $W = V$.*

Definition 2.1.12 (Coordinates). Let $\mathcal{B} = \{\mathbf{v}_1, \ldots, \mathbf{v}_n\}$ be an ordered basis for V. For $\mathbf{v} = \sum a_i \mathbf{v}_i$, the **coordinate vector** of $\mathbf{v}$ with respect to $\mathcal{B}$ is

$$[\mathbf{v}]_{\mathcal{B}} := \begin{pmatrix} a_1 \\ \vdots \\ a_n \end{pmatrix} \in \mathbb{F}^n.$$

The coordinate map $\mathbf{v} \mapsto [\mathbf{v}]_{\mathcal{B}}$ is an isomorphism from V to $\mathbb{F}^n$.

2.2 Linear Maps

2.2.1 Definition and Basic Properties

Definition 2.2.1 (Linear map). A **linear map** (or **linear transformation**) from a vector space V to a vector space W (both over $\mathbb{F}$) is a function $T : V \to W$ satisfying:

(i) $T(\mathbf{u} + \mathbf{v}) = T(\mathbf{u}) + T(\mathbf{v})$ for all $\mathbf{u}, \mathbf{v} \in V$

(ii) $T(a\mathbf{v}) = aT(\mathbf{v})$ for all $a \in \mathbb{F}$, $\mathbf{v} \in V$

Equivalently: $T(a\mathbf{u} + b\mathbf{v}) = aT(\mathbf{u}) + bT(\mathbf{v})$ for all $a, b \in \mathbb{F}$ and $\mathbf{u}, \mathbf{v} \in V$.

The set of all linear maps from V to W is denoted $\mathcal{L}(V, W)$ or $\mathrm{Hom}(V, W)$.

Proposition 2.2.1 (Basic properties of linear maps). *Let $T : V \to W$ be linear.*

(i) $T(\mathbf{0}) = \mathbf{0}$.

(ii) $T(-\mathbf{v}) = -T(\mathbf{v})$.

(iii) $T\left(\sum_{i=1}^n a_i \mathbf{v}_i\right) = \sum_{i=1}^n a_i T(\mathbf{v}_i)$.

(iv) T is completely determined by its values on a basis.

Theorem 2.2.2 (Construction of linear maps). *Let $\mathcal{B} = \{\mathbf{v}_1, \ldots, \mathbf{v}_n\}$ be a basis for V, and let $\mathbf{w}_1, \ldots, \mathbf{w}_n$ be any vectors in W. Then there exists a unique linear map $T : V \to W$ such that $T(\mathbf{v}_i) = \mathbf{w}_i$ for each i.*

Example 2.2.1 (Important linear maps).

(a) **Zero map**: $T(\mathbf{v}) = \mathbf{0}$ for all $\mathbf{v}$.

(b) **Identity map**: $I(\mathbf{v}) = \mathbf{v}$.

(c) **Scaling**: $T(\mathbf{v}) = a\mathbf{v}$ for fixed $a \in \mathbb{F}$.

(d) **Projection**: $P : \mathbb{R}^3 \to \mathbb{R}^2$ by $P(x, y, z) = (x, y)$.

(e) **Differentiation**: $D : C^1(\mathbb{R}) \to C(\mathbb{R})$ by $Df = f'$.

(f) **Integration**: $T : C([0, 1]) \to \mathbb{R}$ by $Tf = \int_0^1 f(x)\,dx$.

(g) **Rotation**: $R_\theta : \mathbb{R}^2 \to \mathbb{R}^2$ by angle θ.

2.2.2 Kernel and Image

Definition 2.2.2 (Kernel and image). For a linear map $T : V \to W$:

- The **kernel** (or **null space**) is $\ker T := \{\mathbf{v} \in V : T(\mathbf{v}) = \mathbf{0}\}$.
- The **image** (or **range**) is $\operatorname{im} T := \{T(\mathbf{v}) : \mathbf{v} \in V\}$.

Proposition 2.2.3. $\ker T$ *is a subspace of V, and* $\operatorname{im} T$ *is a subspace of W.*

Definition 2.2.3 (Nullity and rank). The **nullity** of T is $\operatorname{null}(T) := \dim(\ker T)$. The **rank** of T is $\operatorname{rank}(T) := \dim(\operatorname{im} T)$.

Theorem 2.2.4 (Rank-Nullity Theorem). *If $T : V \to W$ is linear and V is finite-dimensional, then*

$$\dim V = \operatorname{null}(T) + \operatorname{rank}(T) = \dim(\ker T) + \dim(\operatorname{im} T).$$

Proof. Let $\{\mathbf{u}_1, \ldots, \mathbf{u}_k\}$ be a basis for $\ker T$. Extend to a basis $\{\mathbf{u}_1, \ldots, \mathbf{u}_k, \mathbf{v}_1, \ldots, \mathbf{v}_r\}$ for V. We claim $\{T(\mathbf{v}_1), \ldots, T(\mathbf{v}_r)\}$ is a basis for $\operatorname{im} T$.

Spanning: Any $\mathbf{w} \in \operatorname{im} T$ equals $T(\mathbf{x})$ for some $\mathbf{x} = \sum a_i\mathbf{u}_i + \sum b_j\mathbf{v}_j$. Then $\mathbf{w} = T(\mathbf{x}) = \sum b_j T(\mathbf{v}_j)$ since $T(\mathbf{u}_i) = \mathbf{0}$.

Independence: If $\sum c_j T(\mathbf{v}_j) = \mathbf{0}$, then $T(\sum c_j\mathbf{v}_j) = \mathbf{0}$, so $\sum c_j\mathbf{v}_j \in \ker T$. Thus $\sum c_j\mathbf{v}_j = \sum d_i\mathbf{u}_i$ for some d_i. By independence of the full basis, all $c_j = 0$.

Hence $\dim(\operatorname{im} T) = r$ and $\dim V = k + r = \dim(\ker T) + \dim(\operatorname{im} T)$. □

Corollary 2.2.5. *Let $T : V \to W$ with* $\dim V = \dim W = n < \infty$. *Then:*

$$T \text{ is injective} \quad \Leftrightarrow \quad T \text{ is surjective} \quad \Leftrightarrow \quad T \text{ is bijective.}$$

2.2.3 Isomorphisms

Definition 2.2.4 (Isomorphism). A linear map $T : V \to W$ is an **isomorphism** if it is bijective (injective and surjective). We say V and W are **isomorphic**, written $V \cong W$.

Proposition 2.2.6 (Characterizations of isomorphism). *For a linear map $T : V \to W$:*

(i) T is injective iff $\ker T = \{\mathbf{0}\}$.

(ii) T is surjective iff $\operatorname{im} T = W$.

(iii) T is an isomorphism iff T has a linear inverse $T^{-1} : W \to V$.

Theorem 2.2.7 (Classification by dimension). *Two finite-dimensional vector spaces over the same field are isomorphic if and only if they have the same dimension. In particular, every n-dimensional space over $\mathbb{F}$ is isomorphic to $\mathbb{F}^n$.*

This theorem justifies using $\mathbb{F}^n$ as the "standard model" for n-dimensional spaces.

2.2.4 The Space of Linear Maps

Proposition 2.2.8. *$\mathcal{L}(V, W)$ is itself a vector space under:*

- $(S + T)(\mathbf{v}) := S(\mathbf{v}) + T(\mathbf{v})$
- $(aT)(\mathbf{v}) := a \cdot T(\mathbf{v})$

Proposition 2.2.9. *If* $\dim V = n$ *and* $\dim W = m$, *then* $\dim \mathcal{L}(V, W) = mn$.

Definition 2.2.5 (Composition). If $T : U \to V$ and $S : V \to W$ are linear, then $S \circ T : U \to W$ is linear.

Composition is associative but generally not commutative.

2.3 Matrices

2.3.1 Matrix Representation of Linear Maps

Let $T : V \to W$ be linear, with ordered bases $\mathcal{B} = \{\mathbf{v}_1, \ldots, \mathbf{v}_n\}$ for V and $\mathcal{C} = \{\mathbf{w}_1, \ldots, \mathbf{w}_m\}$ for W.

Each $T(\mathbf{v}_j)$ can be written uniquely as a linear combination of the $\mathbf{w}_i$:

$$T(\mathbf{v}_j) = \sum_{i=1}^{m} a_{ij} \mathbf{w}_i.$$

Definition 2.3.1 (Matrix of a linear map). The **matrix of** T with respect to $\mathcal{B}$ and $\mathcal{C}$ is the $m \times n$ matrix

$$[T]_{\mathcal{B}}^{\mathcal{C}} := (a_{ij}) = \begin{pmatrix} a_{11} & a_{12} & \cdots & a_{1n} \\ a_{21} & a_{22} & \cdots & a_{2n} \\ \vdots & \vdots & \ddots & \vdots \\ a_{m1} & a_{m2} & \cdots & a_{mn} \end{pmatrix}$$

where the j-th column is $[T(\mathbf{v}_j)]_{\mathcal{C}}$.

Theorem 2.3.1 (Matrix-vector multiplication). *For any* $\mathbf{v} \in V$*:*

$$[T(\mathbf{v})]_{\mathcal{C}} = [T]_{\mathcal{B}}^{\mathcal{C}} \cdot [\mathbf{v}]_{\mathcal{B}}.$$

This explains why matrix multiplication is defined the way it is.

2.3.2 Matrix Operations

Definition 2.3.2 (Matrix multiplication). If A is $m \times n$ and B is $n \times p$, their product AB is the $m \times p$ matrix with entries

$$(AB)_{ij} = \sum_{k=1}^{n} A_{ik} B_{kj}.$$

Theorem 2.3.2 (Correspondence). *The map* $T \mapsto [T]_{\mathcal{B}}^{\mathcal{C}}$ *is an isomorphism* $\mathcal{L}(V, W) \to M_{m \times n}(\mathbb{F})$*. Moreover:*

(i) $[S + T] = [S] + [T]$

(ii) $[aT] = a[T]$

(iii) $[S \circ T] = [S][T]$ *(composition corresponds to matrix multiplication)*

Definition 2.3.3 (Matrix transpose). The **transpose** of $A = (a_{ij})$ is $A^T = (a_{ji})$, obtained by reflecting across the diagonal.

Proposition 2.3.3 (Properties of transpose).

(i) $(A^T)^T = A$

(ii) $(A + B)^T = A^T + B^T$

(iii) $(aA)^T = aA^T$

(iv) $(AB)^T = B^T A^T$ *(note the order reversal)*

2.3.3 Change of Basis

Definition 2.3.4 (Change of basis matrix). Let $\mathcal{B} = \{\mathbf{v}_1, \ldots, \mathbf{v}_n\}$ and $\mathcal{B}' = \{\mathbf{v}'_1, \ldots, \mathbf{v}'_n\}$ be bases for V. The **change of basis matrix** from $\mathcal{B}'$ to $\mathcal{B}$ is

$$P_{\mathcal{B}' \to \mathcal{B}} := [I]^{\mathcal{B}}_{\mathcal{B}'}$$

whose j-th column is $[\mathbf{v}'_j]_{\mathcal{B}}$.

Proposition 2.3.4. *For any* $\mathbf{v} \in V$*:*

$$[\mathbf{v}]_{\mathcal{B}} = P_{\mathcal{B}' \to \mathcal{B}} \cdot [\mathbf{v}]_{\mathcal{B}'}.$$

Theorem 2.3.5 (Similarity). *Let $T : V \to V$ be a linear operator, with $[T]_{\mathcal{B}} = A$ and $[T]_{\mathcal{B}'} = A'$. If $P = P_{\mathcal{B}' \to \mathcal{B}}$, then*

$$A' = P^{-1}AP.$$

*Matrices related this way are called **similar**.*

Similar matrices represent the same linear transformation in different bases. Many important matrix properties—determinant, trace, eigenvalues, rank—are preserved under similarity.

2.3.4 Elementary Operations and Row Reduction

Definition 2.3.5 (Elementary row operations). The three types are:

(i) Swap two rows: $R_i \leftrightarrow R_j$

(ii) Multiply a row by a nonzero scalar: $R_i \to cR_i$

(iii) Add a multiple of one row to another: $R_i \to R_i + cR_j$

Each elementary operation corresponds to left-multiplication by an **elementary matrix**. Elementary matrices are invertible, and every invertible matrix is a product of elementary matrices.

Definition 2.3.6 (Row echelon form). A matrix is in **row echelon form** (REF) if:

(i) All zero rows are at the bottom.

(ii) The leading entry (pivot) of each nonzero row is to the right of the pivot above it.

It is in **reduced row echelon form** (RREF) if additionally:

(iii) Each pivot is 1.

(iv) Each pivot is the only nonzero entry in its column.

Theorem 2.3.6. *Every matrix can be reduced to a unique RREF by elementary row operations (**Gaussian elimination**).*

Proposition 2.3.7 (Rank from RREF). *The rank of a matrix equals the number of pivots in its RREF.*

2.4 Determinants

2.4.1 Definition and Properties

The determinant is a function $\det : M_{n\times n}(\mathbb{F}) \to \mathbb{F}$ characterized by several equivalent definitions.

Definition 2.4.1 (Determinant via permutations)**.** For an $n \times n$ matrix $A = (a_{ij})$:

$$\det A = \sum_{\sigma \in S_n} \operatorname{sgn}(\sigma) \prod_{i=1}^{n} a_{i,\sigma(i)}$$

where S_n is the symmetric group (all permutations of $\{1, \ldots, n\}$) and $\operatorname{sgn}(\sigma) = \pm 1$ is the sign of the permutation.

Example 2.4.1. For $n = 2$: $\det \begin{pmatrix} a & b \\ c & d \end{pmatrix} = ad - bc$.

For $n = 3$: $\det \begin{pmatrix} a & b & c \\ d & e & f \\ g & h & i \end{pmatrix} = aei + bfg + cdh - ceg - bdi - afh.$

Theorem 2.4.1 (Properties of determinant)**.** *(i)* $\det I = 1$.

(ii) Swapping two rows multiplies $\det$ *by* -1.

(iii) Multiplying a row by c *multiplies* $\det$ *by* c.

(iv) Adding a multiple of one row to another leaves $\det$ *unchanged.*

(v) $\det(AB) = \det A \cdot \det B$.

(vi) $\det A^T = \det A$.

(vii) A is invertible iff $\det A \neq 0$*, in which case* $\det(A^{-1}) = (\det A)^{-1}$.

Properties (i)–(iv) can alternatively be taken as axioms defining the determinant.

Theorem 2.4.2 (Cofactor expansion). *Let A_{ij} denote the $(n-1)\times(n-1)$ matrix obtained by deleting row i and column j. The **cofactor** is $C_{ij} = (-1)^{i+j}\det A_{ij}$. Then:*

$$\det A = \sum_{j=1}^{n} a_{ij}C_{ij} \quad \textit{(expansion along row } i\textit{)}$$

$$\det A = \sum_{i=1}^{n} a_{ij}C_{ij} \quad \textit{(expansion along column } j\textit{)}$$

2.4.2 Geometric Interpretation

Theorem 2.4.3 (Geometric meaning). *If the columns of A are $\mathbf{v}_1,\dots,\mathbf{v}_n \in \mathbb{R}^n$, then $|\det A|$ equals the n-dimensional volume of the parallelepiped spanned by these vectors. The sign indicates orientation.*

For $n=2$: $|\det A|$ is the area of the parallelogram. For $n=3$: $|\det A|$ is the volume of the parallelepiped.

Corollary 2.4.4. *A linear map $T:\mathbb{R}^n\to\mathbb{R}^n$ scales volumes by the factor $|\det[T]|$.*

2.4.3 Cramer's Rule and the Adjugate

Definition 2.4.2 (Adjugate matrix). The **adjugate** (or **classical adjoint**) of A is the matrix $\operatorname{adj}(A)$ whose (i,j) entry is the cofactor C_{ji} (note the transposition).

Theorem 2.4.5. $A\cdot\operatorname{adj}(A) = \operatorname{adj}(A)\cdot A = (\det A)I$.
Hence, if $\det A\neq 0$: $A^{-1} = \frac{1}{\det A}\operatorname{adj}(A)$.

Theorem 2.4.6 (Cramer's rule). *If A is invertible and $A\mathbf{x}=\mathbf{b}$, then*

$$x_j = \frac{\det A_j}{\det A}$$

where A_j is A with column j replaced by $\mathbf{b}$.

Cramer's rule is elegant but computationally inefficient for large systems.

2.5 Inner Product Spaces

2.5.1 Inner Products

Definition 2.5.1 (Inner product). An **inner product** on a real vector space V is a function $\langle \cdot, \cdot \rangle : V \times V \to \mathbb{R}$ satisfying:

(i) $\langle \mathbf{u}, \mathbf{v} \rangle = \langle \mathbf{v}, \mathbf{u} \rangle$ (symmetry)

(ii) $\langle a\mathbf{u} + b\mathbf{v}, \mathbf{w} \rangle = a\langle \mathbf{u}, \mathbf{w} \rangle + b\langle \mathbf{v}, \mathbf{w} \rangle$ (linearity)

(iii) $\langle \mathbf{v}, \mathbf{v} \rangle \geq 0$, with equality iff $\mathbf{v} = \mathbf{0}$ (positive definiteness)

Definition 2.5.2 (Inner product on complex spaces). For a complex vector space, an inner product satisfies:

(i) $\langle \mathbf{u}, \mathbf{v} \rangle = \overline{\langle \mathbf{v}, \mathbf{u} \rangle}$ (conjugate symmetry)

(ii) $\langle a\mathbf{u} + b\mathbf{v}, \mathbf{w} \rangle = a\langle \mathbf{u}, \mathbf{w} \rangle + b\langle \mathbf{v}, \mathbf{w} \rangle$ (linearity in first argument)

(iii) $\langle \mathbf{v}, \mathbf{v} \rangle \geq 0$ real, with equality iff $\mathbf{v} = \mathbf{0}$

Note: $\langle \mathbf{u}, c\mathbf{v} \rangle = \bar{c}\langle \mathbf{u}, \mathbf{v} \rangle$ (conjugate-linearity in second argument).

A vector space with an inner product is called an **inner product space**.

Example 2.5.1 (Standard inner products).

(a) On $\mathbb{R}^n$: $\langle \mathbf{x}, \mathbf{y} \rangle = \sum_{i=1}^n x_i y_i = \mathbf{x}^T \mathbf{y}$ (dot product)

(b) On $\mathbb{C}^n$: $\langle \mathbf{x}, \mathbf{y} \rangle = \sum_{i=1}^n x_i \bar{y}_i = \mathbf{x}^\dagger \mathbf{y}$

(c) On $C([a, b])$: $\langle f, g \rangle = \int_a^b f(x)g(x)\, dx$

(d) On $C([a, b])$ with weight: $\langle f, g \rangle_w = \int_a^b f(x)g(x)w(x)\, dx$ where $w > 0$

2.5.2 Norms and Distance

Definition 2.5.3 (Norm). The **norm** induced by an inner product is

$$\|\mathbf{v}\| := \sqrt{\langle \mathbf{v}, \mathbf{v} \rangle}.$$

Theorem 2.5.1 (Cauchy-Schwarz inequality). *For any* $\mathbf{u}, \mathbf{v}$ *in an inner product space:*

$$|\langle \mathbf{u}, \mathbf{v} \rangle| \leq \|\mathbf{u}\| \cdot \|\mathbf{v}\|$$

with equality iff $\mathbf{u}$ *and* $\mathbf{v}$ *are linearly dependent.*

Proof. If $\mathbf{v} = \mathbf{0}$, both sides are zero. Otherwise, for any $t \in \mathbb{R}$ (or $\mathbb{C}$):

$$0 \leq \|\mathbf{u} - t\mathbf{v}\|^2 = \|\mathbf{u}\|^2 - 2t\,\mathrm{Re}\langle \mathbf{u}, \mathbf{v}\rangle + t^2\|\mathbf{v}\|^2.$$

This quadratic in t is nonnegative, so its discriminant is ≤ 0:

$$4(\mathrm{Re}\langle \mathbf{u}, \mathbf{v}\rangle)^2 \leq 4\|\mathbf{u}\|^2\|\mathbf{v}\|^2.$$

Applying this to $e^{i\theta}\mathbf{u}$ where θ is chosen so $e^{i\theta}\langle \mathbf{u}, \mathbf{v}\rangle$ is real and positive gives the result. □

Corollary 2.5.2 (Triangle inequality). $\|\mathbf{u} + \mathbf{v}\| \leq \|\mathbf{u}\| + \|\mathbf{v}\|$.

Proposition 2.5.3 (Norm properties). *Any norm satisfies:*

(i) $\|\mathbf{v}\| \geq 0$, *with equality iff* $\mathbf{v} = \mathbf{0}$

(ii) $\|a\mathbf{v}\| = |a|\|\mathbf{v}\|$

(iii) $\|\mathbf{u} + \mathbf{v}\| \leq \|\mathbf{u}\| + \|\mathbf{v}\|$

Definition 2.5.4 (Distance). The **distance** between $\mathbf{u}$ and $\mathbf{v}$ is $d(\mathbf{u}, \mathbf{v}) := \|\mathbf{u} - \mathbf{v}\|$.

Theorem 2.5.4 (Parallelogram law).

$$\|\mathbf{u} + \mathbf{v}\|^2 + \|\mathbf{u} - \mathbf{v}\|^2 = 2\|\mathbf{u}\|^2 + 2\|\mathbf{v}\|^2.$$

A norm satisfies the parallelogram law iff it comes from an inner product.

2.5.3 Orthogonality

Definition 2.5.5 (Orthogonality). Vectors $\mathbf{u}$ and $\mathbf{v}$ are **orthogonal**, written $\mathbf{u} \perp \mathbf{v}$, if $\langle \mathbf{u}, \mathbf{v}\rangle = 0$.

A set S is **orthogonal** if any two distinct vectors in S are orthogonal.

A set S is **orthonormal** if it is orthogonal and every vector has norm 1.

Theorem 2.5.5 (Pythagorean theorem). *If* $\mathbf{u} \perp \mathbf{v}$, *then* $\|\mathbf{u} + \mathbf{v}\|^2 = \|\mathbf{u}\|^2 + \|\mathbf{v}\|^2$.

Proposition 2.5.6. *An orthogonal set of nonzero vectors is linearly independent.*

Definition 2.5.6 (Orthonormal basis). An **orthonormal basis** (ONB) is a basis that is an orthonormal set.

Proposition 2.5.7. *If $\{\mathbf{e}_1, \ldots, \mathbf{e}_n\}$ is an orthonormal basis, then for any $\mathbf{v}$:*

$$\mathbf{v} = \sum_{i=1}^{n} \langle \mathbf{v}, \mathbf{e}_i \rangle \mathbf{e}_i, \qquad \|\mathbf{v}\|^2 = \sum_{i=1}^{n} |\langle \mathbf{v}, \mathbf{e}_i \rangle|^2.$$

The second formula is **Parseval's identity**.

Theorem 2.5.8 (Gram-Schmidt process). *Given any basis $\{\mathbf{v}_1, \ldots, \mathbf{v}_n\}$, there exists an orthonormal basis $\{\mathbf{e}_1, \ldots, \mathbf{e}_n\}$ with* $\mathrm{span}\{\mathbf{e}_1, \ldots, \mathbf{e}_k\} = \mathrm{span}\{\mathbf{v}_1, \ldots, \mathbf{v}_k\}$ *for each k.*

Construction. Set $\mathbf{u}_1 = \mathbf{v}_1$. For $k \geq 2$:

$$\mathbf{u}_k = \mathbf{v}_k - \sum_{j=1}^{k-1} \frac{\langle \mathbf{v}_k, \mathbf{u}_j \rangle}{\langle \mathbf{u}_j, \mathbf{u}_j \rangle} \mathbf{u}_j.$$

Then $\mathbf{e}_k = \mathbf{u}_k / \|\mathbf{u}_k\|$. □

Corollary 2.5.9. *Every finite-dimensional inner product space has an orthonormal basis.*

2.5.4 Orthogonal Complements and Projections

Definition 2.5.7 (Orthogonal complement). The **orthogonal complement** of a subset $S \subseteq V$ is

$$S^\perp := \{\mathbf{v} \in V : \langle \mathbf{v}, \mathbf{s} \rangle = 0 \text{ for all } \mathbf{s} \in S\}.$$

Proposition 2.5.10. *$S^\perp$ is always a subspace. If W is a subspace, then $W \cap W^\perp = \{\mathbf{0}\}$.*

Theorem 2.5.11 (Orthogonal decomposition). *If V is finite-dimensional and W is a subspace, then $V = W \oplus W^\perp$. That is, every $\mathbf{v} \in V$ can be written uniquely as $\mathbf{v} = \mathbf{w} + \mathbf{w}^\perp$ with $\mathbf{w} \in W$ and $\mathbf{w}^\perp \in W^\perp$.*

Definition 2.5.8 (Orthogonal projection). The **orthogonal projection** of $\mathbf{v}$ onto W is the component $\mathbf{w}$ in the above decomposition. We write $P_W(\mathbf{v}) = \mathbf{w}$.

Proposition 2.5.12. *If $\{\mathbf{e}_1, \ldots, \mathbf{e}_k\}$ is an orthonormal basis for W, then*

$$P_W(\mathbf{v}) = \sum_{i=1}^{k} \langle \mathbf{v}, \mathbf{e}_i \rangle \mathbf{e}_i.$$

Theorem 2.5.13 (Best approximation). *$P_W(\mathbf{v})$ is the unique element of W closest to $\mathbf{v}$:*

$$\|\mathbf{v} - P_W(\mathbf{v})\| < \|\mathbf{v} - \mathbf{w}\| \quad \textit{for all } \mathbf{w} \in W, \mathbf{w} \neq P_W(\mathbf{v}).$$

This theorem underlies least-squares approximation and Fourier series.

2.6 Eigenvalues and Eigenvectors

2.6.1 Definitions

Let $T : V \to V$ be a linear operator on a finite-dimensional space, or equivalently, let A be an $n \times n$ matrix.

Definition 2.6.1 (Eigenvalue and eigenvector). A scalar $\lambda \in \mathbb{F}$ is an **eigenvalue** of T if there exists a nonzero vector $\mathbf{v}$ such that

$$T(\mathbf{v}) = \lambda\mathbf{v}.$$

Such a nonzero $\mathbf{v}$ is called an **eigenvector** corresponding to λ.

Equivalently, λ is an eigenvalue of A iff $A\mathbf{v} = \lambda\mathbf{v}$ for some $\mathbf{v} \neq \mathbf{0}$.

Definition 2.6.2 (Eigenspace). The **eigenspace** of λ is

$$E_\lambda := \ker(T - \lambda I) = \{\mathbf{v} \in V : T(\mathbf{v}) = \lambda\mathbf{v}\}.$$

This includes the zero vector.

Proposition 2.6.1. *E_λ is a subspace of V. Its dimension is the* ***geometric multiplicity*** *of λ.*

2.6.2 The Characteristic Polynomial

Theorem 2.6.2. *λ is an eigenvalue of A iff* $\det(A - \lambda I) = 0$.

Proof. λ is an eigenvalue iff $(A - \lambda I)\mathbf{v} = \mathbf{0}$ has a nonzero solution iff $A - \lambda I$ is singular iff $\det(A - \lambda I) = 0$. □

Definition 2.6.3 (Characteristic polynomial). The **characteristic polynomial** of A is

$$p_A(\lambda) := \det(\lambda I - A).$$

(Some texts use $\det(A - \lambda I)$, which differs by a sign.)

The characteristic polynomial has degree n and leading coefficient 1:

$$p_A(\lambda) = \lambda^n - (\operatorname{tr} A)\lambda^{n-1} + \cdots + (-1)^n \det A.$$

Theorem 2.6.3 (Eigenvalues are roots). *The eigenvalues of A are exactly the roots of $p_A(\lambda)$.*

Definition 2.6.4 (Algebraic multiplicity). The **algebraic multiplicity** of an eigenvalue λ is its multiplicity as a root of $p_A(\lambda)$.

Proposition 2.6.4. *For any eigenvalue* λ*:*

$$1 \leq \textit{geometric multiplicity} \leq \textit{algebraic multiplicity}.$$

Theorem 2.6.5 (Invariance under similarity)**.** *Similar matrices have the same characteristic polynomial, hence the same eigenvalues with the same algebraic multiplicities.*

Example 2.6.1. For $A = \begin{pmatrix} 4 & -2 \\ 1 & 1 \end{pmatrix}$:

$$p_A(\lambda) = \det \begin{pmatrix} \lambda - 4 & 2 \\ -1 & \lambda - 1 \end{pmatrix} = (\lambda-4)(\lambda-1)+2 = \lambda^2-5\lambda+6 = (\lambda-2)(\lambda-3).$$

Eigenvalues: $\lambda_1 = 2$, $\lambda_2 = 3$.

For $\lambda = 2$: solve $(A - 2I)\mathbf{v} = \mathbf{0}$, i.e., $\begin{pmatrix} 2 & -2 \\ 1 & -1 \end{pmatrix} \mathbf{v} = \mathbf{0}$. Solution: $\mathbf{v} = t\begin{pmatrix} 1 \\ 1 \end{pmatrix}$.

For $\lambda = 3$: solve $(A - 3I)\mathbf{v} = \mathbf{0}$, i.e., $\begin{pmatrix} 1 & -2 \\ 1 & -2 \end{pmatrix} \mathbf{v} = \mathbf{0}$. Solution: $\mathbf{v} = t\begin{pmatrix} 2 \\ 1 \end{pmatrix}$.

2.6.3 Diagonalization

Definition 2.6.5 (Diagonalizable)**.** A matrix A (or operator T) is **diagonalizable** if it is similar to a diagonal matrix, i.e., there exists an invertible P such that $P^{-1}AP$ is diagonal.

Theorem 2.6.6 (Diagonalization criterion)**.** *A is diagonalizable iff V has a basis of eigenvectors of A. In this case, if P has these eigenvectors as columns, then $P^{-1}AP = \operatorname{diag}(\lambda_1, \ldots, \lambda_n)$.*

Proposition 2.6.7 (Independence of eigenvectors)**.** *Eigenvectors corresponding to distinct eigenvalues are linearly independent.*

Corollary 2.6.8. *If A has n distinct eigenvalues (where $n = \dim V$), then A is diagonalizable.*

Theorem 2.6.9 (Diagonalizability conditions)**.** *A is diagonalizable iff for each eigenvalue, the geometric multiplicity equals the algebraic multiplicity.*

Example 2.6.2 (Non-diagonalizable matrix)**.** $A = \begin{pmatrix} 1 & 1 \\ 0 & 1 \end{pmatrix}$ has $p_A(\lambda) = (\lambda - 1)^2$, so $\lambda = 1$ with algebraic multiplicity 2. But $E_1 = \ker(A - I) = \ker \begin{pmatrix} 0 & 1 \\ 0 & 0 \end{pmatrix} = \operatorname{span}\{(1, 0)\}$ has dimension 1. So A is not diagonalizable.

2.6.4 The Cayley-Hamilton Theorem

Theorem 2.6.10 (Cayley-Hamilton). *Every matrix satisfies its own characteristic polynomial: if $p_A(\lambda) = \det(\lambda I - A)$, then $p_A(A) = 0$.*

Example 2.6.3. For $A = \begin{pmatrix} 1 & 2 \\ 3 & 4 \end{pmatrix}$: $p_A(\lambda) = \lambda^2 - 5\lambda - 2$.

Then $A^2 - 5A - 2I = \begin{pmatrix} 7 & 10 \\ 15 & 22 \end{pmatrix} - \begin{pmatrix} 5 & 10 \\ 15 & 20 \end{pmatrix} - \begin{pmatrix} 2 & 0 \\ 0 & 2 \end{pmatrix} = \begin{pmatrix} 0 & 0 \\ 0 & 0 \end{pmatrix}$.

Corollary 2.6.11. *A^{-1} (if it exists) can be expressed as a polynomial in A of degree at most $n - 1$.*

2.7 Special Classes of Matrices and Operators

2.7.1 Self-Adjoint and Hermitian Operators

Definition 2.7.1 (Adjoint operator). Let V be an inner product space and $T : V \to V$ linear. The **adjoint** of T is the unique operator $T^* : V \to V$ satisfying

$$\langle T\mathbf{u}, \mathbf{v} \rangle = \langle \mathbf{u}, T^*\mathbf{v} \rangle \quad \text{for all } \mathbf{u}, \mathbf{v} \in V.$$

Proposition 2.7.1. *In terms of matrices with respect to an orthonormal basis:*

- *Over $\mathbb{R}$: $[T^*] = [T]^T$ (transpose)*
- *Over $\mathbb{C}$: $[T^*] = [T]^\dagger = \overline{[T]}^T$ (conjugate transpose)*

Definition 2.7.2 (Special matrices).

- A is **symmetric** if $A^T = A$ (real case).
- A is **Hermitian** (or **self-adjoint**) if $A^\dagger = A$ (complex case).
- A is **orthogonal** if $A^T A = I$ (real case).
- A is **unitary** if $A^\dagger A = I$ (complex case).
- A is **normal** if $A^\dagger A = AA^\dagger$.

Proposition 2.7.2 (Properties).

(i) Hermitian matrices have real eigenvalues.

(ii) Eigenvectors of a Hermitian matrix corresponding to distinct eigenvalues are orthogonal.

(iii) Unitary matrices preserve inner products: $\langle U\mathbf{u}, U\mathbf{v}\rangle = \langle \mathbf{u}, \mathbf{v}\rangle$.

(iv) Eigenvalues of unitary matrices have absolute value 1.

Theorem 2.7.3 (Spectral theorem for Hermitian matrices). *If A is Hermitian (or real symmetric), then:*

(i) All eigenvalues are real.

(ii) A is diagonalizable.

(iii) There exists an orthonormal basis of eigenvectors.

(iv) $A = UDU^\dagger$ where U is unitary and D is real diagonal.

This theorem is fundamental to quantum mechanics, where observables are Hermitian operators.

Theorem 2.7.4 (Spectral theorem for normal matrices). *A matrix is unitarily diagonalizable (i.e., $A = UDU^\dagger$ with U unitary) iff A is normal.*

2.7.2 Positive Definite Matrices

Definition 2.7.3 (Positive definite). A Hermitian matrix A is:

- **Positive definite** if $\mathbf{x}^\dagger A\mathbf{x} > 0$ for all $\mathbf{x} \neq \mathbf{0}$.
- **Positive semidefinite** if $\mathbf{x}^\dagger A\mathbf{x} \geq 0$ for all $\mathbf{x}$.
- **Negative definite/semidefinite**: reverse the inequalities.

Theorem 2.7.5 (Characterizations of positive definiteness). *For Hermitian A, the following are equivalent:*

(i) A is positive definite.

(ii) All eigenvalues of A are positive.

(iii) $A = B^\dagger B$ for some invertible B.

(iv) All leading principal minors are positive (Sylvester's criterion).

2.8 Jordan Normal Form

Not every matrix is diagonalizable, but every matrix over $\mathbb{C}$ can be brought to a nearly-diagonal form.

Definition 2.8.1 (Jordan block). A **Jordan block** $J_k(\lambda)$ is the $k \times k$ matrix

$$J_k(\lambda) = \begin{pmatrix} \lambda & 1 & 0 & \cdots & 0 \\ 0 & \lambda & 1 & \cdots & 0 \\ \vdots & & \ddots & \ddots & \vdots \\ 0 & 0 & \cdots & \lambda & 1 \\ 0 & 0 & \cdots & 0 & \lambda \end{pmatrix}.$$

Theorem 2.8.1 (Jordan normal form). *Every matrix A over $\mathbb{C}$ is similar to a block-diagonal matrix*

$$J = \begin{pmatrix} J_{k_1}(\lambda_1) & & 0 \\ & \ddots & \\ 0 & & J_{k_r}(\lambda_r) \end{pmatrix}$$

where each λ_i is an eigenvalue of A. This J is unique up to reordering of blocks.

The Jordan form reveals the complete structure of a linear operator. The matrix A is diagonalizable iff all Jordan blocks are 1×1.

Remark 2.8.1 (Physical significance). Jordan blocks arise in systems with "generalized eigenvectors." In physics, this occurs in degenerate perturbation theory and in systems at exceptional points.

2.9 Bilinear and Quadratic Forms

2.9.1 Bilinear Forms

Definition 2.9.1 (Bilinear form). A **bilinear form** on a real vector space V is a function $B : V \times V \to \mathbb{R}$ that is linear in each argument:

$$\begin{aligned} B(a\mathbf{u} + b\mathbf{v}, \mathbf{w}) &= aB(\mathbf{u}, \mathbf{w}) + bB(\mathbf{v}, \mathbf{w}) \\ B(\mathbf{u}, a\mathbf{v} + b\mathbf{w}) &= aB(\mathbf{u}, \mathbf{v}) + bB(\mathbf{u}, \mathbf{w}) \end{aligned}$$

With respect to a basis, $B(\mathbf{u}, \mathbf{v}) = \mathbf{u}^T M \mathbf{v}$ for some matrix M.

Definition 2.9.2 (Properties of bilinear forms). A bilinear form B is:

- **Symmetric** if $B(\mathbf{u}, \mathbf{v}) = B(\mathbf{v}, \mathbf{u})$.

- **Antisymmetric** (or **skew-symmetric**) if $B(\mathbf{u}, \mathbf{v}) = -B(\mathbf{v}, \mathbf{u})$.
- **Non-degenerate** if $B(\mathbf{u}, \mathbf{v}) = 0$ for all $\mathbf{v}$ implies $\mathbf{u} = \mathbf{0}$.

An inner product is a positive definite symmetric bilinear form.

2.9.2 Quadratic Forms

Definition 2.9.3 (Quadratic form)**.** A **quadratic form** on V is a function $Q : V \to \mathbb{R}$ of the form $Q(\mathbf{v}) = B(\mathbf{v}, \mathbf{v})$ for some symmetric bilinear form B.

In coordinates: $Q(\mathbf{x}) = \mathbf{x}^T A \mathbf{x} = \sum_{i,j} a_{ij} x_i x_j$ where A is symmetric.

Theorem 2.9.1 (Classification of quadratic forms)**.** *Every real quadratic form can be reduced by a change of basis to*

$$Q(\mathbf{y}) = y_1^2 + \cdots + y_p^2 - y_{p+1}^2 - \cdots - y_{p+q}^2$$

where $p + q \leq n$. The pair (p, q), called the ***signature****, is an invariant.*

This is **Sylvester's law of inertia**. The form is:

- Positive definite iff $p = n$, $q = 0$
- Negative definite iff $p = 0$, $q = n$
- Indefinite if $p > 0$ and $q > 0$

Example 2.9.1 (Minkowski metric)**.** The metric of special relativity has signature $(1, 3)$ or $(3, 1)$ depending on convention:

$$ds^2 = c^2 dt^2 - dx^2 - dy^2 - dz^2 \quad \text{or} \quad ds^2 = -c^2 dt^2 + dx^2 + dy^2 + dz^2.$$

2.10 Tensor Products and Multilinear Algebra

2.10.1 Motivation

In quantum mechanics, the state space of a composite system is not the direct sum but the *tensor product* of component state spaces. Understanding tensor products is essential for quantum information, particle physics, and differential geometry.

2.10.2 The Tensor Product of Vector Spaces

Definition 2.10.1 (Tensor product). The **tensor product** of vector spaces V and W, denoted $V \otimes W$, is a vector space together with a bilinear map $\otimes : V \times W \to V \otimes W$ satisfying the universal property: for any bilinear map $B : V \times W \to U$, there exists a unique linear map $\tilde{B} : V \otimes W \to U$ such that $B = \tilde{B} \circ \otimes$.

Theorem 2.10.1 (Existence and basis). *$V \otimes W$ exists and is unique up to isomorphism. If $\{\mathbf{e}_i\}$ is a basis for V and $\{\mathbf{f}_j\}$ is a basis for W, then $\{\mathbf{e}_i \otimes \mathbf{f}_j\}$ is a basis for $V \otimes W$.*

Corollary 2.10.2. $\dim(V \otimes W) = \dim V \cdot \dim W$.

Proposition 2.10.3 (Properties).

(i) $(a\mathbf{v}) \otimes \mathbf{w} = \mathbf{v} \otimes (a\mathbf{w}) = a(\mathbf{v} \otimes \mathbf{w})$

(ii) $(\mathbf{v}_1 + \mathbf{v}_2) \otimes \mathbf{w} = \mathbf{v}_1 \otimes \mathbf{w} + \mathbf{v}_2 \otimes \mathbf{w}$

(iii) $\mathbf{v} \otimes (\mathbf{w}_1 + \mathbf{w}_2) = \mathbf{v} \otimes \mathbf{w}_1 + \mathbf{v} \otimes \mathbf{w}_2$

(iv) $V \otimes W \cong W \otimes V$

(v) $(U \otimes V) \otimes W \cong U \otimes (V \otimes W)$

(vi) $V \otimes (W_1 \oplus W_2) \cong (V \otimes W_1) \oplus (V \otimes W_2)$

Remark 2.10.1. Not every element of $V \otimes W$ is a simple tensor $\mathbf{v} \otimes \mathbf{w}$. General elements are sums of simple tensors. Elements that cannot be written as a single $\mathbf{v} \otimes \mathbf{w}$ are called **entangled** in quantum mechanics.

2.10.3 Tensors

Definition 2.10.2 (Dual space). The **dual space** of V is $V^* := \mathcal{L}(V, \mathbb{F})$, the space of linear functionals on V.

If $\{\mathbf{e}_i\}$ is a basis for V, the **dual basis** $\{\mathbf{e}^i\}$ for V^* is defined by $\mathbf{e}^i(\mathbf{e}_j) = \delta^i_j$.

Definition 2.10.3 (Tensors). A **tensor of type** (r, s) on V is an element of

$$T^r_s(V) := \underbrace{V \otimes \cdots \otimes V}_{r} \otimes \underbrace{V^* \otimes \cdots \otimes V^*}_{s}.$$

Equivalently, it is a multilinear map

$$T : \underbrace{V^* \times \cdots \times V^*}_{r} \times \underbrace{V \times \cdots \times V}_{s} \to \mathbb{F}.$$

Example 2.10.1.

- Type $(0,0)$: scalars.
- Type $(1,0)$: vectors.
- Type $(0,1)$: covectors (linear functionals).
- Type $(1,1)$: linear maps $V \to V$.
- Type $(0,2)$: bilinear forms.
- Type $(2,0)$: contravariant 2-tensors.

In components with respect to a basis:

$$T = T^{i_1 \cdots i_r}_{j_1 \cdots j_s} \mathbf{e}_{i_1} \otimes \cdots \otimes \mathbf{e}_{i_r} \otimes \mathbf{e}^{j_1} \otimes \cdots \otimes \mathbf{e}^{j_s}.$$

2.10.4 Symmetric and Exterior Algebras

Definition 2.10.4 (Symmetric tensors)**.** A tensor $T \in V^{\otimes k}$ is **symmetric** if it is invariant under all permutations of its factors. The space of symmetric k-tensors is $S^k(V)$.

Definition 2.10.5 (Alternating tensors)**.** A tensor $T \in V^{\otimes k}$ is **alternating** (or **antisymmetric**) if it changes sign under transposition of any two factors. The space of alternating k-tensors is $\Lambda^k(V)$.

Definition 2.10.6 (Exterior product)**.** The **exterior product** (or **wedge product**) $\wedge : \Lambda^k(V) \times \Lambda^l(V) \to \Lambda^{k+l}(V)$ is the antisymmetrized tensor product.

Proposition 2.10.4. *For $\omega \in \Lambda^k(V)$ and $\eta \in \Lambda^l(V)$:*

$$\omega \wedge \eta = (-1)^{kl} \eta \wedge \omega.$$

In particular, $\omega \wedge \omega = 0$ for odd k.

Theorem 2.10.5. *If $\dim V = n$, then $\dim \Lambda^k(V) = \binom{n}{k}$. In particular:*

- *$\Lambda^0(V) = \mathbb{F}$ (scalars)*
- *$\Lambda^1(V) = V$*
- *$\Lambda^n(V)$ is 1-dimensional*
- *$\Lambda^k(V) = \{0\}$ for $k > n$*

The **exterior algebra** $\Lambda(V) := \bigoplus_{k=0}^{n} \Lambda^k(V)$ is fundamental to differential forms and integration on manifolds.

2.11 Summary and Physical Applications

Linear algebra pervades theoretical physics:

Quantum mechanics: States are vectors in a Hilbert space. Observables are Hermitian operators. Measurement outcomes are eigenvalues; post-measurement states are eigenvectors. The spectral theorem guarantees real eigenvalues and orthogonal eigenstates.

Classical mechanics: The equations of motion are linear(ized) near equilibria. Normal modes are eigenvectors of the dynamical matrix. Symplectic structure involves antisymmetric bilinear forms.

Special and general relativity: Lorentz transformations are linear maps preserving the Minkowski metric (indefinite quadratic form). Tensors describe geometric objects invariantly.

Quantum field theory: Fock space is built from tensor products and symmetric/antisymmetric subspaces (bosons/fermions). Lie algebra representations classify particles.

Data analysis and computation: Principal component analysis uses eigendecomposition. Singular value decomposition underlies many numerical methods.

Exercises

Exercise 2.11.1. Prove that if $\mathbf{v}_1, \ldots, \mathbf{v}_n$ span V and $\mathbf{v}_1, \ldots, \mathbf{v}_{n-1}$ are linearly independent, then either $\{\mathbf{v}_1, \ldots, \mathbf{v}_{n-1}\}$ spans V or $\{\mathbf{v}_1, \ldots, \mathbf{v}_n\}$ is a basis.

Exercise 2.11.2. Let $T : V \to W$ be linear with $\dim V > \dim W$. Prove that T cannot be injective.

Exercise 2.11.3. Prove that similar matrices have the same trace, determinant, and eigenvalues.

Exercise 2.11.4. Show that if A and B are $n \times n$ matrices, then AB and BA have the same eigenvalues. (Hint: consider the case where one is invertible, then use continuity.)

Exercise 2.11.5. Let P be a projection ($P^2 = P$). Prove that the only possible eigenvalues of P are 0 and 1.

Exercise 2.11.6. Prove that an orthogonal matrix has determinant ± 1.

Exercise 2.11.7. Use the Gram-Schmidt process to find an orthonormal basis for $\mathbb{R}^3$ starting from $\{(1, 1, 0), (1, 0, 1), (0, 1, 1)\}$.

Exercise 2.11.8. Prove that $\operatorname{tr}(AB) = \operatorname{tr}(BA)$ for any matrices A, B of compatible sizes.

Exercise 2.11.9. Let A be a real $n \times n$ matrix. Prove that if λ is a complex eigenvalue of A, then $\bar{\lambda}$ is also an eigenvalue.

Exercise 2.11.10. Find the Jordan normal form of $A = \begin{pmatrix} 2 & 1 & 0 \\ 0 & 2 & 1 \\ 0 & 0 & 2 \end{pmatrix}$.

Exercise 2.11.11. In $\mathbb{C}^2$, write $(1,0) \otimes (1,1) + (0,1) \otimes (1,-1)$ in terms of the standard basis for $\mathbb{C}^2 \otimes \mathbb{C}^2$. Is this tensor entangled?

Exercise 2.11.12. Prove that a real symmetric matrix is positive definite if and only if all its eigenvalues are positive.

Chapter 3

Real Analysis

Real analysis provides the rigorous foundation for calculus and its extensions. For physicists, it underpins the theory of differential equations, Fourier analysis, and the functional analysis needed for quantum mechanics. This chapter develops the essential concepts: limits, continuity, differentiation, integration, and series.

3.1 The Real Number System

3.1.1 Axiomatic Characterization

The real numbers $\mathbb{R}$ can be characterized as the unique *complete ordered field.*

Definition 3.1.1 (Ordered field)**.** An **ordered field** is a field $\mathbb{F}$ with a total order $\leq$ compatible with the field operations:

(i) $x \leq y \Rightarrow x + z \leq y + z$ for all z

(ii) $x \geq 0$ and $y \geq 0 \Rightarrow xy \geq 0$

Both $\mathbb{Q}$ and $\mathbb{R}$ are ordered fields, but they differ crucially in completeness.

Definition 3.1.2 (Completeness axiom)**.** An ordered field is **complete** if every nonempty subset that is bounded above has a least upper bound (supremum).

Theorem 3.1.1. *$\mathbb{R}$ is the unique complete ordered field (up to isomorphism). $\mathbb{Q}$ is not complete.*

Example 3.1.1 ($\mathbb{Q}$ is not complete)**.** The set $S = \{x \in \mathbb{Q} : x^2 < 2\}$ is bounded above in $\mathbb{Q}$ (e.g., by 2), but has no supremum in $\mathbb{Q}$. In $\mathbb{R}$, $\sup S = \sqrt{2}$.

3.1.2 Properties of Suprema and Infima

Proposition 3.1.2 (Characterization of supremum). *Let $S \subseteq \mathbb{R}$ be nonempty and bounded above. Then $s = \sup S$ if and only if:*

(i) $x \leq s$ for all $x \in S$ (upper bound)

(ii) For every $\epsilon > 0$, there exists $x \in S$ with $x > s - \epsilon$ (least such bound)

Theorem 3.1.3 (Archimedean property). *For any $x, y \in \mathbb{R}$ with $x > 0$, there exists $n \in \mathbb{N}$ such that $nx > y$.*

Corollary 3.1.4. *(i) For any $\epsilon > 0$, there exists $n \in \mathbb{N}$ with $1/n < \epsilon$.*

(ii) $\mathbb{Q}$ is dense in $\mathbb{R}$: between any two reals lies a rational.

(iii) $\mathbb{R} \setminus \mathbb{Q}$ is dense in $\mathbb{R}$: between any two reals lies an irrational.

3.1.3 The Extended Real Line

Definition 3.1.3 (Extended reals). The **extended real numbers** $\overline{\mathbb{R}} = \mathbb{R} \cup \{-\infty, +\infty\}$ extend $\mathbb{R}$ with:

- $-\infty < x < +\infty$ for all $x \in \mathbb{R}$
- $x + \infty = +\infty$ for $x \neq -\infty$; $\quad x \cdot (+\infty) = +\infty$ for $x > 0$
- $\sup \varnothing = -\infty$; $\quad \inf \varnothing = +\infty$

This allows us to write $\sup S$ and $\inf S$ for any subset $S \subseteq \mathbb{R}$.

3.2 Sequences and Limits

3.2.1 Convergence of Sequences

Definition 3.2.1 (Sequence). A **sequence** in $\mathbb{R}$ is a function $a : \mathbb{N} \to \mathbb{R}$, written $(a_n)_{n=1}^{\infty}$ or (a_n).

Definition 3.2.2 (Limit of a sequence). A sequence (a_n) **converges** to $L \in \mathbb{R}$, written $\lim_{n\to\infty} a_n = L$ or $a_n \to L$, if:

$$\forall \epsilon > 0,\ \exists N \in \mathbb{N} : n > N \Rightarrow |a_n - L| < \epsilon.$$

A sequence that converges is **convergent**; otherwise it is **divergent**.

Proposition 3.2.1 (Uniqueness). *A sequence can have at most one limit.*

Proposition 3.2.2 (Boundedness). *Every convergent sequence is bounded.*

Theorem 3.2.3 (Algebraic properties of limits). *If $a_n \to a$ and $b_n \to b$, then:*

(i) $a_n + b_n \to a + b$

(ii) $a_n - b_n \to a - b$

(iii) $a_n b_n \to ab$

(iv) $a_n / b_n \to a/b$ *(provided $b \neq 0$ and $b_n \neq 0$)*

(v) $|a_n| \to |a|$

(vi) If $a_n \leq b_n$ for all n, then $a \leq b$

Theorem 3.2.4 (Squeeze theorem). *If $a_n \leq b_n \leq c_n$ for all n, and $a_n \to L$ and $c_n \to L$, then $b_n \to L$.*

Example 3.2.1 (Standard limits).

(a) $\lim_{n\to\infty} \frac{1}{n^p} = 0$ for $p > 0$

(b) $\lim_{n\to\infty} r^n = 0$ for $|r| < 1$

(c) $\lim_{n\to\infty} n^{1/n} = 1$

(d) $\lim_{n\to\infty} \left(1 + \frac{1}{n}\right)^n = e$

(e) $\lim_{n\to\infty} \frac{n!}{n^n} = 0$

3.2.2 Monotone Sequences

Definition 3.2.3. A sequence (a_n) is **increasing** if $a_n \leq a_{n+1}$ for all n, **strictly increasing** if $a_n < a_{n+1}$, and similarly for **decreasing**. A sequence is **monotone** if it is either increasing or decreasing.

Theorem 3.2.5 (Monotone convergence theorem). *Every bounded monotone sequence converges. Specifically:*

- *If (a_n) is increasing and bounded above, then $a_n \to \sup\{a_n\}$.*
- *If (a_n) is decreasing and bounded below, then $a_n \to \inf\{a_n\}$.*

This theorem is equivalent to the completeness of $\mathbb{R}$.

3.2.3 Subsequences and Limit Points

Definition 3.2.4 (Subsequence). A **subsequence** of (a_n) is a sequence (a_{n_k}) where $n_1 < n_2 < n_3 < \cdots$.

Proposition 3.2.6. *If $a_n \to L$, then every subsequence also converges to L.*

Definition 3.2.5 (Limit point). L is a **limit point** (or **cluster point**) of (a_n) if some subsequence converges to L.

Theorem 3.2.7 (Bolzano-Weierstrass). *Every bounded sequence has a convergent subsequence.*

Definition 3.2.6 (Limit superior and inferior). For a bounded sequence (a_n):

$$\limsup_{n\to\infty} a_n := \lim_{n\to\infty} \left(\sup_{k\geq n} a_k \right) = \inf_n \sup_{k\geq n} a_k$$

$$\liminf_{n\to\infty} a_n := \lim_{n\to\infty} \left(\inf_{k\geq n} a_k \right) = \sup_n \inf_{k\geq n} a_k$$

Proposition 3.2.8. $\liminf a_n \leq \limsup a_n$, *with equality iff (a_n) converges (to their common value).*

3.2.4 Cauchy Sequences

Definition 3.2.7 (Cauchy sequence). A sequence (a_n) is a **Cauchy sequence** if:

$$\forall \epsilon > 0,\ \exists N : m, n > N \Rightarrow |a_m - a_n| < \epsilon.$$

Theorem 3.2.9 (Cauchy criterion for convergence). *A sequence in $\mathbb{R}$ converges if and only if it is Cauchy.*

The Cauchy criterion is valuable because it characterizes convergence without knowing the limit. It is also the key to defining completeness in metric spaces.

3.3 Series

3.3.1 Definition and Convergence

Definition 3.3.1 (Series). A **series** is a formal sum $\sum_{n=1}^{\infty} a_n$. The **partial sums** are $S_N := \sum_{n=1}^{N} a_n$. The series **converges** to S if $S_N \to S$; we write $\sum_{n=1}^{\infty} a_n = S$.

Proposition 3.3.1 (Necessary condition). *If $\sum a_n$ converges, then $a_n \to 0$.*

The converse is false: $\sum 1/n$ diverges even though $1/n \to 0$.

Theorem 3.3.2 (Cauchy criterion for series). *$\sum a_n$ converges iff for every $\epsilon > 0$, there exists N such that $m > n > N$ implies $|a_{n+1} + \cdots + a_m| < \epsilon$.*

3.3.2 Series with Nonnegative Terms

Theorem 3.3.3 (Comparison test). *Let $0 \leq a_n \leq b_n$ for all n.*

(i) If $\sum b_n$ converges, so does $\sum a_n$.

(ii) If $\sum a_n$ diverges, so does $\sum b_n$.

Theorem 3.3.4 (Limit comparison test). *Let $a_n, b_n > 0$ and suppose $a_n/b_n \to L$.*

(i) If $0 < L < \infty$, then $\sum a_n$ and $\sum b_n$ both converge or both diverge.

(ii) If $L = 0$ and $\sum b_n$ converges, then $\sum a_n$ converges.

(iii) If $L = \infty$ and $\sum b_n$ diverges, then $\sum a_n$ diverges.

Theorem 3.3.5 (Ratio test). *Let $a_n > 0$ and suppose $a_{n+1}/a_n \to L$.*

- *If $L < 1$, the series converges.*
- *If $L > 1$, the series diverges.*
- *If $L = 1$, the test is inconclusive.*

Theorem 3.3.6 (Root test). *Let (a_n) be nonnegative and suppose $(a_n)^{1/n} \to L$.*

- *If $L < 1$, the series converges.*
- *If $L > 1$, the series diverges.*
- *If $L = 1$, the test is inconclusive.*

Theorem 3.3.7 (Integral test). *Let $f : [1, \infty) \to \mathbb{R}$ be positive, continuous, and decreasing. Then $\sum_{n=1}^{\infty} f(n)$ converges iff $\int_1^{\infty} f(x)\,dx$ converges.*

Example 3.3.1 (p-series). $\displaystyle\sum_{n=1}^{\infty} \frac{1}{n^p}$ converges iff $p > 1$. In particular:

- $\sum 1/n$ (harmonic series) diverges.
- $\sum 1/n^2 = \pi^2/6$ (Basel problem).

3.3.3 Absolute and Conditional Convergence

Definition 3.3.2 (Absolute convergence). A series $\sum a_n$ **converges absolutely** if $\sum |a_n|$ converges. It **converges conditionally** if it converges but not absolutely.

Theorem 3.3.8. *Absolute convergence implies convergence.*

Theorem 3.3.9 (Alternating series test). *If (b_n) is positive, decreasing, and $b_n \to 0$, then $\sum_{n=1}^{\infty}(-1)^{n+1}b_n$ converges.*

Example 3.3.2. $\sum_{n=1}^{\infty} \frac{(-1)^{n+1}}{n} = \ln 2$ converges conditionally.

Theorem 3.3.10 (Riemann rearrangement theorem). *If $\sum a_n$ converges conditionally, then for any $L \in \overline{\mathbb{R}}$, there exists a rearrangement of the series that converges to L.*

This remarkable theorem shows that conditionally convergent series are highly sensitive to the order of terms.

Theorem 3.3.11. *If $\sum a_n$ converges absolutely, then every rearrangement converges to the same sum.*

3.3.4 Power Series

Definition 3.3.3 (Power series). A **power series** centered at c is a series of the form

$$\sum_{n=0}^{\infty} a_n(x-c)^n = a_0 + a_1(x-c) + a_2(x-c)^2 + \cdots$$

Theorem 3.3.12 (Radius of convergence). *For any power series $\sum a_n(x-c)^n$, there exists $R \in [0, \infty]$ (the **radius of convergence**) such that:*

- *The series converges absolutely for $|x-c| < R$.*
- *The series diverges for $|x-c| > R$.*
- *At $|x-c| = R$, convergence must be checked separately.*

Proposition 3.3.13 (Formulas for radius of convergence).

$$\frac{1}{R} = \limsup_{n\to\infty} |a_n|^{1/n} = \lim_{n\to\infty} \left|\frac{a_{n+1}}{a_n}\right| \quad \textit{(if the limit exists).}$$

Theorem 3.3.14 (Properties of power series). *Within the radius of convergence:*

(i) The sum $f(x) = \sum a_n(x-c)^n$ *is continuous.*

(ii) The series can be differentiated term by term: $f'(x) = \sum na_n(x-c)^{n-1}$.

(iii) The series can be integrated term by term.

(iv) The radii of convergence of f, f', *and* $\int f$ *are all equal.*

Example 3.3.3 (Important power series).

$$e^x = \sum_{n=0}^{\infty} \frac{x^n}{n!} \qquad R = \infty$$

$$\sin x = \sum_{n=0}^{\infty} \frac{(-1)^n x^{2n+1}}{(2n+1)!} \qquad R = \infty$$

$$\cos x = \sum_{n=0}^{\infty} \frac{(-1)^n x^{2n}}{(2n)!} \qquad R = \infty$$

$$\ln(1+x) = \sum_{n=1}^{\infty} \frac{(-1)^{n+1} x^n}{n} \qquad R = 1$$

$$\frac{1}{1-x} = \sum_{n=0}^{\infty} x^n \qquad R = 1$$

$$(1+x)^\alpha = \sum_{n=0}^{\infty} \binom{\alpha}{n} x^n \qquad R = 1 \text{ (binomial series)}$$

3.4 Continuity

3.4.1 Limits of Functions

Definition 3.4.1 (Limit of a function). Let $f : D \to \mathbb{R}$ and let c be a limit point of D. We say $\lim_{x\to c} f(x) = L$ if:

$$\forall \epsilon > 0,\ \exists \delta > 0 : 0 < |x - c| < \delta \text{ and } x \in D \Rightarrow |f(x) - L| < \epsilon.$$

Proposition 3.4.1 (Sequential characterization). $\lim_{x\to c} f(x) = L$ *iff for every sequence* (x_n) *in* $D \setminus \{c\}$ *with* $x_n \to c$, *we have* $f(x_n) \to L$.

This characterization is often useful for proving limits exist (or don't exist).

Definition 3.4.2 (One-sided limits).

$$\lim_{x\to c^+} f(x) = L \quad \text{if } \forall \epsilon > 0, \exists \delta > 0 : 0 < x - c < \delta \Rightarrow |f(x) - L| < \epsilon$$

$$\lim_{x\to c^-} f(x) = L \quad \text{if } \forall \epsilon > 0, \exists \delta > 0 : 0 < c - x < \delta \Rightarrow |f(x) - L| < \epsilon$$

Proposition 3.4.2. $\lim_{x\to c} f(x) = L$ *iff both one-sided limits exist and equal* L.

3.4.2 Continuous Functions

Definition 3.4.3 (Continuity)**.** $f : D \to \mathbb{R}$ is **continuous at** $c \in D$ if:

$$\forall \epsilon > 0,\ \exists \delta > 0 : |x - c| < \delta \text{ and } x \in D \Rightarrow |f(x) - f(c)| < \epsilon.$$

Equivalently, if c is a limit point of D: $\lim_{x\to c} f(x) = f(c)$.

f is **continuous on** D if it is continuous at every point of D.

Proposition 3.4.3 (Sequential characterization)**.** *f is continuous at c iff for every sequence (x_n) in D with $x_n \to c$, we have $f(x_n) \to f(c)$.*

Theorem 3.4.4 (Algebra of continuous functions)**.** *If f and g are continuous at c, then so are:*

(i) $f + g$, $f - g$, fg, $|f|$

(ii) f/g *(provided $g(c) \neq 0$)*

(iii) $g \circ f$ *(if f is continuous at c and g is continuous at $f(c)$)*

Example 3.4.1. Polynomials are continuous everywhere. Rational functions are continuous on their domain. Trigonometric functions, exponentials, and logarithms are continuous on their domains.

3.4.3 Properties of Continuous Functions

Theorem 3.4.5 (Extreme value theorem)**.** *If $f : [a, b] \to \mathbb{R}$ is continuous, then f attains its maximum and minimum: there exist $c, d \in [a, b]$ with $f(c) \leq f(x) \leq f(d)$ for all $x \in [a, b]$.*

Theorem 3.4.6 (Intermediate value theorem)**.** *If $f : [a, b] \to \mathbb{R}$ is continuous and y lies between $f(a)$ and $f(b)$, then $y = f(c)$ for some $c \in (a, b)$.*

Corollary 3.4.7. *The continuous image of an interval is an interval.*

Corollary 3.4.8 (Root existence)**.** *If $f : [a, b] \to \mathbb{R}$ is continuous with $f(a)$ and $f(b)$ of opposite signs, then $f(c) = 0$ for some $c \in (a, b)$.*

Theorem 3.4.9 (Continuous bijections)**.** *If $f : [a, b] \to \mathbb{R}$ is continuous and strictly monotone, then $f^{-1} : f([a, b]) \to [a, b]$ exists and is continuous.*

3.4.4 Uniform Continuity

Definition 3.4.4 (Uniform continuity). $f : D \to \mathbb{R}$ is **uniformly continuous** on D if:

$$\forall \epsilon > 0,\ \exists \delta > 0 : |x - y| < \delta \text{ and } x, y \in D \Rightarrow |f(x) - f(y)| < \epsilon.$$

The key difference from ordinary continuity: the same δ works for all points.

Example 3.4.2. $f(x) = x^2$ is continuous but not uniformly continuous on $\mathbb{R}$. However, it is uniformly continuous on any bounded interval.

Theorem 3.4.10 (Heine-Cantor). *If $f : [a, b] \to \mathbb{R}$ is continuous, then f is uniformly continuous.*

Proposition 3.4.11. *If $f : D \to \mathbb{R}$ is uniformly continuous and (x_n) is Cauchy in D, then $(f(x_n))$ is Cauchy.*

This allows extending uniformly continuous functions to the closure of their domain.

3.5 Differentiation

3.5.1 The Derivative

Definition 3.5.1 (Derivative). The **derivative** of f at c is

$$f'(c) := \lim_{h \to 0} \frac{f(c+h) - f(c)}{h} = \lim_{x \to c} \frac{f(x) - f(c)}{x - c}$$

if this limit exists. We say f is **differentiable** at c.

Proposition 3.5.1. *If f is differentiable at c, then f is continuous at c.*

The converse is false: $f(x) = |x|$ is continuous but not differentiable at 0.

Theorem 3.5.2 (Differentiation rules). *If f and g are differentiable at c:*

(i) $(f + g)'(c) = f'(c) + g'(c)$

(ii) $(fg)'(c) = f'(c)g(c) + f(c)g'(c)$ *(product rule)*

(iii) $(f/g)'(c) = \frac{f'(c)g(c) - f(c)g'(c)}{g(c)^2}$ *if* $g(c) \neq 0$ *(quotient rule)*

(iv) $(g \circ f)'(c) = g'(f(c)) \cdot f'(c)$ *(chain rule)*

Theorem 3.5.3 (Inverse function theorem). *If f is differentiable at c with $f'(c) \neq 0$, and f^{-1} exists near $f(c)$, then f^{-1} is differentiable at $f(c)$ with*

$$(f^{-1})'(f(c)) = \frac{1}{f'(c)}.$$

3.5.2 Mean Value Theorems

Theorem 3.5.4 (Fermat's theorem). *If f has a local extremum at an interior point c and $f'(c)$ exists, then $f'(c) = 0$.*

Theorem 3.5.5 (Rolle's theorem). *If $f : [a, b] \to \mathbb{R}$ is continuous on $[a, b]$, differentiable on (a, b), and $f(a) = f(b)$, then $f'(c) = 0$ for some $c \in (a, b)$.*

Theorem 3.5.6 (Mean value theorem). *If $f : [a, b] \to \mathbb{R}$ is continuous on $[a, b]$ and differentiable on (a, b), then*

$$f(b) - f(a) = f'(c)(b - a)$$

for some $c \in (a, b)$.

Corollary 3.5.7 (Monotonicity criterion). *Let f be continuous on $[a, b]$ and differentiable on (a, b).*

(i) If $f'(x) > 0$ for all $x \in (a, b)$, then f is strictly increasing.

(ii) If $f'(x) < 0$ for all $x \in (a, b)$, then f is strictly decreasing.

(iii) If $f'(x) = 0$ for all $x \in (a, b)$, then f is constant.

Theorem 3.5.8 (Cauchy's mean value theorem). *If $f, g : [a, b] \to \mathbb{R}$ are continuous on $[a, b]$ and differentiable on (a, b), then for some $c \in (a, b)$:*

$$[f(b) - f(a)]g'(c) = [g(b) - g(a)]f'(c).$$

Theorem 3.5.9 (L'Hôpital's rule). *Suppose f and g are differentiable near c (except possibly at c), with $g'(x) \neq 0$ near c. If $\lim_{x\to c} f(x) = \lim_{x\to c} g(x) = 0$ (or both $\pm\infty$), and $\lim_{x\to c} f'(x)/g'(x)$ exists, then:*

$$\lim_{x\to c} \frac{f(x)}{g(x)} = \lim_{x\to c} \frac{f'(x)}{g'(x)}.$$

3.5.3 Higher Derivatives and Taylor's Theorem

Definition 3.5.2 (Higher derivatives). $f''(x) := (f')'(x)$, $f'''(x) := (f'')'(x)$, and in general $f^{(n)}(x)$ is the n-th derivative.

Definition 3.5.3 (C^k functions). $f \in C^k(D)$ if $f^{(k)}$ exists and is continuous on D. $f \in C^\infty(D)$ if $f^{(n)}$ exists for all n.

Theorem 3.5.10 (Taylor's theorem). *If $f \in C^{n+1}([a, b])$ and $c \in [a, b]$, then for any $x \in [a, b]$:*

$$f(x) = \sum_{k=0}^{n} \frac{f^{(k)}(c)}{k!}(x - c)^k + R_n(x)$$

where the remainder $R_n(x)$ can be expressed as:

- ***Lagrange form:*** *$R_n(x) = \frac{f^{(n+1)}(\xi)}{(n+1)!}(x-c)^{n+1}$ for some ξ between c and x.*
- ***Integral form:*** *$R_n(x) = \frac{1}{n!}\int_c^x (x-t)^n f^{(n+1)}(t)\,dt$.*

Definition 3.5.4 (Taylor series)**.** The **Taylor series** of f at c is $\sum_{n=0}^{\infty} \frac{f^{(n)}(c)}{n!}(x-c)^n$. When $c = 0$, this is the **Maclaurin series**.

Remark 3.5.1. The Taylor series may not converge to $f(x)$ even if $f \in C^\infty$. The function $f(x) = e^{-1/x^2}$ (with $f(0) = 0$) has all derivatives zero at 0, so its Maclaurin series is identically zero.

Definition 3.5.5 (Analytic function)**.** f is **(real) analytic** at c if f equals its Taylor series in some neighborhood of c.

3.5.4 Convexity

Definition 3.5.6 (Convex function)**.** $f : I \to \mathbb{R}$ (where I is an interval) is **convex** if for all $x, y \in I$ and $t \in [0,1]$:

$$f(tx + (1-t)y) \le tf(x) + (1-t)f(y).$$

f is **concave** if $-f$ is convex.

Geometrically, the graph lies below (or on) the chord connecting any two points.

Theorem 3.5.11 (Characterization of convexity)**.** *Let $f : I \to \mathbb{R}$.*

(i) If f is differentiable, f is convex iff f' is increasing.

(ii) If f is twice differentiable, f is convex iff $f''(x) \ge 0$ for all x.

Theorem 3.5.12 (Jensen's inequality)**.** *If f is convex and $t_1, \dots, t_n \ge 0$ with $\sum t_i = 1$, then*

$$f\left(\sum_{i=1}^{n} t_i x_i\right) \le \sum_{i=1}^{n} t_i f(x_i).$$

3.6 The Riemann Integral

3.6.1 Definition

Definition 3.6.1 (Partition)**.** A **partition** of $[a,b]$ is a finite set $P = \{x_0, x_1, \dots, x_n\}$ with $a = x_0 < x_1 < \cdots < x_n = b$. The **mesh** (or **norm**) is $\|P\| := \max_i (x_i - x_{i-1})$.

Definition 3.6.2 (Riemann sums and integral). For a bounded function $f : [a, b] \to \mathbb{R}$ and partition P:

$$L(f, P) := \sum_{i=1}^{n} m_i(x_i - x_{i-1}) \quad \text{where } m_i = \inf_{[x_{i-1}, x_i]} f \quad \text{(lower sum)}$$

$$U(f, P) := \sum_{i=1}^{n} M_i(x_i - x_{i-1}) \quad \text{where } M_i = \sup_{[x_{i-1}, x_i]} f \quad \text{(upper sum)}$$

The **lower and upper integrals** are:

$$\underline{\int_a^b} f := \sup_P L(f, P), \qquad \overline{\int_a^b} f := \inf_P U(f, P).$$

f is **Riemann integrable** if $\underline{\int_a^b} f = \overline{\int_a^b} f$, and this common value is $\int_a^b f(x)\, dx$.

Theorem 3.6.1 (Riemann criterion). *f is Riemann integrable on $[a, b]$ iff for every $\epsilon > 0$, there exists a partition P with $U(f, P) - L(f, P) < \epsilon$.*

Theorem 3.6.2 (Integrability of continuous functions). *Every continuous function on $[a, b]$ is Riemann integrable.*

Theorem 3.6.3 (Integrability of monotone functions). *Every monotone function on $[a, b]$ is Riemann integrable.*

More generally:

Theorem 3.6.4 (Lebesgue criterion). *A bounded function $f : [a, b] \to \mathbb{R}$ is Riemann integrable iff the set of discontinuities of f has measure zero.*

3.6.2 Properties of the Integral

Theorem 3.6.5 (Properties of the Riemann integral). *Let f, g be integrable on $[a, b]$.*

(i) ***Linearity:*** $\int_a^b (\alpha f + \beta g) = \alpha \int_a^b f + \beta \int_a^b g$

(ii) ***Monotonicity:*** *If $f \leq g$, then $\int_a^b f \leq \int_a^b g$*

(iii) ***Additivity:*** $\int_a^b f = \int_a^c f + \int_c^b f$ *for $a < c < b$*

(iv) ***Absolute value:*** $\left|\int_a^b f\right| \leq \int_a^b |f|$

(v) ***Product:*** *fg is integrable*

Theorem 3.6.6 (Mean value theorem for integrals). *If f is continuous on $[a,b]$, then for some $c \in [a,b]$:*

$$\int_a^b f(x)\,dx = f(c)(b-a).$$

3.6.3 The Fundamental Theorem of Calculus

Theorem 3.6.7 (Fundamental theorem of calculus, Part I). *If f is integrable on $[a,b]$ and $F(x) := \int_a^x f(t)\,dt$, then F is continuous on $[a,b]$. Moreover, if f is continuous at $c \in (a,b)$, then F is differentiable at c with $F'(c) = f(c)$.*

Theorem 3.6.8 (Fundamental theorem of calculus, Part II). *If f is continuous on $[a,b]$ and F is any antiderivative of f (i.e., $F' = f$), then:*

$$\int_a^b f(x)\,dx = F(b) - F(a).$$

Theorem 3.6.9 (Integration by parts). *If $f, g \in C^1([a,b])$, then:*

$$\int_a^b f(x)g'(x)\,dx = [f(x)g(x)]_a^b - \int_a^b f'(x)g(x)\,dx.$$

Theorem 3.6.10 (Change of variables). *If $\phi : [\alpha,\beta] \to [a,b]$ is C^1 with $\phi(\alpha) = a$ and $\phi(\beta) = b$, and f is continuous on $[a,b]$, then:*

$$\int_a^b f(x)\,dx = \int_\alpha^\beta f(\phi(t))\phi'(t)\,dt.$$

3.6.4 Improper Integrals

Definition 3.6.3 (Improper integrals). For unbounded intervals or unbounded functions:

$$\begin{aligned}
\int_a^\infty f(x)\,dx &:= \lim_{R\to\infty} \int_a^R f(x)\,dx \\
\int_{-\infty}^b f(x)\,dx &:= \lim_{R\to-\infty} \int_R^b f(x)\,dx \\
\int_a^b f(x)\,dx &:= \lim_{\epsilon\to 0^+} \int_{a+\epsilon}^b f(x)\,dx \quad \text{(if } f \text{ unbounded near } a\text{)}
\end{aligned}$$

Theorem 3.6.11 (Comparison for improper integrals). *If $0 \le f(x) \le g(x)$ for $x \ge a$:*

(i) *If $\int_a^\infty g$ converges, so does $\int_a^\infty f$.*

(ii) If $\int_a^\infty f$ diverges, so does $\int_a^\infty g$.

Example 3.6.1. $\int_1^\infty x^{-p}\,dx$ converges iff $p > 1$. $\int_0^1 x^{-p}\,dx$ converges iff $p < 1$.

3.7 Sequences and Series of Functions

3.7.1 Pointwise and Uniform Convergence

Definition 3.7.1 (Pointwise convergence). A sequence of functions (f_n) **converges pointwise** to f on D if for each $x \in D$:

$$\lim_{n\to\infty} f_n(x) = f(x).$$

Pointwise convergence does not preserve continuity, differentiability, or integrability in general.

Definition 3.7.2 (Uniform convergence). (f_n) **converges uniformly** to f on D if:

$$\forall \epsilon > 0,\ \exists N : n > N \Rightarrow |f_n(x) - f(x)| < \epsilon \text{ for all } x \in D.$$

Equivalently, $\sup_{x\in D} |f_n(x) - f(x)| \to 0$.

Theorem 3.7.1 (Uniform limit of continuous functions). *If each f_n is continuous on D and $f_n \to f$ uniformly, then f is continuous on D.*

Theorem 3.7.2 (Interchange of limit and integral). *If $f_n \to f$ uniformly on $[a, b]$ and each f_n is integrable, then f is integrable and:*

$$\lim_{n\to\infty} \int_a^b f_n(x)\,dx = \int_a^b f(x)\,dx.$$

Theorem 3.7.3 (Interchange of limit and derivative). *Let (f_n) be a sequence of C^1 functions on $[a, b]$. If:*

(i) $(f_n(c))$ converges for some $c \in [a, b]$, and

(ii) (f_n') converges uniformly on $[a, b]$,

then (f_n) converges uniformly to some f, and $f' = \lim f_n'$.

3.7.2 Uniform Convergence of Series

Definition 3.7.3. A series $\sum f_n$ of functions **converges uniformly** if the partial sums $S_N = \sum_{n=1}^{N} f_n$ converge uniformly.

Theorem 3.7.4 (Weierstrass M-test). *If $|f_n(x)| \leq M_n$ for all $x \in D$ and $\sum M_n$ converges, then $\sum f_n$ converges uniformly and absolutely on D.*

Example 3.7.1. $\sum_{n=1}^{\infty} \frac{\sin(nx)}{n^2}$ converges uniformly on $\mathbb{R}$ since $|\sin(nx)/n^2| \leq 1/n^2$ and $\sum 1/n^2$ converges.

Corollary 3.7.5. *A power series $\sum a_n(x-c)^n$ converges uniformly on any closed interval $[c-r, c+r]$ with r less than the radius of convergence.*

3.8 Metric Spaces (Preview)

Many concepts of real analysis generalize to **metric spaces**, which will be essential for functional analysis.

Definition 3.8.1 (Metric space). A **metric space** is a set X with a function $d : X \times X \to [0, \infty)$ satisfying:

(i) $d(x, y) = 0 \Leftrightarrow x = y$ (identity)

(ii) $d(x, y) = d(y, x)$ (symmetry)

(iii) $d(x, z) \leq d(x, y) + d(y, z)$ (triangle inequality)

Example 3.8.1 (Standard metrics).

(a) $\mathbb{R}^n$ with $d(\mathbf{x}, \mathbf{y}) = \|\mathbf{x} - \mathbf{y}\|_2 = \sqrt{\sum (x_i - y_i)^2}$

(b) Any normed vector space with $d(\mathbf{x}, \mathbf{y}) = \|\mathbf{x} - \mathbf{y}\|$

(c) $C([a, b])$ with $d(f, g) = \sup_{x \in [a,b]} |f(x) - g(x)|$

(d) Discrete metric: $d(x, y) = 1$ if $x \neq y$, $d(x, x) = 0$

Limits, continuity, Cauchy sequences, and completeness all generalize naturally. A metric space is **complete** if every Cauchy sequence converges.

Theorem 3.8.1 (Fixed point theorem). *Let (X, d) be a complete metric space and $T : X \to X$ a **contraction**: $d(Tx, Ty) \leq k \cdot d(x, y)$ for some $k < 1$. Then T has a unique fixed point $x^* = Tx^*$.*

This theorem has wide applications: existence and uniqueness for differential equations, iterative numerical methods, and more.

3.9 Asymptotic Notation

Asymptotic notation describes limiting behavior and is essential for approximations in physics.

Definition 3.9.1 (Big-O notation). $f(x) = O(g(x))$ as $x \to a$ means there exist constants $C > 0$ and $\delta > 0$ such that $|f(x)| \leq C|g(x)|$ for $|x - a| < \delta$.

Definition 3.9.2 (Little-o notation). $f(x) = o(g(x))$ as $x \to a$ means $\lim_{x\to a} f(x)/g(x) = 0$.

Definition 3.9.3 (Asymptotic equivalence). $f(x) \sim g(x)$ as $x \to a$ means $\lim_{x\to a} f(x)/g(x) = 1$.

Example 3.9.1. As $x \to 0$:

- $\sin x = x - \frac{x^3}{6} + O(x^5) = x + o(x)$
- $\sin x \sim x$
- $e^x = 1 + x + \frac{x^2}{2} + O(x^3)$
- $\ln(1 + x) = x - \frac{x^2}{2} + O(x^3)$

Example 3.9.2 (Stirling's formula). As $n \to \infty$:

$$n! \sim \sqrt{2\pi n}\left(\frac{n}{e}\right)^n.$$

More precisely: $\ln(n!) = n \ln n - n + \frac{1}{2}\ln(2\pi n) + O(1/n)$.

3.10 Summary and Physical Applications

Real analysis provides the rigorous foundation for the calculations of physics:

Limits and continuity: Physical quantities vary continuously (in classical physics). The epsilon-delta definition makes precise the notion of arbitrarily small changes.

Differentiation: Rates of change—velocity, acceleration, field gradients—are derivatives. Taylor series give local approximations essential for perturbation theory.

Integration: Accumulated quantities—work, probability, charge—are integrals. Improper integrals arise naturally (infinite domains, singular integrands).

Series: Solutions to differential equations often come as power series. Fourier series decompose periodic functions. Asymptotic series give useful approximations even when divergent.

Uniform convergence: Essential for interchanging limits with integrals and derivatives—operations routinely performed in physics. The Weierstrass M-test justifies term-by-term integration of power series.

Metric spaces: The abstraction to metric spaces leads to functional analysis, where state spaces are infinite-dimensional and operators replace matrices.

Exercises

Exercise 3.10.1. Prove that between any two distinct real numbers there exist infinitely many rationals and infinitely many irrationals.

Exercise 3.10.2. Let $a_1 = 1$ and $a_{n+1} = \sqrt{2 + a_n}$. Prove that (a_n) converges and find its limit.

Exercise 3.10.3. Determine whether the following series converge:

(a) $\displaystyle\sum_{n=1}^{\infty} \frac{n!}{n^n}$

(b) $\displaystyle\sum_{n=2}^{\infty} \frac{1}{n(\ln n)^2}$

(c) $\displaystyle\sum_{n=1}^{\infty} \frac{(-1)^n}{\sqrt{n}}$

Exercise 3.10.4. Prove that $f(x) = \sum_{n=1}^{\infty} \frac{\cos(nx)}{n^2}$ is continuous on $\mathbb{R}$.

Exercise 3.10.5. Let $f(x) = x^2 \sin(1/x)$ for $x \neq 0$ and $f(0) = 0$. Show that f is differentiable everywhere but f' is not continuous at 0.

Exercise 3.10.6. Use Taylor's theorem to show that e is irrational. (Hint: assume $e = p/q$ and derive a contradiction from the remainder formula.)

Exercise 3.10.7. Evaluate $\displaystyle\int_0^1 \frac{\ln(1+x)}{x}\,dx$ by expanding the integrand as a power series and integrating term by term.

Exercise 3.10.8. Prove that $\int_0^{\infty} e^{-x^2}\,dx = \frac{\sqrt{\pi}}{2}$. (Hint: consider I^2 as a double integral and switch to polar coordinates.)

Exercise 3.10.9. Let $f_n(x) = x^n$ on $[0, 1]$. Find the pointwise limit. Does the convergence uniform? What about on $[0, r]$ for $r < 1$?

Exercise 3.10.10. Prove that if f is convex on (a, b), then f is continuous on (a, b).

Exercise 3.10.11. Use the contraction mapping theorem to prove that $x = \cos x$ has a unique solution in $[0, 1]$.

Exercise 3.10.12. Find the first four terms of the asymptotic expansion of $\int_x^\infty e^{-t^2}\, dt$ as $x \to \infty$.

Chapter 4

Complex Analysis

Complex analysis extends calculus to functions of a complex variable. The theory is remarkably rich: complex differentiability imposes such strong constraints that complex-differentiable functions have properties far beyond their real counterparts. For physics, complex analysis provides powerful computational tools—contour integration, residue calculus, conformal mapping—and deep insights into analyticity, causality, and dispersion relations.

4.1 The Complex Number System

4.1.1 Algebraic Structure

Definition 4.1.1 (Complex numbers). The **complex numbers** $\mathbb{C}$ consist of expressions $z = x + iy$ where $x, y \in \mathbb{R}$ and $i^2 = -1$. We call $x = \operatorname{Re}(z)$ the **real part** and $y = \operatorname{Im}(z)$ the **imaginary part**.

Proposition 4.1.1 (Field structure). *$\mathbb{C}$ is a field under the operations:*

$$(x_1 + iy_1) + (x_2 + iy_2) = (x_1 + x_2) + i(y_1 + y_2)$$
$$(x_1 + iy_1)(x_2 + iy_2) = (x_1x_2 - y_1y_2) + i(x_1y_2 + x_2y_1)$$

Definition 4.1.2 (Complex conjugate). The **complex conjugate** of $z = x + iy$ is $\bar{z} = x - iy$.

Proposition 4.1.2 (Properties of conjugation). *(i)* $\bar{\bar{z}} = z$

(ii) $\overline{z + w} = \bar{z} + \bar{w}, \quad \overline{zw} = \bar{z}\bar{w}$

(iii) $z + \bar{z} = 2\operatorname{Re}(z), \quad z - \bar{z} = 2i\operatorname{Im}(z)$

(iv) $z\bar{z} = x^2 + y^2 \geq 0$

(v) $z \in \mathbb{R} \Leftrightarrow z = \bar{z}$

Definition 4.1.3 (Modulus). The **modulus** (or **absolute value**) of $z = x + iy$ is $|z| = \sqrt{z\bar{z}} = \sqrt{x^2 + y^2}$.

Proposition 4.1.3 (Properties of modulus). *(i)* $|z| \geq 0$, *with equality iff* $z = 0$

(ii) $|zw| = |z||w|$

(iii) $|z + w| \leq |z| + |w|$ *(triangle inequality)*

(iv) $||z| - |w|| \leq |z - w|$ *(reverse triangle inequality)*

(v) $|\mathrm{Re}(z)| \leq |z|, \quad |\mathrm{Im}(z)| \leq |z|$

4.1.2 Geometric Representation

The complex plane (or **Argand diagram**) identifies $z = x + iy$ with the point $(x, y) \in \mathbb{R}^2$.

Definition 4.1.4 (Polar form). Every nonzero $z \in \mathbb{C}$ can be written as

$$z = r(\cos\theta + i\sin\theta) = re^{i\theta}$$

where $r = |z|$ and $\theta = \arg(z)$ is the **argument** (angle from positive real axis).

Theorem 4.1.4 (Euler's formula). *For all* $\theta \in \mathbb{R}$*:*

$$e^{i\theta} = \cos\theta + i\sin\theta.$$

Corollary 4.1.5.

$$\cos\theta = \frac{e^{i\theta} + e^{-i\theta}}{2}, \quad \sin\theta = \frac{e^{i\theta} - e^{-i\theta}}{2i}$$

$$e^{i\pi} + 1 = 0 \quad \textit{(Euler's identity)}$$

Proposition 4.1.6 (Multiplication in polar form). *If* $z_1 = r_1e^{i\theta_1}$ *and* $z_2 = r_2e^{i\theta_2}$*, then:*

$$z_1z_2 = r_1r_2e^{i(\theta_1+\theta_2)}, \quad \frac{z_1}{z_2} = \frac{r_1}{r_2}e^{i(\theta_1-\theta_2)}.$$

Multiplication rotates and scales.

Theorem 4.1.7 (De Moivre's formula). *For any* $n \in \mathbb{Z}$*:*

$$(\cos\theta + i\sin\theta)^n = \cos(n\theta) + i\sin(n\theta).$$

Proposition 4.1.8 (Roots of unity). *The* n*-th roots of unity are* $e^{2\pi ik/n}$ *for* $k = 0, 1, \ldots, n-1$*. They form the vertices of a regular* n*-gon inscribed in the unit circle.*

4.1.3 The Extended Complex Plane

Definition 4.1.5 (Riemann sphere). The **extended complex plane** (or **Riemann sphere**) is $\hat{\mathbb{C}} = \mathbb{C} \cup \{\infty\}$, where we adjoin a single "point at infinity."

The Riemann sphere can be visualized by stereographic projection: project $\mathbb{C}$ onto the unit sphere in $\mathbb{R}^3$ with the north pole corresponding to ∞.

Proposition 4.1.9 (Arithmetic with infinity). *For $z \in \mathbb{C}$, $z \neq 0$:*

$$z + \infty = \infty, \quad z \cdot \infty = \infty, \quad \frac{z}{\infty} = 0, \quad \frac{z}{0} = \infty.$$

Expressions like $\infty - \infty$, $0 \cdot \infty$, ∞/∞ are undefined.

4.2 Complex Functions

4.2.1 Basic Definitions

Definition 4.2.1 (Complex function). A **complex function** is a function $f : D \to \mathbb{C}$ where $D \subseteq \mathbb{C}$. Writing $z = x + iy$ and $f(z) = u(x, y) + iv(x, y)$, we identify f with a pair of real-valued functions $u, v : D \to \mathbb{R}$.

Definition 4.2.2 (Limit and continuity). $\lim_{z \to z_0} f(z) = L$ if for every $\epsilon > 0$, there exists $\delta > 0$ such that $|f(z) - L| < \epsilon$ whenever $0 < |z - z_0| < \delta$.

f is **continuous** at z_0 if $\lim_{z \to z_0} f(z) = f(z_0)$.

Since $|z - z_0|$ measures distance in $\mathbb{R}^2$, complex continuity is equivalent to continuity of (u, v) as a map $\mathbb{R}^2 \to \mathbb{R}^2$.

4.2.2 Elementary Functions

Definition 4.2.3 (Complex exponential).

$$e^z := e^x(\cos y + i \sin y) \quad \text{for } z = x + iy.$$

Proposition 4.2.1 (Properties of e^z).

(i) $e^{z_1 + z_2} = e^{z_1} e^{z_2}$

(ii) $e^z \neq 0$ *for all* z

(iii) $|e^z| = e^x, \quad \arg(e^z) = y$

(iv) e^z *is periodic:* $e^{z + 2\pi i} = e^z$

Definition 4.2.4 (Complex logarithm). For $z \neq 0$:

$$\log z := \ln |z| + i \arg(z).$$

Since $\arg(z)$ is multi-valued (determined only modulo 2π), $\log z$ is multi-valued.

Definition 4.2.5 (Principal value). The **principal value** $\operatorname{Log} z$ uses the principal argument $\operatorname{Arg}(z) \in (-\pi, \pi]$:

$$\operatorname{Log} z = \ln |z| + i \operatorname{Arg}(z).$$

Definition 4.2.6 (Complex powers). For $z \neq 0$ and $\alpha \in \mathbb{C}$:

$$z^\alpha := e^{\alpha \log z}.$$

This is generally multi-valued unless $\alpha \in \mathbb{Z}$.

Example 4.2.1. $i^i = e^{i \log i} = e^{i(i\pi/2 + 2\pi i k)} = e^{-\pi/2 - 2\pi k}$ for $k \in \mathbb{Z}$. The principal value is $e^{-\pi/2} \approx 0.208$.

Definition 4.2.7 (Complex trigonometric functions).

$$\cos z := \frac{e^{iz} + e^{-iz}}{2}, \quad \sin z := \frac{e^{iz} - e^{-iz}}{2i}$$
$$\cosh z := \frac{e^{z} + e^{-z}}{2}, \quad \sinh z := \frac{e^{z} - e^{-z}}{2}$$

Proposition 4.2.2. $\cos(iz) = \cosh z, \quad \sin(iz) = i \sinh z$. *The identity* $\cos^2 z + \sin^2 z = 1$ *holds for all* $z \in \mathbb{C}$.

4.3 Analytic Functions

4.3.1 Complex Differentiability

Definition 4.3.1 (Complex derivative). f is **complex differentiable** (or **holomorphic**) at z_0 if

$$f'(z_0) := \lim_{h \to 0} \frac{f(z_0 + h) - f(z_0)}{h}$$

exists. Here $h \in \mathbb{C}$ approaches 0 from any direction.

Definition 4.3.2 (Holomorphic and analytic). f is **holomorphic** (or **analytic**) on an open set D if $f'(z)$ exists for all $z \in D$. A function holomorphic on all of $\mathbb{C}$ is called **entire**.

The usual differentiation rules hold: sum, product, quotient, and chain rules.

Example 4.3.1. Polynomials and e^z are entire. Rational functions are holomorphic except at poles. $\bar{z}$ is *not* holomorphic anywhere.

4.3.2 The Cauchy-Riemann Equations

The key to complex analysis is that complex differentiability is much more restrictive than real differentiability.

Theorem 4.3.1 (Cauchy-Riemann equations). *Let $f = u + iv$ with $u, v : \mathbb{R}^2 \to \mathbb{R}$. If f is holomorphic at $z_0 = x_0 + iy_0$, then the partial derivatives exist and satisfy:*

$$\frac{\partial u}{\partial x} = \frac{\partial v}{\partial y}, \quad \frac{\partial u}{\partial y} = -\frac{\partial v}{\partial x}$$

at (x_0, y_0).

Conversely, if u and v have continuous first partial derivatives satisfying these equations in a neighborhood of (x_0, y_0), then f is holomorphic at z_0.

Proof of necessity. Suppose $f'(z_0)$ exists. Taking h real ($h = t$):

$$f'(z_0) = \lim_{t \to 0} \frac{f(z_0 + t) - f(z_0)}{t} = \frac{\partial u}{\partial x} + i\frac{\partial v}{\partial x}.$$

Taking h purely imaginary ($h = it$):

$$f'(z_0) = \lim_{t \to 0} \frac{f(z_0 + it) - f(z_0)}{it} = \frac{1}{i}\left(\frac{\partial u}{\partial y} + i\frac{\partial v}{\partial y}\right) = \frac{\partial v}{\partial y} - i\frac{\partial u}{\partial y}.$$

Equating real and imaginary parts gives the Cauchy-Riemann equations. □

Proposition 4.3.2 (Derivative formula). *If $f = u + iv$ is holomorphic:*

$$f'(z) = \frac{\partial u}{\partial x} + i\frac{\partial v}{\partial x} = \frac{\partial v}{\partial y} - i\frac{\partial u}{\partial y}.$$

Corollary 4.3.3 (Harmonic functions). *If $f = u + iv$ is holomorphic on D, then u and v are **harmonic**: $\nabla^2 u = \nabla^2 v = 0$, where $\nabla^2 = \partial^2/\partial x^2 + \partial^2/\partial y^2$ is the Laplacian.*

Proof. From the Cauchy-Riemann equations:

$$\frac{\partial^2 u}{\partial x^2} = \frac{\partial}{\partial x}\frac{\partial v}{\partial y} = \frac{\partial^2 v}{\partial x \partial y}, \quad \frac{\partial^2 u}{\partial y^2} = -\frac{\partial}{\partial y}\frac{\partial v}{\partial x} = -\frac{\partial^2 v}{\partial y \partial x}.$$

Adding (assuming mixed partials are equal): $\nabla^2 u = 0$. Similarly for v. □

Definition 4.3.3 (Harmonic conjugate). If u is harmonic on a simply connected domain, there exists a harmonic function v (unique up to a constant) such that $f = u + iv$ is holomorphic. We call v a **harmonic conjugate** of u.

4.3.3 Wirtinger Derivatives

Definition 4.3.4 (Wirtinger derivatives).

$$\frac{\partial}{\partial z} := \frac{1}{2}\left(\frac{\partial}{\partial x} - i\frac{\partial}{\partial y}\right), \quad \frac{\partial}{\partial \bar{z}} := \frac{1}{2}\left(\frac{\partial}{\partial x} + i\frac{\partial}{\partial y}\right).$$

Proposition 4.3.4. *f is holomorphic iff $\partial f/\partial \bar{z} = 0$. In this case, $f'(z) = \partial f/\partial z$.*

This elegant formulation treats z and $\bar{z}$ as independent variables.

4.4 Complex Integration

4.4.1 Contour Integrals

Definition 4.4.1 (Contour). A **contour** (or **path**) is a piecewise smooth curve $\gamma : [a, b] \to \mathbb{C}$. It is **closed** if $\gamma(a) = \gamma(b)$, and **simple** if it doesn't self-intersect (except possibly at endpoints).

Definition 4.4.2 (Contour integral). For a continuous f and a smooth contour γ:

$$\int_\gamma f(z)\, dz := \int_a^b f(\gamma(t))\gamma'(t)\, dt.$$

Proposition 4.4.1 (Properties).

(i) $\int_\gamma (af + bg)\, dz = a\int_\gamma f\, dz + b\int_\gamma g\, dz$

(ii) $\int_{-\gamma} f\, dz = -\int_\gamma f\, dz$ *(reversing orientation)*

(iii) $\left|\int_\gamma f\, dz\right| \leq \max_{z\in\gamma} |f(z)| \cdot \text{length}(\gamma)$ *(ML inequality)*

Example 4.4.1. Let γ be the unit circle traversed counterclockwise: $\gamma(t) = e^{it}$, $t \in [0, 2\pi]$.

$$\int_\gamma \frac{dz}{z} = \int_0^{2\pi} \frac{ie^{it}}{e^{it}}\, dt = \int_0^{2\pi} i\, dt = 2\pi i.$$

$$\int_\gamma z^n\, dz = \int_0^{2\pi} e^{int} \cdot ie^{it}\, dt = i\int_0^{2\pi} e^{i(n+1)t}\, dt = \begin{cases} 0 & n \neq -1 \\ 2\pi i & n = -1 \end{cases}.$$

4.4.2 Cauchy's Theorem

Theorem 4.4.2 (Cauchy's theorem). *If f is holomorphic on a simply connected domain D, then for any closed contour γ in D:*

$$\oint_\gamma f(z)\,dz = 0.$$

Proof sketch using Green's theorem. Write $f\,dz = (u+iv)(dx+i\,dy) = (u\,dx-v\,dy) + i(v\,dx + u\,dy)$. By Green's theorem:

$$\oint_\gamma u\,dx - v\,dy = \iint_D \left(-\frac{\partial v}{\partial x} - \frac{\partial u}{\partial y}\right) dx\,dy = 0$$

by the Cauchy-Riemann equations. Similarly for the imaginary part. □

Corollary 4.4.3 (Path independence). *If f is holomorphic on a simply connected domain, then $\int_\gamma f\,dz$ depends only on the endpoints of γ.*

Corollary 4.4.4 (Antiderivative existence). *If f is holomorphic on a simply connected domain D, then f has an antiderivative: there exists holomorphic F on D with $F' = f$.*

Theorem 4.4.5 (Cauchy's theorem for multiply connected domains). *If f is holomorphic on a domain D bounded by an outer contour γ_0 and inner contours $\gamma_1, \ldots, \gamma_n$ (all oriented appropriately), then:*

$$\oint_{\gamma_0} f\,dz = \sum_{k=1}^{n} \oint_{\gamma_k} f\,dz.$$

4.4.3 Cauchy's Integral Formula

Theorem 4.4.6 (Cauchy's integral formula). *If f is holomorphic inside and on a simple closed contour γ, and z_0 is inside γ, then:*

$$f(z_0) = \frac{1}{2\pi i}\oint_\gamma \frac{f(z)}{z-z_0}\,dz.$$

Proof. Let C_ϵ be a small circle of radius ϵ centered at z_0. By Cauchy's theorem for multiply connected domains:

$$\oint_\gamma \frac{f(z)}{z-z_0}\,dz = \oint_{C_\epsilon} \frac{f(z)}{z-z_0}\,dz.$$

On C_ϵ, write $z = z_0 + \epsilon e^{i\theta}$:

$$\oint_{C_\epsilon} \frac{f(z)}{z-z_0}\,dz = \int_0^{2\pi} \frac{f(z_0+\epsilon e^{i\theta})}{\epsilon e^{i\theta}} \cdot i\epsilon e^{i\theta}\,d\theta = i\int_0^{2\pi} f(z_0+\epsilon e^{i\theta})\,d\theta.$$

As $\epsilon \to 0$, this approaches $i \cdot 2\pi f(z_0) = 2\pi i f(z_0)$. □

Theorem 4.4.7 (Cauchy's formula for derivatives). *Under the same hypotheses:*

$$f^{(n)}(z_0) = \frac{n!}{2\pi i} \oint_\gamma \frac{f(z)}{(z - z_0)^{n+1}} \, dz.$$

Corollary 4.4.8 (Holomorphic implies infinitely differentiable). *If f is holomorphic on D, then f has derivatives of all orders on D, and each derivative is holomorphic.*

This is strikingly different from real analysis, where differentiability does not imply twice differentiability.

Theorem 4.4.9 (Morera's theorem). *If f is continuous on a domain D and $\oint_\gamma f \, dz = 0$ for every closed contour γ in D, then f is holomorphic on D.*

Theorem 4.4.10 (Liouville's theorem). *A bounded entire function is constant.*

Proof. Let $|f(z)| \leq M$ for all z. By Cauchy's formula for derivatives, with γ a circle of radius R centered at any point a:

$$|f'(a)| = \left| \frac{1}{2\pi i} \oint_\gamma \frac{f(z)}{(z - a)^2} \, dz \right| \leq \frac{1}{2\pi} \cdot \frac{M}{R^2} \cdot 2\pi R = \frac{M}{R}.$$

As $R \to \infty$, we get $f'(a) = 0$ for all a. Thus f is constant. □

Corollary 4.4.11 (Fundamental theorem of algebra). *Every non-constant polynomial has a root in $\mathbb{C}$.*

Proof. If $p(z)$ has no root, then $1/p(z)$ is entire. Since $|p(z)| \to \infty$ as $|z| \to \infty$, $1/p$ is bounded, hence constant by Liouville's theorem. Contradiction. □

4.5 Power Series and Taylor Series

4.5.1 Power Series

Theorem 4.5.1 (Convergence of power series). *A power series $\sum_{n=0}^{\infty} a_n (z - z_0)^n$ has a radius of convergence $R \in [0, \infty]$ such that:*

- *The series converges absolutely for $|z - z_0| < R$.*
- *The series diverges for $|z - z_0| > R$.*
- *The convergence is uniform on any closed disk $|z - z_0| \leq r$ with $r < R$.*

Theorem 4.5.2 (Holomorphy of power series). *A power series defines a holomorphic function inside its disk of convergence, and can be differentiated term by term:*

$$f(z) = \sum_{n=0}^{\infty} a_n (z - z_0)^n \quad \Rightarrow \quad f'(z) = \sum_{n=1}^{\infty} n a_n (z - z_0)^{n-1}.$$

The derivative has the same radius of convergence.

4.5.2 Taylor Series

Theorem 4.5.3 (Taylor series). *If f is holomorphic on a disk $|z - z_0| < R$, then f equals its Taylor series:*

$$f(z) = \sum_{n=0}^{\infty} \frac{f^{(n)}(z_0)}{n!} (z - z_0)^n$$

for all z in the disk.

This is a profound difference from real analysis: complex-holomorphic functions are automatically analytic (equal to their Taylor series).

Theorem 4.5.4 (Identity theorem). *If f and g are holomorphic on a connected domain D and $f(z_n) = g(z_n)$ for a sequence (z_n) with a limit point in D, then $f = g$ on all of D.*

Corollary 4.5.5. *If f is holomorphic and not identically zero, its zeros are isolated.*

Example 4.5.1 (Standard Taylor series).

$$\begin{aligned}
e^z &= \sum_{n=0}^{\infty} \frac{z^n}{n!} & |z| &< \infty \\
\frac{1}{1-z} &= \sum_{n=0}^{\infty} z^n & |z| &< 1 \\
\log(1+z) &= \sum_{n=1}^{\infty} \frac{(-1)^{n+1}}{n} z^n & |z| &< 1 \\
\sin z &= \sum_{n=0}^{\infty} \frac{(-1)^n z^{2n+1}}{(2n+1)!} & |z| &< \infty \\
\cos z &= \sum_{n=0}^{\infty} \frac{(-1)^n z^{2n}}{(2n)!} & |z| &< \infty
\end{aligned}$$

4.6 Laurent Series and Singularities

4.6.1 Laurent Series

Theorem 4.6.1 (Laurent series). *If f is holomorphic in an annulus $r < |z - z_0| < R$, then:*

$$f(z) = \sum_{n=-\infty}^{\infty} a_n (z - z_0)^n$$

where

$$a_n = \frac{1}{2\pi i} \oint_\gamma \frac{f(\zeta)}{(\zeta - z_0)^{n+1}} \, d\zeta$$

for any circle γ in the annulus centered at z_0.

Definition 4.6.1 (Principal part). The **principal part** of the Laurent series is $\sum_{n=1}^{\infty} a_{-n}(z - z_0)^{-n}$, the terms with negative powers.

4.6.2 Classification of Singularities

Definition 4.6.2 (Isolated singularity). z_0 is an **isolated singularity** of f if f is holomorphic in some punctured disk $0 < |z - z_0| < R$.

Definition 4.6.3 (Types of singularities). An isolated singularity z_0 is:

- A **removable singularity** if f can be extended to a holomorphic function at z_0. Equivalently, the principal part is zero.
- A **pole of order** m if the principal part has finitely many terms, with $(z - z_0)^{-m}$ the lowest power. Equivalently, $\lim_{z \to z_0} (z - z_0)^m f(z)$ exists and is nonzero.
- An **essential singularity** if the principal part has infinitely many terms.

Theorem 4.6.2 (Riemann's removable singularity theorem). *If f is holomorphic in $0 < |z - z_0| < R$ and bounded near z_0, then z_0 is a removable singularity.*

Theorem 4.6.3 (Casorati-Weierstrass). *If z_0 is an essential singularity of f, then for any $w \in \mathbb{C}$ and any $\epsilon, \delta > 0$, there exists z with $0 < |z - z_0| < \delta$ and $|f(z) - w| < \epsilon$. That is, f comes arbitrarily close to every complex value in any neighborhood of z_0.*

Theorem 4.6.4 (Picard's theorem). *In any neighborhood of an essential singularity, f takes every complex value infinitely often, with at most one exception.*

Example 4.6.1.

- $f(z) = \sin z/z$ has a removable singularity at $z = 0$.
- $f(z) = 1/(z-1)^3$ has a pole of order 3 at $z = 1$.
- $f(z) = e^{1/z}$ has an essential singularity at $z = 0$.

4.6.3 Zeros and Poles

Definition 4.6.4 (Order of a zero). If $f(z_0) = 0$ and f is holomorphic near z_0, the **order** of the zero is the smallest $m \geq 1$ such that $f^{(m)}(z_0) \neq 0$. Equivalently, $f(z) = (z - z_0)^m g(z)$ where $g(z_0) \neq 0$.

Proposition 4.6.5. *f has a pole of order m at z_0 iff $1/f$ has a zero of order m at z_0.*

Definition 4.6.5 (Meromorphic function). A function is **meromorphic** on D if it is holomorphic except for isolated poles.

Example 4.6.2. Rational functions are meromorphic on $\mathbb{C}$. $\tan z = \sin z/\cos z$ is meromorphic with poles at $z = (n + 1/2)\pi$.

4.7 The Residue Theorem

4.7.1 Residues

Definition 4.7.1 (Residue). The **residue** of f at an isolated singularity z_0 is the coefficient a_{-1} in the Laurent expansion:

$$\operatorname{Res}(f, z_0) := a_{-1} = \frac{1}{2\pi i} \oint_\gamma f(z)\, dz$$

where γ is a small circle around z_0.

Proposition 4.7.1 (Residue formulas). *(i) If z_0 is a simple pole:* $\operatorname{Res}(f, z_0) = \lim_{z\to z_0}(z - z_0)f(z)$.

(ii) If z_0 is a pole of order m:

$$\operatorname{Res}(f, z_0) = \frac{1}{(m-1)!} \lim_{z\to z_0} \frac{d^{m-1}}{dz^{m-1}} \left[(z - z_0)^m f(z)\right].$$

(iii) If $f = g/h$ where $g(z_0) \neq 0$, $h(z_0) = 0$, and $h'(z_0) \neq 0$ (simple zero):

$$\operatorname{Res}(f, z_0) = \frac{g(z_0)}{h'(z_0)}.$$

4.7.2 The Residue Theorem

Theorem 4.7.2 (Residue theorem). *Let f be holomorphic inside and on a simple closed contour γ, except for isolated singularities $z_1, \dots, z_n$ inside γ. Then:*

$$\oint_\gamma f(z)\, dz = 2\pi i \sum_{k=1}^{n} \operatorname{Res}(f, z_k).$$

Proof. Surround each singularity z_k by a small circle C_k. By Cauchy's theorem for multiply connected domains:

$$\oint_\gamma f\, dz = \sum_{k=1}^{n} \oint_{C_k} f\, dz = \sum_{k=1}^{n} 2\pi i \cdot \operatorname{Res}(f, z_k). \qquad \square$$

Example 4.7.1. Evaluate $\oint_{|z|=2} \frac{z}{z^2-1}\, dz$.

The integrand has simple poles at $z = \pm 1$, both inside $|z| = 2$.

$$\begin{aligned} \operatorname{Res}(f, 1) &= \lim_{z\to 1}(z-1) \cdot \frac{z}{(z-1)(z+1)} = \frac{1}{2} \\ \operatorname{Res}(f, -1) &= \lim_{z\to -1}(z+1) \cdot \frac{z}{(z-1)(z+1)} = \frac{-1}{-2} = \frac{1}{2} \end{aligned}$$

Thus $\oint_{|z|=2} \frac{z}{z^2-1}\, dz = 2\pi i \left(\frac{1}{2} + \frac{1}{2}\right) = 2\pi i$.

4.7.3 Argument Principle and Rouché's Theorem

Theorem 4.7.3 (Argument principle). *Let f be meromorphic inside and on a simple closed contour γ, with no zeros or poles on γ. Let N be the number of zeros and P the number of poles inside γ (counted with multiplicity). Then:*

$$\frac{1}{2\pi i} \oint_\gamma \frac{f'(z)}{f(z)}\, dz = N - P.$$

Theorem 4.7.4 (Rouché's theorem). *If f and g are holomorphic inside and on a simple closed contour γ, and $|g(z)| < |f(z)|$ on γ, then f and $f+g$ have the same number of zeros inside γ.*

Example 4.7.2 (Application of Rouché's theorem). Show that $p(z) = z^5 + 3z + 1$ has exactly one root in $|z| < 1$.

On $|z| = 1$: Let $f(z) = 3z$ and $g(z) = z^5 + 1$. Then $|f(z)| = 3$ and $|g(z)| \le |z|^5 + 1 = 2 < 3 = |f(z)|$. By Rouché, $p(z) = f(z) + g(z)$ has the same number of zeros as $f(z) = 3z$, which has exactly one zero (at $z = 0$) in $|z| < 1$.

4.8 Evaluation of Real Integrals

The residue theorem provides powerful techniques for evaluating real integrals.

4.8.1 Trigonometric Integrals

Proposition 4.8.1. *For integrals $\int_0^{2\pi} R(\cos\theta, \sin\theta)\, d\theta$ where R is a rational function:*

Substitute $z = e^{i\theta}$, so $\cos\theta = (z + z^{-1})/2$, $\sin\theta = (z - z^{-1})/(2i)$, $d\theta = dz/(iz)$. The integral becomes a contour integral over $|z| = 1$.

Example 4.8.1. $\displaystyle\int_0^{2\pi} \frac{d\theta}{2 + \cos\theta}$.

With $z = e^{i\theta}$:

$$\int_0^{2\pi} \frac{d\theta}{2 + \cos\theta} = \oint_{|z|=1} \frac{1}{2 + \frac{z+z^{-1}}{2}} \cdot \frac{dz}{iz} = \oint_{|z|=1} \frac{2\, dz}{i(z^2 + 4z + 1)}.$$

The denominator has roots $z = -2 \pm \sqrt{3}$. Only $z = -2 + \sqrt{3}$ lies inside $|z| = 1$.

$$\text{Res}\left(\frac{2}{i(z^2 + 4z + 1)}, -2 + \sqrt{3}\right) = \frac{2}{i \cdot 2\sqrt{3}} = \frac{1}{i\sqrt{3}}.$$

Thus the integral equals $2\pi i \cdot \frac{1}{i\sqrt{3}} = \frac{2\pi}{\sqrt{3}}$.

4.8.2 Integrals over $(-\infty, \infty)$

Proposition 4.8.2 (Semicircular contour)**.** *For $\int_{-\infty}^{\infty} f(x)\, dx$ where $f(z)$ is rational with no real poles:*

If $f(z) \to 0$ fast enough as $|z| \to \infty$ in the upper half-plane, close the contour with a large semicircle in the upper half-plane. The semicircle contribution vanishes, giving:

$$\int_{-\infty}^{\infty} f(x)\, dx = 2\pi i \sum_{\text{Im}(z_k)>0} \text{Res}(f, z_k).$$

Theorem 4.8.3 (Jordan's lemma)**.** *If $f(z) \to 0$ uniformly as $|z| \to \infty$ in the upper half-plane, and $a > 0$, then:*

$$\lim_{R\to\infty} \int_{C_R} f(z) e^{iaz}\, dz = 0$$

where C_R is the semicircle $|z| = R$, $\text{Im}(z) \geq 0$.

Example 4.8.2. $\int_{-\infty}^{\infty} \frac{x^2}{x^4+1}\,dx.$

The poles of $f(z) = z^2/(z^4+1)$ are at $z^4 = -1 = e^{i\pi}$, i.e., $z = e^{i\pi(2k+1)/4}$ for $k = 0, 1, 2, 3$. The poles in the upper half-plane are $z_1 = e^{i\pi/4}$ and $z_2 = e^{3i\pi/4}$.

Computing residues and summing:

$$\int_{-\infty}^{\infty} \frac{x^2}{x^4+1}\,dx = \frac{\pi}{\sqrt{2}}.$$

4.8.3 Integrals with Branch Cuts

For integrands involving x^α or $\log x$, we use contours that avoid branch cuts.

Example 4.8.3. $\int_0^{\infty} \frac{x^{a-1}}{1+x}\,dx$ for $0 < a < 1$.

Use a keyhole contour around the branch cut on $[0, \infty)$. Setting $f(z) = z^{a-1}/(1+z)$ with branch cut on the positive real axis:

$$\int_0^{\infty} \frac{x^{a-1}}{1+x}\,dx - e^{2\pi i(a-1)} \int_0^{\infty} \frac{x^{a-1}}{1+x}\,dx = 2\pi i \cdot \operatorname{Res}(f, -1).$$

The residue at $z = -1$ is $(-1)^{a-1} = e^{i\pi(a-1)}$. Solving:

$$\int_0^{\infty} \frac{x^{a-1}}{1+x}\,dx = \frac{2\pi i \cdot e^{i\pi(a-1)}}{1 - e^{2\pi i(a-1)}} = \frac{\pi}{\sin(\pi a)}.$$

4.8.4 Principal Value Integrals

Definition 4.8.1 (Cauchy principal value)**.** If f has a singularity at $x_0 \in (a, b)$:

$$\text{P.V.} \int_a^b f(x)\,dx := \lim_{\epsilon \to 0+} \left(\int_a^{x_0-\epsilon} f(x)\,dx + \int_{x_0+\epsilon}^b f(x)\,dx \right).$$

Proposition 4.8.4. *If f has a simple pole at $x_0 \in \mathbb{R}$ and we indent the contour with a small semicircle of radius ϵ:*

$$\lim_{\epsilon \to 0} \int_{semicircle} f(z)\,dz = \pm \pi i \cdot \operatorname{Res}(f, x_0)$$

where + for upper semicircle (indent into upper half-plane) and − for lower.

4.9 Conformal Mappings

4.9.1 Conformal Maps

Definition 4.9.1 (Conformal map). A **conformal map** is a function that preserves angles (in both magnitude and orientation) between curves.

Theorem 4.9.1. *A holomorphic function f is conformal at any point where $f'(z) \neq 0$.*

Proof. If $\gamma(t)$ is a curve through z_0, then $(f \circ \gamma)'(t_0) = f'(z_0) \cdot \gamma'(t_0)$. Multiplication by $f'(z_0) \neq 0$ rotates by $\arg f'(z_0)$ and scales by $|f'(z_0)|$—the same transformation for all curves through z_0, hence angles are preserved. □

Definition 4.9.2 (Biholomorphism). A **biholomorphism** (or **conformal equivalence**) between domains D_1 and D_2 is a bijective holomorphic map $f : D_1 \to D_2$ (whose inverse is automatically holomorphic).

4.9.2 Möbius Transformations

Definition 4.9.3 (Möbius transformation). A **Möbius transformation** (or **linear fractional transformation**) is a map of the form:

$$f(z) = \frac{az+b}{cz+d}, \quad ad - bc \neq 0.$$

Proposition 4.9.2 (Properties of Möbius transformations).

(i) They form a group under composition.

(ii) They map circles and lines to circles and lines.

(iii) They are the only conformal automorphisms of $\hat{\mathbb{C}}$.

(iv) Any three distinct points can be mapped to any other three distinct points.

(v) They preserve the cross-ratio: $(z_1, z_2; z_3, z_4) = (f(z_1), f(z_2); f(z_3), f(z_4))$.

Definition 4.9.4 (Cross-ratio). $(z_1, z_2; z_3, z_4) := \frac{(z_1-z_3)(z_2-z_4)}{(z_1-z_4)(z_2-z_3)}$.

Example 4.9.1 (Standard Möbius transformations).

- $f(z) = z + b$: translation
- $f(z) = az$, $|a| = 1$: rotation
- $f(z) = rz$, $r > 0$: dilation
- $f(z) = 1/z$: inversion
- $f(z) = \frac{z-i}{z+i}$: maps upper half-plane to unit disk

4.9.3 The Riemann Mapping Theorem

Theorem 4.9.3 (Riemann mapping theorem). *Every simply connected proper subset of $\mathbb{C}$ is conformally equivalent to the open unit disk $\mathbb{D} = \{z : |z| < 1\}$.*

This is a profound existence theorem—it guarantees conformal maps exist between vastly different-looking regions—but it is non-constructive.

Example 4.9.2 (Useful conformal maps).

- $w = z^2$: maps first quadrant to upper half-plane
- $w = e^z$: maps horizontal strip $0 < \text{Im}(z) < \pi$ to upper half-plane
- $w = \sin z$: maps strip $-\pi/2 < \text{Re}(z) < \pi/2$ to $\mathbb{C} \setminus ((-\infty, -1] \cup [1, \infty))$
- $w = \frac{1+z}{1-z}$: maps unit disk to right half-plane

4.9.4 Applications to Physics

Conformal maps are used extensively in:

Electrostatics and fluid dynamics: Solutions to Laplace's equation in one domain transform to solutions in another. Complex potential $\Phi = \phi + i\psi$ where ϕ is the electrostatic potential and ψ is the stream function.

Heat conduction: Steady-state temperature distributions satisfy Laplace's equation.

Aerodynamics: The Joukowsky transformation $w = z + 1/z$ maps circles to airfoil shapes, enabling calculation of lift.

String theory: Conformal field theory relies on conformal symmetry in two dimensions.

4.10 Special Functions

4.10.1 The Gamma Function

Definition 4.10.1 (Gamma function). For $\text{Re}(z) > 0$:

$$\Gamma(z) := \int_0^\infty t^{z-1} e^{-t}\, dt.$$

Theorem 4.10.1 (Properties of Γ).

(i) $\Gamma(z+1) = z\Gamma(z)$ *(functional equation)*

(ii) $\Gamma(n+1) = n!$ *for* $n \in \mathbb{N}$ *(extends factorial)*

(iii) $\Gamma(1/2) = \sqrt{\pi}$

(iv) $\Gamma(z)\Gamma(1-z) = \frac{\pi}{\sin(\pi z)}$ *(reflection formula)*

(v) Γ *extends to a meromorphic function on* $\mathbb{C}$ *with simple poles at* $z = 0, -1, -2, \ldots$

Proposition 4.10.2 (Stirling's asymptotic formula). *As* $|z| \to \infty$ *with* $|\arg z| < \pi$*:*

$$\Gamma(z) \sim \sqrt{2\pi} z^{z-1/2} e^{-z}.$$

4.10.2 The Beta Function

Definition 4.10.2 (Beta function). For $\mathrm{Re}(a), \mathrm{Re}(b) > 0$:

$$B(a,b) := \int_0^1 t^{a-1}(1-t)^{b-1}\, dt.$$

Theorem 4.10.3. $B(a,b) = \frac{\Gamma(a)\Gamma(b)}{\Gamma(a+b)}$.

4.10.3 The Riemann Zeta Function

Definition 4.10.3 (Riemann zeta function). For $\mathrm{Re}(s) > 1$:

$$\zeta(s) := \sum_{n=1}^{\infty} \frac{1}{n^s} = \prod_p \frac{1}{1-p^{-s}}$$

where the product is over all primes p.

Theorem 4.10.4 (Analytic continuation). $\zeta(s)$ *extends to a meromorphic function on* $\mathbb{C}$ *with a single simple pole at* $s = 1$ *with residue* 1.

Theorem 4.10.5 (Functional equation).

$$\zeta(s) = 2^s \pi^{s-1} \sin\left(\frac{\pi s}{2}\right) \Gamma(1-s)\zeta(1-s).$$

Proposition 4.10.6 (Special values).

$$\zeta(2) = \frac{\pi^2}{6}, \quad \zeta(4) = \frac{\pi^4}{90}, \quad \zeta(-1) = -\frac{1}{12}, \quad \zeta(0) = -\frac{1}{2}.$$

The Riemann Hypothesis—that all non-trivial zeros of $\zeta(s)$ have $\mathrm{Re}(s) = 1/2$—remains one of the most important unsolved problems in mathematics.

4.11 Summary and Physical Applications

Complex analysis is indispensable in theoretical physics:

Electromagnetism: Complex potential methods solve 2D electrostatics and magnetostatics. The electric field $\mathbf{E} = -\nabla\phi$ where ϕ is the real part of an analytic function.

Fluid mechanics: Potential flow is described by analytic functions. The complex velocity potential enables exact solutions for flow around obstacles.

Quantum mechanics: The Schrödinger equation solutions often extend naturally to complex energies. Scattering amplitudes are analytic functions of momentum.

Quantum field theory: Propagators are boundary values of analytic functions. Dispersion relations connect real and imaginary parts. Contour integration evaluates Feynman diagrams.

Statistical mechanics: Partition functions are analytic except at phase transitions. Lee-Yang theory relates zeros to critical behavior.

Signal processing: The Laplace transform $\mathcal{L}\{f\}(s) = \int_0^\infty f(t)e^{-st}\,dt$ is analytic in a half-plane. Poles determine system behavior.

Dispersion relations: Causality implies analyticity in the upper half-plane for response functions. The Kramers-Kronig relations connect real and imaginary parts:

$$\operatorname{Re}\chi(\omega) = \frac{1}{\pi}\text{P.V.}\int_{-\infty}^{\infty} \frac{\operatorname{Im}\chi(\omega')}{\omega' - \omega}\,d\omega'.$$

Exercises

Exercise 4.11.1. Find all values of $\log(-1)$, $\sqrt{i}$, and $(1+i)^{1-i}$.

Exercise 4.11.2. Show that $\sin z$ and $\cos z$ are unbounded on $\mathbb{C}$. Find all zeros of $\sin z$.

Exercise 4.11.3. Verify the Cauchy-Riemann equations for $f(z) = e^z$ and find $f'(z)$ directly from the definition.

Exercise 4.11.4. If $f(z)$ is entire and $\operatorname{Re}(f(z)) \leq M$ for all z, prove that f is constant. (Hint: consider $e^{f(z)}$.)

Exercise 4.11.5. Find the Laurent series of $f(z) = \frac{1}{z(z-1)}$ valid for $0 < |z| < 1$ and for $|z| > 1$.

Exercise 4.11.6. Classify the singularities of $f(z) = \frac{\sin z}{z^3}$, $g(z) = \frac{e^z - 1}{z^2}$, and $h(z) = e^{1/(z-1)}$.

Exercise 4.11.7. Evaluate $\oint_{|z|=2} \frac{e^z}{z^2(z-1)}\,dz$ using residues.

Exercise 4.11.8. Evaluate $\int_{-\infty}^{\infty} \frac{\cos x}{x^2+1}\,dx$ and $\int_{-\infty}^{\infty} \frac{x \sin x}{x^2+1}\,dx$.

Exercise 4.11.9. Evaluate $\int_0^{2\pi} \frac{d\theta}{a+b\cos\theta}$ for $a > b > 0$.

Exercise 4.11.10. Use the keyhole contour to evaluate $\int_0^{\infty} \frac{\ln x}{x^2+1}\,dx$.

Exercise 4.11.11. Find a conformal map from the upper half-plane to the infinite strip $\{w : 0 < \mathrm{Im}(w) < 1\}$.

Exercise 4.11.12. Show that $\Gamma(z)$ has no zeros in $\mathbb{C}$.

Part II

Calculus on Manifolds

Chapter 5

Multivariable Calculus

Multivariable calculus extends the ideas of differentiation and integration to functions of several variables. This machinery is essential for describing physical systems in three-dimensional space—electromagnetic fields, fluid flow, gravitational potentials—and provides the foundation for the more abstract differential geometry needed in general relativity and gauge theories.

5.1 Topology of $\mathbb{R}^n$

5.1.1 Metric Structure

Definition 5.1.1 (Euclidean space). $\mathbb{R}^n$ is the set of n-tuples $\mathbf{x} = (x_1, \ldots, x_n)$ with the **Euclidean norm**:

$$\|\mathbf{x}\| := \sqrt{\sum_{i=1}^{n} x_i^2} = \sqrt{\mathbf{x} \cdot \mathbf{x}}$$

and **Euclidean distance** $d(\mathbf{x}, \mathbf{y}) := \|\mathbf{x} - \mathbf{y}\|$.

Proposition 5.1.1 (Norm properties).

(i) $\|\mathbf{x}\| \geq 0$, *with equality iff* $\mathbf{x} = \mathbf{0}$

(ii) $\|c\mathbf{x}\| = |c|\|\mathbf{x}\|$

(iii) $\|\mathbf{x} + \mathbf{y}\| \leq \|\mathbf{x}\| + \|\mathbf{y}\|$ *(triangle inequality)*

(iv) $|\mathbf{x} \cdot \mathbf{y}| \leq \|\mathbf{x}\|\|\mathbf{y}\|$ *(Cauchy-Schwarz)*

5.1.2 Open and Closed Sets

Definition 5.1.2 (Ball). The **open ball** of radius $r > 0$ centered at $\mathbf{a}$ is:

$$B_r(\mathbf{a}) := \{\mathbf{x} \in \mathbb{R}^n : \|\mathbf{x} - \mathbf{a}\| < r\}.$$

The **closed ball** uses $\leq$ instead of $<$.

Definition 5.1.3 (Open set). A set $U \subseteq \mathbb{R}^n$ is **open** if for every $\mathbf{a} \in U$, there exists $r > 0$ such that $B_r(\mathbf{a}) \subseteq U$.

Definition 5.1.4 (Closed set). A set $F \subseteq \mathbb{R}^n$ is **closed** if its complement $\mathbb{R}^n \setminus F$ is open. Equivalently, F contains all its limit points.

Proposition 5.1.2 (Properties).

(i) $\varnothing$ and $\mathbb{R}^n$ are both open and closed.

(ii) Arbitrary unions of open sets are open.

(iii) Finite intersections of open sets are open.

(iv) Arbitrary intersections of closed sets are closed.

(v) Finite unions of closed sets are closed.

Definition 5.1.5 (Boundary, interior, closure). For $S \subseteq \mathbb{R}^n$:

- The **interior** S° is the largest open set contained in S.
- The **closure** $\bar{S}$ is the smallest closed set containing S.
- The **boundary** $\partial S := \bar{S} \setminus S^\circ$.

5.1.3 Compactness and Connectedness

Definition 5.1.6 (Compact set). $K \subseteq \mathbb{R}^n$ is **compact** if every open cover has a finite subcover.

Theorem 5.1.3 (Heine-Borel). *A subset of $\mathbb{R}^n$ is compact iff it is closed and bounded.*

Theorem 5.1.4 (Sequential compactness). *$K \subseteq \mathbb{R}^n$ is compact iff every sequence in K has a subsequence converging to a point in K.*

Definition 5.1.7 (Connected set). $S \subseteq \mathbb{R}^n$ is **connected** if it cannot be written as $S = A \cup B$ where A and B are nonempty, disjoint, and both relatively open in S.

Definition 5.1.8 (Path-connected). S is **path-connected** if for any $\mathbf{a}, \mathbf{b} \in S$, there exists a continuous path $\gamma : [0,1] \to S$ with $\gamma(0) = \mathbf{a}$ and $\gamma(1) = \mathbf{b}$.

Proposition 5.1.5. *Path-connected implies connected. In $\mathbb{R}^n$, open connected sets are path-connected.*

Definition 5.1.9 (Simply connected). A path-connected set S is **simply connected** if every closed loop in S can be continuously contracted to a point within S.

Example 5.1.1. $\mathbb{R}^n$ is simply connected. $\mathbb{R}^n \setminus \{\mathbf{0}\}$ is simply connected for $n \geq 3$, but not for $n = 2$. A torus is connected but not simply connected.

5.2 Limits and Continuity

5.2.1 Limits of Functions

Definition 5.2.1 (Limit). Let $f : D \subseteq \mathbb{R}^n \to \mathbb{R}^m$ and let $\mathbf{a}$ be a limit point of D. We say $\lim_{\mathbf{x}\to\mathbf{a}} f(\mathbf{x}) = \mathbf{L}$ if:

$$\forall \epsilon > 0,\ \exists \delta > 0 : 0 < \|\mathbf{x} - \mathbf{a}\| < \delta \text{ and } \mathbf{x} \in D \Rightarrow \|f(\mathbf{x}) - \mathbf{L}\| < \epsilon.$$

Proposition 5.2.1 (Sequential characterization). $\lim_{\mathbf{x}\to\mathbf{a}} f(\mathbf{x}) = \mathbf{L}$ *iff for every sequence* $(\mathbf{x}_k)$ *in* $D \setminus \{\mathbf{a}\}$ *with* $\mathbf{x}_k \to \mathbf{a}$, *we have* $f(\mathbf{x}_k) \to \mathbf{L}$.

Remark 5.2.1. In multiple dimensions, $\mathbf{x}$ can approach $\mathbf{a}$ along infinitely many paths. For a limit to exist, all paths must yield the same value.

Example 5.2.1. Consider $f(x,y) = \frac{xy}{x^2+y^2}$ for $(x,y) \neq (0,0)$.
Along $y = 0$: $f(x,0) = 0 \to 0$.
Along $y = x$: $f(x,x) = \frac{x^2}{2x^2} = \frac{1}{2} \to \frac{1}{2}$.
Different paths give different limits, so $\lim_{(x,y)\to(0,0)} f(x,y)$ does not exist.

5.2.2 Continuous Functions

Definition 5.2.2 (Continuity). $f : D \to \mathbb{R}^m$ is **continuous at $\mathbf{a} \in D$** if $\lim_{\mathbf{x}\to\mathbf{a}} f(\mathbf{x}) = f(\mathbf{a})$. It is **continuous on** D if continuous at every point of D.

Proposition 5.2.2 (Component functions). $f = (f_1, \ldots, f_m) : D \to \mathbb{R}^m$ *is continuous iff each component* $f_i : D \to \mathbb{R}$ *is continuous.*

Theorem 5.2.3 (Properties of continuous functions).

(i) Sums, products, and compositions of continuous functions are continuous.

(ii) If K is compact and $f : K \to \mathbb{R}$ is continuous, then f attains its maximum and minimum.

(iii) Continuous image of a compact set is compact.

(iv) Continuous image of a connected set is connected.

Definition 5.2.3 (Uniform continuity). $f : D \to \mathbb{R}^m$ is **uniformly continuous** if:

$$\forall \epsilon > 0,\ \exists \delta > 0 : \|\mathbf{x} - \mathbf{y}\| < \delta \Rightarrow \|f(\mathbf{x}) - f(\mathbf{y})\| < \epsilon.$$

Theorem 5.2.4. *A continuous function on a compact set is uniformly continuous.*

5.3 Differentiation

5.3.1 Partial Derivatives

Definition 5.3.1 (Partial derivative). The **partial derivative** of $f : \mathbb{R}^n \to \mathbb{R}$ with respect to x_i at $\mathbf{a}$ is:

$$\frac{\partial f}{\partial x_i}(\mathbf{a}) := \lim_{h \to 0} \frac{f(\mathbf{a} + h\mathbf{e}_i) - f(\mathbf{a})}{h}$$

where $\mathbf{e}_i$ is the i-th standard basis vector.

Alternative notations: $\partial_i f$, f_{x_i}, $D_i f$, $\partial f / \partial x_i$.

Definition 5.3.2 (Gradient). If all partial derivatives exist, the **gradient** of $f : \mathbb{R}^n \to \mathbb{R}$ is:

$$\nabla f := \left(\frac{\partial f}{\partial x_1}, \ldots, \frac{\partial f}{\partial x_n} \right).$$

Remark 5.3.1. Existence of partial derivatives does not imply continuity. The function $f(x, y) = \frac{xy}{x^2+y^2}$ (with $f(0, 0) = 0$) has partial derivatives at the origin but is not continuous there.

5.3.2 The Total Derivative

Definition 5.3.3 (Differentiability). $f : U \subseteq \mathbb{R}^n \to \mathbb{R}^m$ (where U is open) is **differentiable** at $\mathbf{a} \in U$ if there exists a linear map $L : \mathbb{R}^n \to \mathbb{R}^m$ such that:

$$\lim_{\mathbf{h} \to \mathbf{0}} \frac{\|f(\mathbf{a} + \mathbf{h}) - f(\mathbf{a}) - L(\mathbf{h})\|}{\|\mathbf{h}\|} = 0.$$

The map L is the **total derivative** (or **differential**) of f at $\mathbf{a}$, denoted $Df(\mathbf{a})$ or $df_{\mathbf{a}}$.

Equivalently: $f(\mathbf{a}+\mathbf{h}) = f(\mathbf{a}) + L(\mathbf{h}) + o(\|\mathbf{h}\|)$ as $\mathbf{h} \to \mathbf{0}$.

Theorem 5.3.1 (Differentiability implies continuity). *If f is differentiable at $\mathbf{a}$, then f is continuous at $\mathbf{a}$.*

Theorem 5.3.2 (Jacobian matrix). *If $f = (f_1, \ldots, f_m) : \mathbb{R}^n \to \mathbb{R}^m$ is differentiable at $\mathbf{a}$, then all partial derivatives exist and $Df(\mathbf{a})$ is represented by the $m \times n$ **Jacobian matrix**:*

$$J_f(\mathbf{a}) = \begin{pmatrix} \frac{\partial f_1}{\partial x_1} & \cdots & \frac{\partial f_1}{\partial x_n} \\ \vdots & \ddots & \vdots \\ \frac{\partial f_m}{\partial x_1} & \cdots & \frac{\partial f_m}{\partial x_n} \end{pmatrix}_{\mathbf{a}}$$

For $f : \mathbb{R}^n \to \mathbb{R}$, the Jacobian is the row vector $(\nabla f)^T$.

Theorem 5.3.3 (Sufficient condition for differentiability). *If all partial derivatives of f exist in a neighborhood of $\mathbf{a}$ and are continuous at $\mathbf{a}$, then f is differentiable at $\mathbf{a}$.*

Definition 5.3.4 (C^k functions). f is of class C^k if all partial derivatives up to order k exist and are continuous. C^∞ means smooth (infinitely differentiable).

5.3.3 Chain Rule

Theorem 5.3.4 (Chain rule). *If $f : U \subseteq \mathbb{R}^n \to \mathbb{R}^m$ is differentiable at $\mathbf{a}$ and $g : V \subseteq \mathbb{R}^m \to \mathbb{R}^p$ is differentiable at $f(\mathbf{a})$, then $g \circ f$ is differentiable at $\mathbf{a}$ with:*

$$D(g \circ f)(\mathbf{a}) = Dg(f(\mathbf{a})) \circ Df(\mathbf{a}).$$

In matrix form: $J_{g\circ f}(\mathbf{a}) = J_g(f(\mathbf{a})) \cdot J_f(\mathbf{a})$.

Example 5.3.1. If $z = f(x, y)$ and $x = x(s, t)$, $y = y(s, t)$, then:

$$\frac{\partial z}{\partial s} = \frac{\partial f}{\partial x}\frac{\partial x}{\partial s} + \frac{\partial f}{\partial y}\frac{\partial y}{\partial s}.$$

5.3.4 Directional Derivatives

Definition 5.3.5 (Directional derivative). The **directional derivative** of $f : \mathbb{R}^n \to \mathbb{R}$ at $\mathbf{a}$ in the direction $\mathbf{v}$ (with $\|\mathbf{v}\| = 1$) is:

$$D_{\mathbf{v}} f(\mathbf{a}) := \lim_{t\to 0} \frac{f(\mathbf{a}+t\mathbf{v}) - f(\mathbf{a})}{t} = \left.\frac{d}{dt}\right|_{t=0} f(\mathbf{a}+t\mathbf{v}).$$

Theorem 5.3.5. *If f is differentiable at* $\mathbf{a}$*, then:*

$$D_{\mathbf{v}} f(\mathbf{a}) = \nabla f(\mathbf{a}) \cdot \mathbf{v}.$$

Corollary 5.3.6 (Gradient as direction of steepest ascent). *The gradient $\nabla f(\mathbf{a})$ points in the direction of maximum rate of increase of f. The magnitude $\|\nabla f(\mathbf{a})\|$ is that maximum rate.*

Corollary 5.3.7 (Level sets). *∇f is perpendicular to level sets of f. If $f(\mathbf{x}) = c$ defines a surface, then ∇f is normal to the surface.*

5.3.5 Higher-Order Derivatives

Definition 5.3.6 (Second partial derivatives).

$$\frac{\partial^2 f}{\partial x_j \partial x_i} := \frac{\partial}{\partial x_j}\left(\frac{\partial f}{\partial x_i}\right).$$

Theorem 5.3.8 (Symmetry of mixed partials (Schwarz/Clairaut)). *If f is C^2 on an open set, then:*

$$\frac{\partial^2 f}{\partial x_i \partial x_j} = \frac{\partial^2 f}{\partial x_j \partial x_i}.$$

Definition 5.3.7 (Hessian matrix). For $f : \mathbb{R}^n \to \mathbb{R}$, the **Hessian** is the $n \times n$ matrix of second partial derivatives:

$$H_f(\mathbf{a}) = \left(\frac{\partial^2 f}{\partial x_i \partial x_j}(\mathbf{a})\right).$$

The Hessian is symmetric for C^2 functions and represents the second derivative $D^2 f$.

5.3.6 Taylor's Theorem

Theorem 5.3.9 (Taylor's theorem in $\mathbb{R}^n$). *If $f : \mathbb{R}^n \to \mathbb{R}$ is C^{k+1} near* $\mathbf{a}$*, then:*

$$f(\mathbf{a}+\mathbf{h}) = \sum_{|\alpha|\le k} \frac{1}{\alpha!} D^\alpha f(\mathbf{a})\mathbf{h}^\alpha + R_k(\mathbf{h})$$

where $\alpha = (\alpha_1, \ldots, \alpha_n)$ is a multi-index, $|\alpha| = \sum \alpha_i$, $\alpha! = \prod \alpha_i!$, $\mathbf{h}^\alpha = \prod h_i^{\alpha_i}$, $D^\alpha = \prod \partial^{\alpha_i}/\partial x_i^{\alpha_i}$, and $R_k = O(\|\mathbf{h}\|^{k+1})$.

Proposition 5.3.10 (Second-order Taylor expansion).

$$f(\mathbf{a}+\mathbf{h}) = f(\mathbf{a}) + \nabla f(\mathbf{a}) \cdot \mathbf{h} + \frac{1}{2}\mathbf{h}^T H_f(\mathbf{a})\mathbf{h} + O(\|\mathbf{h}\|^3).$$

5.4 Inverse and Implicit Function Theorems

5.4.1 The Inverse Function Theorem

Theorem 5.4.1 (Inverse function theorem). *Let $f : U \subseteq \mathbb{R}^n \to \mathbb{R}^n$ be C^1 with $\det J_f(\mathbf{a}) \neq 0$. Then there exist open neighborhoods V of $\mathbf{a}$ and W of $f(\mathbf{a})$ such that $f : V \to W$ is a bijection with C^1 inverse $f^{-1} : W \to V$. Moreover:*

$$J_{f^{-1}}(f(\mathbf{a})) = [J_f(\mathbf{a})]^{-1}.$$

Remark 5.4.1. The condition $\det J_f(\mathbf{a}) \neq 0$ means $Df(\mathbf{a})$ is invertible. The theorem is local: f need not be globally invertible.

Example 5.4.1. The polar coordinate map $(r, \theta) \mapsto (r\cos\theta, r\sin\theta)$ has Jacobian determinant r. It is locally invertible for $r > 0$, but not globally one-to-one (periodic in θ).

5.4.2 The Implicit Function Theorem

Theorem 5.4.2 (Implicit function theorem). *Let $F : \mathbb{R}^n \times \mathbb{R}^m \to \mathbb{R}^m$ be C^1 near $(\mathbf{a}, \mathbf{b})$ with $F(\mathbf{a}, \mathbf{b}) = \mathbf{0}$. Write points as $(\mathbf{x}, \mathbf{y})$ with $\mathbf{x} \in \mathbb{R}^n$, $\mathbf{y} \in \mathbb{R}^m$.*

If the $m \times m$ matrix $\frac{\partial F}{\partial \mathbf{y}}(\mathbf{a}, \mathbf{b})$ is invertible, then there exist neighborhoods U of $\mathbf{a}$ and V of $\mathbf{b}$, and a C^1 function $g : U \to V$ such that:

$$F(\mathbf{x}, \mathbf{y}) = \mathbf{0} \text{ in } U \times V \quad \Leftrightarrow \quad \mathbf{y} = g(\mathbf{x}).$$

Example 5.4.2. The equation $F(x, y) = x^2 + y^2 - 1 = 0$ defines a circle. At $(0, 1)$: $\partial F/\partial y = 2y = 2 \neq 0$. So near $(0, 1)$, we can solve for $y = g(x) = \sqrt{1 - x^2}$.

At $(1, 0)$: $\partial F/\partial y = 0$, but $\partial F/\partial x = 2 \neq 0$. Here we solve for x as a function of y.

Corollary 5.4.3 (Implicit differentiation). *If $F(\mathbf{x}, g(\mathbf{x})) = \mathbf{0}$, then:*

$$\frac{\partial g}{\partial \mathbf{x}} = -\left(\frac{\partial F}{\partial \mathbf{y}}\right)^{-1} \frac{\partial F}{\partial \mathbf{x}}.$$

5.5 Optimization

5.5.1 Local Extrema

Definition 5.5.1 (Local extremum). $f : D \to \mathbb{R}$ has a **local maximum** at $\mathbf{a}$ if $f(\mathbf{x}) \leq f(\mathbf{a})$ for all $\mathbf{x}$ in some neighborhood of $\mathbf{a}$. Similarly for local minimum.

Theorem 5.5.1 (First-order necessary condition). *If f has a local extremum at an interior point $\mathbf{a}$ and f is differentiable there, then $\nabla f(\mathbf{a}) = \mathbf{0}$.*

Definition 5.5.2 (Critical point). A **critical point** (or **stationary point**) is a point where $\nabla f = \mathbf{0}$ or ∇f does not exist.

Theorem 5.5.2 (Second derivative test). *Let $f : \mathbb{R}^n \to \mathbb{R}$ be C^2 with $\nabla f(\mathbf{a}) = \mathbf{0}$. Let $H = H_f(\mathbf{a})$ be the Hessian.*

(i) If H is positive definite, $\mathbf{a}$ is a local minimum.

(ii) If H is negative definite, $\mathbf{a}$ is a local maximum.

(iii) If H is indefinite (has both positive and negative eigenvalues), $\mathbf{a}$ is a saddle point.

(iv) If H is semidefinite (has zero eigenvalues), the test is inconclusive.

Example 5.5.1. For $f(x, y) = x^2 - y^2$: $\nabla f = (2x, -2y) = \mathbf{0}$ at origin. $H = \begin{pmatrix} 2 & 0 \\ 0 & -2 \end{pmatrix}$ is indefinite, so $(0, 0)$ is a saddle point.

5.5.2 Constrained Optimization: Lagrange Multipliers

Theorem 5.5.3 (Lagrange multipliers). *To find extrema of $f(\mathbf{x})$ subject to constraints $g_1(\mathbf{x}) = 0, \ldots, g_k(\mathbf{x}) = 0$:*

If $\mathbf{a}$ is an extremum and the gradients $\nabla g_1(\mathbf{a}), \ldots, \nabla g_k(\mathbf{a})$ are linearly independent, then there exist scalars $\lambda_1, \ldots, \lambda_k$ (Lagrange multipliers) such that:

$$\nabla f(\mathbf{a}) = \sum_{i=1}^{k} \lambda_i \nabla g_i(\mathbf{a}).$$

Definition 5.5.3 (Lagrangian). The **Lagrangian** is $\mathcal{L}(\mathbf{x}, \boldsymbol{\lambda}) := f(\mathbf{x}) - \sum_{i=1}^{k} \lambda_i g_i(\mathbf{x})$. Critical points satisfy $\nabla_{\mathbf{x}} \mathcal{L} = \mathbf{0}$ and $\nabla_{\boldsymbol{\lambda}} \mathcal{L} = \mathbf{0}$.

Example 5.5.2. Maximize $f(x, y) = xy$ subject to $g(x, y) = x^2 + y^2 - 1 = 0$.
$\nabla f = (y, x)$, $\nabla g = (2x, 2y)$.
$\nabla f = \lambda \nabla g$ gives $y = 2\lambda x$ and $x = 2\lambda y$.
If $\lambda \neq 0$: $y = 2\lambda x$ and $x = 2\lambda \cdot 2\lambda x = 4\lambda^2 x$. So $4\lambda^2 = 1$, giving $\lambda = \pm 1/2$.
For $\lambda = 1/2$: $y = x$. From constraint: $2x^2 = 1$, so $x = \pm 1/\sqrt{2}$.
Maximum value: $f = xy = 1/2$ at $(1/\sqrt{2}, 1/\sqrt{2})$ and $(-1/\sqrt{2}, -1/\sqrt{2})$.

Remark 5.5.1 (Physical interpretation). The Lagrange multiplier λ measures the sensitivity of the optimum to the constraint: $\lambda = df^*/dc$ where f^* is the optimal value and c is the constraint level.

5.6 Multiple Integrals

5.6.1 Double Integrals

Definition 5.6.1 (Double integral over rectangles). For $f : R = [a, b] \times [c, d] \to \mathbb{R}$, the **double integral** is defined via Riemann sums:

$$\iint_R f(x, y)\, dA := \lim_{\|P\| \to 0} \sum_{i,j} f(x_{ij}^*, y_{ij}^*) \Delta x_i \Delta y_j.$$

Theorem 5.6.1 (Fubini's theorem). *If f is continuous on $R = [a, b] \times [c, d]$, then:*

$$\iint_R f(x, y)\, dA = \int_a^b \left(\int_c^d f(x, y)\, dy \right) dx = \int_c^d \left(\int_a^b f(x, y)\, dx \right) dy.$$

Fubini's theorem allows us to compute double integrals as iterated single integrals.

Definition 5.6.2 (Integral over general regions). For a bounded region D:

$$\iint_D f\, dA := \iint_R f \cdot \chi_D\, dA$$

where R is a rectangle containing D and χ_D is the characteristic function.

Proposition 5.6.2 (Type I and Type II regions).

- ***Type I:*** $D = \{(x, y) : a \le x \le b,\ g_1(x) \le y \le g_2(x)\}$

$$\iint_D f\, dA = \int_a^b \int_{g_1(x)}^{g_2(x)} f(x, y)\, dy\, dx.$$

- ***Type II:*** $D = \{(x, y) : c \le y \le d,\ h_1(y) \le x \le h_2(y)\}$

$$\iint_D f\, dA = \int_c^d \int_{h_1(y)}^{h_2(y)} f(x, y)\, dx\, dy.$$

5.6.2 Triple Integrals

Definition 5.6.3 (Triple integral). For $f : E \subseteq \mathbb{R}^3 \to \mathbb{R}$:

$$\iiint_E f(x, y, z)\, dV = \iiint_E f\, dx\, dy\, dz.$$

Fubini's theorem extends: for suitable regions, triple integrals reduce to iterated single integrals.

5.6.3 Change of Variables

Theorem 5.6.3 (Change of variables). *Let $T : U \to V$ be a C^1 bijection between open sets in $\mathbb{R}^n$, and let f be integrable on V. Then:*

$$\int_V f(\mathbf{x})\,d\mathbf{x} = \int_U f(T(\mathbf{u}))\,|\det J_T(\mathbf{u})|\,d\mathbf{u}.$$

The factor $|\det J_T|$ is the **Jacobian** and accounts for how T stretches or compresses volume.

Example 5.6.1 (Polar coordinates). $T(r, \theta) = (r\cos\theta, r\sin\theta)$.

$$J_T = \begin{pmatrix} \cos\theta & -r\sin\theta \\ \sin\theta & r\cos\theta \end{pmatrix}, \qquad \det J_T = r.$$

Thus $dA = dx\,dy = r\,dr\,d\theta$.

Example 5.6.2 (Cylindrical coordinates). $(x, y, z) = (r\cos\theta, r\sin\theta, z)$. Jacobian: r.

$$dV = r\,dr\,d\theta\,dz.$$

Example 5.6.3 (Spherical coordinates). $(x, y, z) = (\rho\sin\phi\cos\theta, \rho\sin\phi\sin\theta, \rho\cos\phi)$.

$$J = \rho^2\sin\phi, \quad dV = \rho^2\sin\phi\,d\rho\,d\phi\,d\theta.$$

Here $\rho \geq 0$, $0 \leq \phi \leq \pi$, $0 \leq \theta < 2\pi$.

Example 5.6.4 (Gaussian integral).

$$\int_{-\infty}^{\infty} e^{-x^2}\,dx = \sqrt{\pi}.$$

Proof: Let $I = \int_{-\infty}^{\infty} e^{-x^2}dx$. Then:

$$I^2 = \int_{-\infty}^{\infty}\int_{-\infty}^{\infty} e^{-(x^2+y^2)}dx\,dy = \int_0^{2\pi}\int_0^{\infty} e^{-r^2}r\,dr\,d\theta = 2\pi\cdot\frac{1}{2} = \pi.$$

So $I = \sqrt{\pi}$.

5.7 Vector Calculus

5.7.1 Vector Fields

Definition 5.7.1 (Vector field). A **vector field** on $D \subseteq \mathbb{R}^n$ is a function $\mathbf{F} : D \to \mathbb{R}^n$ assigning a vector to each point.

In $\mathbb{R}^3$: $\mathbf{F}(x, y, z) = P(x, y, z)\mathbf{i} + Q(x, y, z)\mathbf{j} + R(x, y, z)\mathbf{k}$.

Example 5.7.1 (Physical vector fields).

- Gravitational field: $\mathbf{F} = -\frac{Gm}{r^3}\mathbf{r}$
- Electric field: $\mathbf{E} = \frac{q}{4\pi\epsilon_0 r^3}\mathbf{r}$
- Velocity field of a fluid

5.7.2 Gradient, Divergence, and Curl

Definition 5.7.2 (Gradient). For $f : \mathbb{R}^3 \to \mathbb{R}$:

$$\nabla f = \operatorname{grad} f = \frac{\partial f}{\partial x}\mathbf{i} + \frac{\partial f}{\partial y}\mathbf{j} + \frac{\partial f}{\partial z}\mathbf{k}.$$

Definition 5.7.3 (Divergence). For $\mathbf{F} = (P, Q, R)$:

$$\nabla \cdot \mathbf{F} = \operatorname{div} \mathbf{F} = \frac{\partial P}{\partial x} + \frac{\partial Q}{\partial y} + \frac{\partial R}{\partial z}.$$

The divergence measures the "outward flux density" or "source strength" at a point.

Definition 5.7.4 (Curl). For $\mathbf{F} = (P, Q, R)$:

$$\nabla \times \mathbf{F} = \operatorname{curl} \mathbf{F} = \begin{vmatrix} \mathbf{i} & \mathbf{j} & \mathbf{k} \\ \partial_x & \partial_y & \partial_z \\ P & Q & R \end{vmatrix} = \left(\frac{\partial R}{\partial y} - \frac{\partial Q}{\partial z}, \frac{\partial P}{\partial z} - \frac{\partial R}{\partial x}, \frac{\partial Q}{\partial x} - \frac{\partial P}{\partial y} \right).$$

The curl measures the "rotation" or "circulation density" of the field.

Definition 5.7.5 (Laplacian). For $f : \mathbb{R}^3 \to \mathbb{R}$:

$$\nabla^2 f = \Delta f = \nabla \cdot (\nabla f) = \frac{\partial^2 f}{\partial x^2} + \frac{\partial^2 f}{\partial y^2} + \frac{\partial^2 f}{\partial z^2}.$$

5.7.3 Vector Identities

Proposition 5.7.1 (Important identities). *For scalar field f and vector fields $\mathbf{F}$, $\mathbf{G}$:*

$$\begin{aligned} \nabla \times (\nabla f) &= \mathbf{0} && \textit{(curl of gradient} \\ \nabla \cdot (\nabla \times \mathbf{F}) &= 0 && \textit{(divergence of cu} \\ \nabla \times (\nabla \times \mathbf{F}) &= \nabla(\nabla \cdot \mathbf{F}) - \nabla^2 \mathbf{F} \\ \nabla \cdot (f\mathbf{F}) &= f(\nabla \cdot \mathbf{F}) + \mathbf{F} \cdot \nabla f \\ \nabla \times (f\mathbf{F}) &= f(\nabla \times \mathbf{F}) + (\nabla f) \times \mathbf{F} \\ \nabla(\mathbf{F} \cdot \mathbf{G}) &= (\mathbf{F} \cdot \nabla)\mathbf{G} + (\mathbf{G} \cdot \nabla)\mathbf{F} + \mathbf{F} \times (\nabla \times \mathbf{G}) + \mathbf{G} \times (\nabla \times \mathbf{F}) \end{aligned}$$

Definition 5.7.6 (Conservative and irrotational fields). A vector field $\mathbf{F}$ is:

- **Conservative** (or **gradient field**) if $\mathbf{F} = \nabla f$ for some scalar f (called a **potential**).
- **Irrotational** if $\nabla \times \mathbf{F} = \mathbf{0}$.
- **Solenoidal** (or **incompressible**) if $\nabla \cdot \mathbf{F} = 0$.

Theorem 5.7.2. *On a simply connected domain, $\mathbf{F}$ is conservative iff $\mathbf{F}$ is irrotational.*

5.8 Line and Surface Integrals

5.8.1 Line Integrals

Definition 5.8.1 (Line integral of a scalar field). For $f : \mathbb{R}^n \to \mathbb{R}$ and curve C parametrized by $\mathbf{r}(t)$, $a \leq t \leq b$:

$$\int_C f\, ds = \int_a^b f(\mathbf{r}(t)) \|\mathbf{r}'(t)\|\, dt.$$

Definition 5.8.2 (Line integral of a vector field). For $\mathbf{F} : \mathbb{R}^n \to \mathbb{R}^n$:

$$\int_C \mathbf{F} \cdot d\mathbf{r} = \int_a^b \mathbf{F}(\mathbf{r}(t)) \cdot \mathbf{r}'(t)\, dt.$$

In $\mathbb{R}^3$ with $\mathbf{F} = (P, Q, R)$:

$$\int_C \mathbf{F} \cdot d\mathbf{r} = \int_C P\, dx + Q\, dy + R\, dz.$$

Proposition 5.8.1 (Physical interpretation). *$\int_C \mathbf{F} \cdot d\mathbf{r}$ represents the work done by force $\mathbf{F}$ along path C.*

Theorem 5.8.2 (Path independence). *The following are equivalent for a vector field $\mathbf{F}$ on a connected open set D:*

(i) $\int_C \mathbf{F} \cdot d\mathbf{r}$ is independent of path (depends only on endpoints).

(ii) $\oint_C \mathbf{F} \cdot d\mathbf{r} = 0$ for every closed curve C.

(iii) $\mathbf{F} = \nabla f$ for some scalar function f.

If D is simply connected, these are also equivalent to:

(iv) $\nabla \times \mathbf{F} = \mathbf{0}$ (in $\mathbb{R}^3$).

Theorem 5.8.3 (Fundamental theorem for line integrals). *If $\mathbf{F} = \nabla f$, then:*

$$\int_C \mathbf{F} \cdot d\mathbf{r} = f(\mathbf{r}(b)) - f(\mathbf{r}(a)).$$

5.8.2 Surface Integrals

Definition 5.8.3 (Parametric surface). A **parametric surface** is a map $\mathbf{r} : D \subseteq \mathbb{R}^2 \to \mathbb{R}^3$, $(u, v) \mapsto \mathbf{r}(u, v)$.

Definition 5.8.4 (Surface integral of a scalar field).

$$\iint_S f\,dS = \iint_D f(\mathbf{r}(u,v)) \left\| \frac{\partial \mathbf{r}}{\partial u} \times \frac{\partial \mathbf{r}}{\partial v} \right\| du\,dv.$$

The factor $\|\mathbf{r}_u \times \mathbf{r}_v\|$ is the area element.

Definition 5.8.5 (Oriented surface and flux). An **oriented surface** has a continuous choice of unit normal $\mathbf{n}$. The **flux** of $\mathbf{F}$ through S is:

$$\iint_S \mathbf{F} \cdot d\mathbf{S} = \iint_S \mathbf{F} \cdot \mathbf{n}\,dS = \iint_D \mathbf{F} \cdot (\mathbf{r}_u \times \mathbf{r}_v)\,du\,dv.$$

Example 5.8.1. For $z = g(x, y)$:

$$\iint_S \mathbf{F} \cdot d\mathbf{S} = \iint_D (-Pg_x - Qg_y + R)\,dx\,dy$$

for upward-pointing normal.

5.9 The Fundamental Theorems of Vector Calculus

5.9.1 Green's Theorem

Theorem 5.9.1 (Green's theorem). *Let D be a region in $\mathbb{R}^2$ bounded by a positively oriented, piecewise smooth, simple closed curve C. If P and Q have continuous partial derivatives on an open region containing D, then:*

$$\oint_C (P\,dx + Q\,dy) = \iint_D \left(\frac{\partial Q}{\partial x} - \frac{\partial P}{\partial y} \right) dA.$$

Corollary 5.9.2 (Area formula).

$$Area(D) = \frac{1}{2} \oint_C (x\,dy - y\,dx) = \oint_C x\,dy = -\oint_C y\,dx.$$

Example 5.9.1. Verify Green's theorem for $\mathbf{F} = (y^2, x)$ on the unit disk.
$\oint_C (y^2\,dx + x\,dy) = \iint_D (1 - 2y)\,dA = \iint_D 1\,dA - 2\iint_D y\,dA = \pi - 0 = \pi$.
Direct computation of the line integral confirms this.

5.9.2 Stokes' Theorem

Theorem 5.9.3 (Stokes' theorem). *Let S be an oriented piecewise smooth surface bounded by a simple, closed, piecewise smooth curve C with compatible orientation. If $\mathbf{F}$ has continuous partial derivatives, then:*

$$\oint_C \mathbf{F} \cdot d\mathbf{r} = \iint_S (\nabla \times \mathbf{F}) \cdot d\mathbf{S}.$$

Stokes' theorem relates circulation around a curve to the flux of curl through any surface bounded by that curve.

Remark 5.9.1. Green's theorem is the special case of Stokes' theorem for a flat surface in the xy-plane.

Corollary 5.9.4. *If $\nabla \times \mathbf{F} = \mathbf{0}$ on a simply connected domain, then $\oint_C \mathbf{F} \cdot d\mathbf{r} = 0$ for any closed curve, confirming that $\mathbf{F}$ is conservative.*

5.9.3 The Divergence Theorem

Theorem 5.9.5 (Divergence theorem (Gauss's theorem)). *Let E be a solid region with outward-oriented boundary surface S. If $\mathbf{F}$ has continuous partial derivatives, then:*

$$\iint_S \mathbf{F} \cdot d\mathbf{S} = \iiint_E (\nabla \cdot \mathbf{F})\, dV.$$

The divergence theorem relates the flux out of a closed surface to the "total source strength" inside.

Example 5.9.2 (Gauss's law in electrostatics). For electric field $\mathbf{E}$:

$$\iint_S \mathbf{E} \cdot d\mathbf{S} = \frac{1}{\epsilon_0} \iiint_E \rho\, dV = \frac{Q_{\text{enc}}}{\epsilon_0}$$

since $\nabla \cdot \mathbf{E} = \rho/\epsilon_0$.

Corollary 5.9.6 (Integration by parts in $\mathbb{R}^n$).

$$\iiint_E f(\nabla \cdot \mathbf{F})\, dV = \iint_S f\mathbf{F} \cdot d\mathbf{S} - \iiint_E \mathbf{F} \cdot \nabla f\, dV.$$

Corollary 5.9.7 (Green's identities). *For scalar functions u and v:*

$$\textit{First:} \quad \iiint_E (u\nabla^2 v + \nabla u \cdot \nabla v)\, dV = \iint_S u\frac{\partial v}{\partial n}\, dS$$

$$\textit{Second:} \quad \iiint_E (u\nabla^2 v - v\nabla^2 u)\, dV = \iint_S \left(u\frac{\partial v}{\partial n} - v\frac{\partial u}{\partial n}\right) dS$$

where $\partial/\partial n$ denotes the normal derivative.

5.9.4 Unified View: The Generalized Stokes' Theorem

All three theorems (Green, Stokes, Divergence) are special cases of the generalized Stokes' theorem:

$$\int_{\partial M} \omega = \int_M d\omega$$

where ω is a differential form, $d\omega$ is its exterior derivative, and ∂M is the boundary of the manifold M. We develop this in the next chapter.

5.10 Curvilinear Coordinates

5.10.1 General Formalism

Definition 5.10.1 (Curvilinear coordinates). A **curvilinear coordinate system** on a region $D \subseteq \mathbb{R}^n$ is a smooth bijection $\mathbf{r} : U \to D$ from an open set $U \subseteq \mathbb{R}^n$.

Definition 5.10.2 (Scale factors and metric). The **scale factors** are $h_i := \left\| \frac{\partial \mathbf{r}}{\partial u_i} \right\|$. The **metric tensor** has components:

$$g_{ij} = \frac{\partial \mathbf{r}}{\partial u_i} \cdot \frac{\partial \mathbf{r}}{\partial u_j}.$$

For orthogonal coordinates: $g_{ij} = h_i^2 \delta_{ij}$.

Proposition 5.10.1 (Differential operators in orthogonal coordinates). *For orthogonal coordinates (u_1, u_2, u_3) with scale factors (h_1, h_2, h_3):*

$$\begin{aligned}
\nabla f &= \sum_{i=1}^{3} \frac{1}{h_i} \frac{\partial f}{\partial u_i} \hat{\mathbf{e}}_i \\
\nabla \cdot \mathbf{F} &= \frac{1}{h_1 h_2 h_3} \left[\frac{\partial}{\partial u_1}(h_2 h_3 F_1) + \frac{\partial}{\partial u_2}(h_1 h_3 F_2) + \frac{\partial}{\partial u_3}(h_1 h_2 F_3) \right] \\
\nabla^2 f &= \frac{1}{h_1 h_2 h_3} \left[\frac{\partial}{\partial u_1} \left(\frac{h_2 h_3}{h_1} \frac{\partial f}{\partial u_1} \right) + \textit{cyclic permutations} \right]
\end{aligned}$$

5.10.2 Common Coordinate Systems

Example 5.10.1 (Cylindrical coordinates (r, θ, z)).

$$x = r\cos\theta, \quad y = r\sin\theta, \quad z = z.$$

Scale factors: $h_r = 1$, $h_\theta = r$, $h_z = 1$.

$$\nabla f = \frac{\partial f}{\partial r}\hat{\mathbf{r}} + \frac{1}{r}\frac{\partial f}{\partial \theta}\hat{\boldsymbol{\theta}} + \frac{\partial f}{\partial z}\hat{\mathbf{z}}$$
$$\nabla^2 f = \frac{1}{r}\frac{\partial}{\partial r}\left(r\frac{\partial f}{\partial r}\right) + \frac{1}{r^2}\frac{\partial^2 f}{\partial \theta^2} + \frac{\partial^2 f}{\partial z^2}$$

Example 5.10.2 (Spherical coordinates (ρ, θ, ϕ))**.** Convention: ρ = radial distance, θ = azimuthal angle (from z-axis), ϕ = polar angle.

$$x = \rho \sin\theta \cos\phi, \quad y = \rho \sin\theta \sin\phi, \quad z = \rho\cos\theta.$$

Scale factors: $h_\rho = 1$, $h_\theta = \rho$, $h_\phi = \rho\sin\theta$.

$$\nabla f = \frac{\partial f}{\partial \rho}\hat{\boldsymbol{\rho}} + \frac{1}{\rho}\frac{\partial f}{\partial \theta}\hat{\boldsymbol{\theta}} + \frac{1}{\rho\sin\theta}\frac{\partial f}{\partial \phi}\hat{\boldsymbol{\phi}}$$
$$\nabla^2 f = \frac{1}{\rho^2}\frac{\partial}{\partial \rho}\left(\rho^2\frac{\partial f}{\partial \rho}\right) + \frac{1}{\rho^2\sin\theta}\frac{\partial}{\partial \theta}\left(\sin\theta\frac{\partial f}{\partial \theta}\right) + \frac{1}{\rho^2\sin^2\theta}\frac{\partial^2 f}{\partial \phi^2}$$

Remark 5.10.1. Convention varies: many physics texts swap the meanings of θ and ϕ. Always check the convention being used.

5.11 Potential Theory

5.11.1 Scalar and Vector Potentials

Theorem 5.11.1 (Helmholtz decomposition)**.** *Any sufficiently smooth vector field* $\mathbf{F}$ *on* $\mathbb{R}^3$ *(decaying at infinity) can be decomposed as:*

$$\mathbf{F} = -\nabla\phi + \nabla \times \mathbf{A}$$

where ϕ *is a scalar potential and* $\mathbf{A}$ *is a vector potential.*

Proposition 5.11.2 (Finding potentials)**.**

- *If* $\nabla \times \mathbf{F} = \mathbf{0}$*, then* $\mathbf{F} = \nabla f$ *where* $f(\mathbf{r}) = -\int_{\mathbf{r}_0}^{\mathbf{r}} \mathbf{F} \cdot d\mathbf{r}'$.
- *If* $\nabla \cdot \mathbf{F} = 0$*, then* $\mathbf{F} = \nabla \times \mathbf{A}$ *for some* $\mathbf{A}$.

5.11.2 Laplace's and Poisson's Equations

Definition 5.11.1 (Harmonic function)**.** A function f is **harmonic** on D if $\nabla^2 f = 0$ on D.

Theorem 5.11.3 (Mean value property). *If f is harmonic on D, then for any ball $B_r(\mathbf{a}) \subseteq D$:*

$$f(\mathbf{a}) = \frac{1}{|S_r|} \iint_{S_r} f\, dS = \frac{1}{|B_r|} \iiint_{B_r} f\, dV$$

where S_r is the sphere of radius r centered at $\mathbf{a}$.

Theorem 5.11.4 (Maximum principle). *If f is harmonic on a bounded domain D and continuous on $\bar{D}$, then f attains its maximum and minimum on ∂D.*

Corollary 5.11.5 (Uniqueness). *A solution to Laplace's equation with given boundary values is unique.*

Definition 5.11.2 (Poisson's equation). $\nabla^2 f = \rho$ where ρ is a given source term.

Theorem 5.11.6 (Solution via Green's function). *The solution to $\nabla^2 f = \rho$ (with suitable boundary conditions) is:*

$$f(\mathbf{r}) = -\frac{1}{4\pi} \iiint \frac{\rho(\mathbf{r}')}{|\mathbf{r} - \mathbf{r}'|} d^3\mathbf{r}'.$$

The function $G(\mathbf{r}, \mathbf{r}') = -1/(4\pi|\mathbf{r} - \mathbf{r}'|)$ is the ***Green's function*** *for the Laplacian.*

5.12 Summary and Physical Applications

Multivariable calculus provides the mathematical language for classical field theories:

Electrostatics: Electric field $\mathbf{E} = -\nabla\phi$ (conservative), with $\nabla \cdot \mathbf{E} = \rho/\epsilon_0$ (Gauss's law) and $\nabla \times \mathbf{E} = \mathbf{0}$ (irrotational).

Magnetostatics: Magnetic field $\nabla \cdot \mathbf{B} = 0$ (solenoidal), $\nabla \times \mathbf{B} = \mu_0 \mathbf{J}$ (Ampère's law), $\mathbf{B} = \nabla \times \mathbf{A}$.

Fluid mechanics: Continuity equation $\partial\rho/\partial t + \nabla \cdot (\rho\mathbf{v}) = 0$. Incompressible flow: $\nabla \cdot \mathbf{v} = 0$. Irrotational flow: $\nabla \times \mathbf{v} = \mathbf{0}$.

Heat conduction: Heat equation $\partial T/\partial t = \alpha\nabla^2 T$. Steady state: Laplace's equation.

Classical mechanics: Lagrangian mechanics on configuration space. Hamilton's equations involve gradients of the Hamiltonian.

General relativity: The mathematical structure generalizes to curved manifolds, with the metric tensor replacing the Euclidean metric.

Exercises

Exercise 5.12.1. Show that $\lim_{(x,y)\to(0,0)} \frac{x^2 y}{x^4+y^2}$ does not exist by approaching along $y = x^2$.

Exercise 5.12.2. If $f(x, y) = \begin{cases} \frac{xy(x^2-y^2)}{x^2+y^2} & (x, y) \neq (0, 0) \\ 0 & (x, y) = (0, 0) \end{cases}$, show that $\frac{\partial^2 f}{\partial x \partial y}(0, 0) \neq \frac{\partial^2 f}{\partial y \partial x}(0, 0)$.

Exercise 5.12.3. Find the extrema of $f(x, y, z) = x + y + z$ subject to $x^2 + y^2 = 1$ and $y + z = 1$.

Exercise 5.12.4. Evaluate $\iint_D e^{-(x^2+y^2)}\, dA$ over the region $D = \{(x, y) : x^2 + y^2 \leq R^2\}$. What happens as $R \to \infty$?

Exercise 5.12.5. Verify Stokes' theorem for $\mathbf{F} = (y, -x, z)$ and S the upper hemisphere $x^2 + y^2 + z^2 = 1$, $z \geq 0$.

Exercise 5.12.6. Verify the divergence theorem for $\mathbf{F} = (x^3, y^3, z^3)$ over the unit ball.

Exercise 5.12.7. Express $\nabla \times \mathbf{F}$ in cylindrical coordinates for a general vector field $\mathbf{F} = F_r\hat{\mathbf{r}} + F_\theta\hat{\boldsymbol{\theta}} + F_z\hat{\mathbf{z}}$.

Exercise 5.12.8. Show that if f is harmonic in $\mathbb{R}^3$ and $f(\mathbf{r}) \to 0$ as $|\mathbf{r}| \to \infty$, then $f \equiv 0$.

Exercise 5.12.9. If $\mathbf{F} = \frac{\mathbf{r}}{r^3}$ for $\mathbf{r} \neq \mathbf{0}$, compute $\nabla \cdot \mathbf{F}$ and explain the apparent paradox with the divergence theorem.

Exercise 5.12.10. Find the potential function for $\mathbf{F} = (2xy + z^3, x^2 + 2yz, 3xz^2 + y^2)$ and verify your answer.

Exercise 5.12.11. Prove that if $\mathbf{F}$ is both irrotational and solenoidal, then each component of $\mathbf{F}$ is harmonic.

Exercise 5.12.12. Use the change of variables formula to derive the volume of an n-dimensional ball of radius R.

Chapter 6

Differential Geometry

Differential geometry studies smooth spaces called manifolds, which generalize curves and surfaces to arbitrary dimensions. This framework is essential for general relativity (where spacetime is a curved 4-manifold), gauge theories (where fields live on fiber bundles), and classical mechanics (where phase space is a symplectic manifold). This chapter develops the core concepts: manifolds, tangent spaces, differential forms, and curvature.

6.1 Manifolds

6.1.1 Topological Manifolds

Definition 6.1.1 (Topological manifold). An **n-dimensional topological manifold** is a topological space M that is:

(i) **Hausdorff**: distinct points have disjoint neighborhoods.

(ii) **Second countable**: has a countable basis of open sets.

(iii) **Locally Euclidean**: every point has a neighborhood homeomorphic to an open subset of $\mathbb{R}^n$.

Definition 6.1.2 (Chart and coordinate system). A **chart** (or **coordinate system**) on M is a pair (U, ϕ) where $U \subseteq M$ is open and $\phi : U \to \phi(U) \subseteq \mathbb{R}^n$ is a homeomorphism. The functions $x^i := \phi^i$ are **local coordinates**.

Definition 6.1.3 (Atlas). An **atlas** on M is a collection of charts $\{(U_\alpha, \phi_\alpha)\}$ such that $\bigcup_\alpha U_\alpha = M$.

6.1.2 Smooth Manifolds

Definition 6.1.4 (Transition maps). If (U_α, ϕ_α) and (U_β, ϕ_β) are charts with $U_\alpha \cap U_\beta \neq \varnothing$, the **transition map** is:

$$\phi_\beta \circ \phi_\alpha^{-1} : \phi_\alpha(U_\alpha \cap U_\beta) \to \phi_\beta(U_\alpha \cap U_\beta).$$

Definition 6.1.5 (Smooth atlas). An atlas is **smooth** (or C^∞) if all transition maps are smooth (infinitely differentiable as maps between open subsets of $\mathbb{R}^n$).

Definition 6.1.6 (Smooth manifold). A **smooth manifold** is a topological manifold equipped with a maximal smooth atlas (one that contains every chart compatible with it).

Example 6.1.1 (Fundamental examples).

(a) $\mathbb{R}^n$ with the identity chart.

(b) The n-sphere $S^n = \{\mathbf{x} \in \mathbb{R}^{n+1} : \|\mathbf{x}\| = 1\}$ with stereographic projections.

(c) The torus $T^n = S^1 \times \cdots \times S^1$.

(d) Real projective space $\mathbb{R}P^n$ (lines through the origin in $\mathbb{R}^{n+1}$).

(e) Matrix groups: $GL(n, \mathbb{R})$, $O(n)$, $SO(n)$, $SL(n, \mathbb{R})$.

(f) Any open subset of a smooth manifold.

Example 6.1.2 (The 2-sphere S^2). Using stereographic projection from the north pole $N = (0, 0, 1)$:

$$\phi_N(x, y, z) = \left(\frac{x}{1 - z}, \frac{y}{1 - z}\right)$$

maps $S^2 \setminus \{N\}$ to $\mathbb{R}^2$. Similarly for the south pole. The transition map is:

$$\phi_S \circ \phi_N^{-1}(u, v) = \frac{1}{u^2 + v^2}(u, v)$$

which is smooth on $\mathbb{R}^2 \setminus \{0\}$.

6.1.3 Smooth Maps

Definition 6.1.7 (Smooth function). A function $f : M \to \mathbb{R}$ is **smooth** if for every chart (U, ϕ), the composition $f \circ \phi^{-1} : \phi(U) \to \mathbb{R}$ is smooth.

Definition 6.1.8 (Smooth map between manifolds). A map $F : M \to N$ between smooth manifolds is **smooth** if for every chart (U, ϕ) on M and (V, ψ) on N with $F(U) \subseteq V$, the map $\psi \circ F \circ \phi^{-1}$ is smooth.

Definition 6.1.9 (Diffeomorphism). A **diffeomorphism** is a smooth bijection with smooth inverse. Manifolds M and N are **diffeomorphic** if such a map exists.

The set of smooth functions $M \to \mathbb{R}$ is denoted $C^\infty(M)$. It is a commutative algebra under pointwise operations.

6.2 Tangent Spaces

6.2.1 Tangent Vectors as Derivations

There are several equivalent ways to define tangent vectors. The most elegant uses derivations.

Definition 6.2.1 (Derivation at a point). A **derivation at** $p \in M$ is a linear map $v : C^\infty(M) \to \mathbb{R}$ satisfying the Leibniz rule:

$$v(fg) = f(p)v(g) + g(p)v(f).$$

Definition 6.2.2 (Tangent space). The **tangent space** to M at p, denoted T_pM, is the set of all derivations at p.

Proposition 6.2.1. *T_pM is a vector space of dimension $n = \dim M$.*

Proposition 6.2.2 (Coordinate basis). *Let (U, ϕ) be a chart with $\phi = (x^1, \ldots, x^n)$. The derivations*

$$\left.\frac{\partial}{\partial x^i}\right|_p : f \mapsto \frac{\partial(f \circ \phi^{-1})}{\partial x^i}(\phi(p))$$

form a basis for T_pM. Any $v \in T_pM$ can be written:

$$v = v^i \left.\frac{\partial}{\partial x^i}\right|_p$$

*where $v^i = v(x^i)$ are the **components** of v.*

6.2.2 Alternative Definitions

Definition 6.2.3 (Tangent vector via curves). A tangent vector at p is an equivalence class of smooth curves $\gamma : (-\epsilon, \epsilon) \to M$ with $\gamma(0) = p$, where $\gamma_1 \sim \gamma_2$ if $(f \circ \gamma_1)'(0) = (f \circ \gamma_2)'(0)$ for all $f \in C^\infty(M)$.

The derivation corresponding to $[\gamma]$ is $v_\gamma(f) := (f \circ \gamma)'(0)$.

6.2.3 The Differential (Pushforward)

Definition 6.2.4 (Differential of a map). Let $F : M \to N$ be smooth and $p \in M$. The **differential** (or **pushforward**) of F at p is the linear map $dF_p : T_pM \to T_{F(p)}N$ defined by:

$$(dF_p(v))(g) := v(g \circ F)$$

for $v \in T_pM$ and $g \in C^\infty(N)$.

Alternative notations: F_{*p}, F_*, DF_p, TF_p.

Proposition 6.2.3 (Chain rule). *If $F : M \to N$ and $G : N \to P$ are smooth, then:*

$$d(G \circ F)_p = dG_{F(p)} \circ dF_p.$$

Proposition 6.2.4 (Coordinate expression). *If $F : M \to N$ with local coordinates x^i on M and y^j on N, then:*

$$dF_p\left(\frac{\partial}{\partial x^i}\right) = \frac{\partial F^j}{\partial x^i}(p)\frac{\partial}{\partial y^j}.$$

The matrix $(\partial F^j/\partial x^i)$ is the Jacobian of F in these coordinates.

6.2.4 The Tangent Bundle

Definition 6.2.5 (Tangent bundle). The **tangent bundle** of M is

$$TM := \bigsqcup_{p \in M} T_pM = \{(p, v) : p \in M, v \in T_pM\}.$$

Theorem 6.2.5. *TM is naturally a smooth $2n$-dimensional manifold. The projection $\pi : TM \to M$, $(p, v) \mapsto p$, is smooth.*

Definition 6.2.6 (Vector field). A **vector field** on M is a smooth section of TM: a smooth map $X : M \to TM$ such that $\pi \circ X = \mathrm{id}_M$. Equivalently, X assigns to each $p \in M$ a tangent vector $X_p \in T_pM$, smoothly.

The set of vector fields on M is denoted $\mathfrak{X}(M)$. It is a $C^\infty(M)$-module and a real vector space.

6.3 Cotangent Spaces and Differential Forms

6.3.1 Cotangent Space

Definition 6.3.1 (Cotangent space). The **cotangent space** at p is the dual space:

$$T_p^*M := (T_pM)^* = \{\omega : T_pM \to \mathbb{R} \text{ linear}\}.$$

Elements are called **covectors** or **1-forms at** p.

Definition 6.3.2 (Differential of a function). For $f \in C^\infty(M)$, the **differential** $df_p \in T_p^*M$ is defined by:

$$df_p(v) := v(f).$$

Proposition 6.3.1 (Dual basis). *If $\{(\partial/\partial x^i)|_p\}$ is the coordinate basis for T_pM, then $\{dx^i|_p\}$ is the dual basis for T_p^*M:*

$$dx^i\left(\frac{\partial}{\partial x^j}\right) = \delta^i_j.$$

Any covector can be written $\omega = \omega_i\, dx^i$.

6.3.2 Pullback

Definition 6.3.3 (Pullback). Let $F : M \to N$ be smooth. The **pullback** $F^* : T^*_{F(p)}N \to T_p^*M$ is the dual of dF_p:

$$(F^*\omega)(v) := \omega(dF_p(v))$$

for $\omega \in T^*_{F(p)}N$ and $v \in T_pM$.

Proposition 6.3.2. *For $g \in C^\infty(N)$: $F^*(dg) = d(g \circ F)$.*

6.3.3 Differential Forms

Definition 6.3.4 (k-form). A **differential k-form** (or just **k-form**) on M is a smooth section of $\Lambda^k(T^*M)$: at each point p, it gives an alternating k-linear map $(T_pM)^k \to \mathbb{R}$.

The space of k-forms is denoted $\Omega^k(M)$. We have $\Omega^0(M) = C^\infty(M)$.

Proposition 6.3.3 (Local expression). *In local coordinates, a k-form can be written:*

$$\omega = \sum_{i_1<\cdots<i_k} \omega_{i_1\cdots i_k}(x)\, dx^{i_1} \wedge \cdots \wedge dx^{i_k}$$

where $\omega_{i_1\cdots i_k}$ are smooth functions.

Example 6.3.1. A 1-form on $\mathbb{R}^3$: $\omega = P\,dx + Q\,dy + R\,dz$.
A 2-form on $\mathbb{R}^3$: $\eta = A\,dy \wedge dz + B\,dz \wedge dx + C\,dx \wedge dy$.
A 3-form on $\mathbb{R}^3$: $\mu = f\,dx \wedge dy \wedge dz$.

6.3.4 The Wedge Product

Definition 6.3.5 (Wedge product). The **wedge product** $\wedge : \Omega^k(M) \times \Omega^l(M) \to \Omega^{k+l}(M)$ is defined by:

$$(\omega\wedge\eta)_p(v_1, \dots, v_{k+l}) = \frac{1}{k!l!} \sum_{\sigma \in S_{k+l}} \operatorname{sgn}(\sigma)\, \omega_p(v_{\sigma(1)}, \dots, v_{\sigma(k)})\eta_p(v_{\sigma(k+1)}, \dots, v_{\sigma(k+l)}).$$

Proposition 6.3.4 (Properties of wedge product).

(i) Bilinear over $C^\infty(M)$.

(ii) Associative: $(\omega \wedge \eta) \wedge \mu = \omega \wedge (\eta \wedge \mu)$.

(iii) Graded commutative: $\omega \wedge \eta = (-1)^{kl}\eta \wedge \omega$ for $\omega \in \Omega^k$, $\eta \in \Omega^l$.

(iv) $dx^i \wedge dx^i = 0$.

The space $\Omega^*(M) := \bigoplus_{k=0}^{n} \Omega^k(M)$ with the wedge product forms a graded algebra.

6.4 The Exterior Derivative

6.4.1 Definition and Properties

Theorem 6.4.1 (Exterior derivative). *There exists a unique family of linear operators $d : \Omega^k(M) \to \Omega^{k+1}(M)$ satisfying:*

(i) On functions ($k = 0$): df is the differential defined earlier.

(ii) $d^2 = 0$: $d(d\omega) = 0$ for all ω.

(iii) Graded Leibniz rule: $d(\omega \wedge \eta) = d\omega \wedge \eta + (-1)^k\omega \wedge d\eta$ for $\omega \in \Omega^k$.

(iv) Locality: $(d\omega)_p$ depends only on ω near p.

Proposition 6.4.2 (Coordinate formula). *If $\omega = \sum_I \omega_I\, dx^I$ (where I is a multi-index), then:*

$$d\omega = \sum_I d\omega_I \wedge dx^I = \sum_I \sum_j \frac{\partial \omega_I}{\partial x^j} dx^j \wedge dx^I.$$

Example 6.4.1. For $f \in C^\infty(\mathbb{R}^3)$:

$$df = \frac{\partial f}{\partial x}dx + \frac{\partial f}{\partial y}dy + \frac{\partial f}{\partial z}dz.$$

For $\omega = P\,dx + Q\,dy + R\,dz$:

$$d\omega = \left(\frac{\partial R}{\partial y} - \frac{\partial Q}{\partial z}\right) dy \wedge dz + \left(\frac{\partial P}{\partial z} - \frac{\partial R}{\partial x}\right) dz \wedge dx + \left(\frac{\partial Q}{\partial x} - \frac{\partial P}{\partial y}\right) dx \wedge dy.$$

This corresponds to the curl!

6.4.2 Correspondence with Vector Calculus

On $\mathbb{R}^3$ with coordinates (x, y, z), we can identify:

Forms	Vector Calculus	Operation
$f \in \Omega^0$	scalar field f	
$\omega = P\,dx + Q\,dy + R\,dz \in \Omega^1$	vector field (P, Q, R)	
$\eta = A\,dy \wedge dz + B\,dz \wedge dx + C\,dx \wedge dy \in \Omega^2$	vector field (A, B, C)	
$\mu = \rho\,dx \wedge dy \wedge dz \in \Omega^3$	scalar field ρ	
df	∇f	gradient
$d\omega$ (for 1-form)	$\nabla \times \mathbf{F}$	curl
$d\eta$ (for 2-form)	$\nabla \cdot \mathbf{G}$	divergence

The identity $d^2 = 0$ encodes both $\nabla \times (\nabla f) = \mathbf{0}$ and $\nabla \cdot (\nabla \times \mathbf{F}) = 0$.

6.4.3 Pullback of Forms

Theorem 6.4.3 (Pullback of differential forms). *If $F : M \to N$ is smooth, there is a pullback $F^* : \Omega^k(N) \to \Omega^k(M)$ satisfying:*

(i) $F^*(g) = g \circ F$ *for* $g \in \Omega^0(N)$.

(ii) $F^*(\omega \wedge \eta) = F^*\omega \wedge F^*\eta$.

(iii) $F^*(d\omega) = d(F^*\omega)$ *(commutes with* d*).*

(iv) $(G \circ F)^* = F^* \circ G^*$.

6.4.4 Closed and Exact Forms

Definition 6.4.1 (Closed and exact forms). A k-form ω is:

- **Closed** if $d\omega = 0$.
- **Exact** if $\omega = d\eta$ for some $(k-1)$-form η.

Since $d^2 = 0$, every exact form is closed. The converse fails in general.

Definition 6.4.2 (de Rham cohomology). The **k-th de Rham cohomology group** is:

$$H^k_{dR}(M) := \frac{\ker(d : \Omega^k \to \Omega^{k+1})}{\operatorname{im}(d : \Omega^{k-1} \to \Omega^k)} = \frac{\text{closed } k\text{-forms}}{\text{exact } k\text{-forms}}.$$

Theorem 6.4.4 (Poincaré lemma). *On a contractible open set (e.g., a ball), every closed form is exact. Thus $H^k_{dR}(\mathbb{R}^n) = 0$ for $k > 0$.*

Example 6.4.2. On $\mathbb{R}^2 \setminus \{0\}$, the 1-form $\omega = \frac{-y\,dx + x\,dy}{x^2+y^2}$ is closed but not exact. This reflects the nontrivial topology: $H^1_{dR}(\mathbb{R}^2 \setminus \{0\}) \cong \mathbb{R}$.

6.5 Integration on Manifolds

6.5.1 Orientation

Definition 6.5.1 (Orientation). An **orientation** on an n-manifold M is a continuous choice of "positive" basis for each T_pM. Equivalently, it is an equivalence class of atlases where all transition maps have positive Jacobian determinant.

Definition 6.5.2 (Volume form). A **volume form** is a nowhere-vanishing n-form on an n-manifold. A volume form determines an orientation.

Proposition 6.5.1. *M is orientable iff it admits a volume form. Non-orientable examples: Möbius strip, Klein bottle, $\mathbb{R}P^{2n}$.*

6.5.2 Integration of Forms

Definition 6.5.3 (Integration of n-forms). Let ω be an n-form on an oriented n-manifold M. In a positively oriented chart (U, ϕ) with $\omega = f\,dx^1 \wedge \cdots \wedge dx^n$:

$$\int_U \omega := \int_{\phi(U)} f(\phi^{-1}(x))\,dx^1 \cdots dx^n.$$

For compact support, extend using a partition of unity.

Theorem 6.5.2 (Change of variables). *If $F : M \to N$ is an orientation-preserving diffeomorphism and $\omega \in \Omega^n(N)$, then:*

$$\int_M F^*\omega = \int_N \omega.$$

6.5.3 Stokes' Theorem

Definition 6.5.4 (Manifold with boundary). An n-**manifold with boundary** is locally modeled on $\mathbb{R}^n$ or the half-space $\{x \in \mathbb{R}^n : x^n \geq 0\}$. The **boundary** ∂M consists of points mapping to $\{x^n = 0\}$.

∂M is an $(n-1)$-manifold without boundary. The orientation of M induces an orientation on ∂M (the "outward normal first" convention).

Theorem 6.5.3 (Generalized Stokes' theorem). *Let M be an oriented n-manifold with boundary, and let $\omega \in \Omega^{n-1}(M)$ have compact support. Then:*

$$\int_M d\omega = \int_{\partial M} \omega.$$

This single theorem encompasses:

- **Fundamental theorem of calculus**: $\int_a^b f'(x)\,dx = f(b) - f(a)$.
- **Green's theorem**: $\iint_D \left(\frac{\partial Q}{\partial x} - \frac{\partial P}{\partial y}\right) dA = \oint_{\partial D} P\,dx + Q\,dy$.
- **Classical Stokes**: $\iint_S (\nabla \times \mathbf{F}) \cdot d\mathbf{S} = \oint_{\partial S} \mathbf{F} \cdot d\mathbf{r}$.
- **Divergence theorem**: $\iiint_E (\nabla \cdot \mathbf{F})\,dV = \iint_{\partial E} \mathbf{F} \cdot d\mathbf{S}$.

6.6 Lie Derivatives and Flows

6.6.1 Integral Curves and Flows

Definition 6.6.1 (Integral curve). An **integral curve** of a vector field X is a curve $\gamma : I \to M$ such that $\gamma'(t) = X_{\gamma(t)}$ for all $t \in I$.

Theorem 6.6.1 (Existence and uniqueness). *For any $p \in M$ and vector field X, there exists a unique maximal integral curve $\gamma : I \to M$ with $\gamma(0) = p$.*

Definition 6.6.2 (Flow). The **flow** of X is the map $\Phi : \mathcal{D} \to M$ (where $\mathcal{D} \subseteq \mathbb{R} \times M$) defined by $\Phi_t(p) := \gamma_p(t)$, where γ_p is the integral curve starting at p.

Proposition 6.6.2 (Flow properties). *(i)* $\Phi_0 = \mathrm{id}_M$.

(ii) $\Phi_s \circ \Phi_t = \Phi_{s+t}$ *(where defined).*

(iii) $\frac{d}{dt}\big|_{t=0} \Phi_t(p) = X_p$.

If X is complete (integral curves exist for all time), the flow is a one-parameter group of diffeomorphisms.

6.6.2 Lie Derivative

Definition 6.6.3 (Lie derivative of a function)**.** For $X \in \mathfrak{X}(M)$ and $f \in C^\infty(M)$:

$$\mathcal{L}_X f := X(f) = \frac{d}{dt}\bigg|_{t=0} (\Phi_t^* f) = \frac{d}{dt}\bigg|_{t=0} (f \circ \Phi_t).$$

Definition 6.6.4 (Lie derivative of a vector field)**.** For $X, Y \in \mathfrak{X}(M)$:

$$\mathcal{L}_X Y := \frac{d}{dt}\bigg|_{t=0} (\Phi_{-t})_* Y = [X, Y]$$

where $[X, Y]$ is the **Lie bracket**.

Definition 6.6.5 (Lie bracket)**.** The **Lie bracket** of vector fields X and Y is the vector field $[X, Y]$ defined by:

$$[X, Y](f) := X(Y(f)) - Y(X(f)).$$

Proposition 6.6.3 (Properties of Lie bracket)**.**

(i) Bilinear over $\mathbb{R}$.

(ii) Antisymmetric: $[X, Y] = -[Y, X]$.

(iii) Jacobi identity: $[[X, Y], Z] + [[Y, Z], X] + [[Z, X], Y] = 0$.

(iv) $[fX, gY] = fg[X, Y] + f(Xg)Y - g(Yf)X$.

In coordinates, if $X = X^i \partial_i$ and $Y = Y^j \partial_j$:

$$[X, Y] = \left(X^i \frac{\partial Y^j}{\partial x^i} - Y^i \frac{\partial X^j}{\partial x^i} \right) \frac{\partial}{\partial x^j}.$$

Definition 6.6.6 (Lie derivative of differential forms)**.** For $\omega \in \Omega^k(M)$:

$$\mathcal{L}_X \omega := \frac{d}{dt}\bigg|_{t=0} \Phi_t^* \omega.$$

Theorem 6.6.4 (Cartan's magic formula).

$$\mathcal{L}_X\omega = d(\iota_X\omega) + \iota_X(d\omega) = (d\iota_X + \iota_X d)\omega$$

where ι_X is the ***interior product*** *(contraction with X).*

Definition 6.6.7 (Interior product). For $\omega \in \Omega^k(M)$ and $X \in \mathfrak{X}(M)$, $\iota_X\omega \in \Omega^{k-1}(M)$ is:

$$(\iota_X\omega)(Y_1, \dots, Y_{k-1}) := \omega(X, Y_1, \dots, Y_{k-1}).$$

6.7 Riemannian Geometry

6.7.1 Riemannian Metrics

Definition 6.7.1 (Riemannian metric). A **Riemannian metric** on M is a smooth assignment of an inner product $g_p : T_pM \times T_pM \to \mathbb{R}$ to each tangent space, i.e., a smooth, symmetric, positive-definite $(0,2)$-tensor field.

In local coordinates:

$$g = g_{ij}(x)\, dx^i \otimes dx^j$$

where (g_{ij}) is a positive-definite symmetric matrix at each point.

Definition 6.7.2 (Riemannian manifold). A **Riemannian manifold** is a smooth manifold equipped with a Riemannian metric.

Example 6.7.1.

(a) $\mathbb{R}^n$ with $g = \sum_i dx^i \otimes dx^i$ (Euclidean metric).

(b) S^2 with the metric induced from $\mathbb{R}^3$.

(c) Upper half-plane $\mathbb{H} = \{(x, y) : y > 0\}$ with $g = (dx^2 + dy^2)/y^2$ (hyperbolic metric).

Definition 6.7.3 (Length and distance). The **length** of a curve $\gamma : [a, b] \to M$ is:

$$L(\gamma) := \int_a^b \sqrt{g_{\gamma(t)}(\gamma'(t), \gamma'(t))}\, dt = \int_a^b \|\gamma'(t)\|\, dt.$$

The **distance** between points is the infimum of lengths of curves connecting them.

6.7.2 The Levi-Civita Connection

Definition 6.7.4 (Connection (covariant derivative)). A **connection** on M is an operator $\nabla : \mathfrak{X}(M) \times \mathfrak{X}(M) \to \mathfrak{X}(M)$, written $(X, Y) \mapsto \nabla_X Y$, satisfying:

(i) $\nabla_{fX+gY} Z = f\nabla_X Z + g\nabla_Y Z$ (linear in X).

(ii) $\nabla_X(Y + Z) = \nabla_X Y + \nabla_X Z$.

(iii) $\nabla_X(fY) = (Xf)Y + f\nabla_X Y$ (Leibniz rule).

Definition 6.7.5 (Christoffel symbols). In coordinates, a connection is determined by the **Christoffel symbols** Γ^k_{ij}:

$$\nabla_{\partial_i} \partial_j = \Gamma^k_{ij} \partial_k.$$

For any vector field $Y = Y^j \partial_j$:

$$\nabla_X Y = X^i \left(\frac{\partial Y^k}{\partial x^i} + \Gamma^k_{ij} Y^j \right) \partial_k.$$

Definition 6.7.6 (Torsion and metric compatibility). A connection is:

- **Torsion-free** if $\nabla_X Y - \nabla_Y X = [X, Y]$, i.e., $\Gamma^k_{ij} = \Gamma^k_{ji}$.
- **Metric-compatible** if $X(g(Y, Z)) = g(\nabla_X Y, Z) + g(Y, \nabla_X Z)$.

Theorem 6.7.1 (Levi-Civita connection). *On a Riemannian manifold, there exists a unique connection that is both torsion-free and metric-compatible. It is called the **Levi-Civita connection**.*

Proposition 6.7.2 (Christoffel symbols for Levi-Civita).

$$\Gamma^k_{ij} = \frac{1}{2} g^{kl} \left(\frac{\partial g_{jl}}{\partial x^i} + \frac{\partial g_{il}}{\partial x^j} - \frac{\partial g_{ij}}{\partial x^l} \right)$$

where (g^{kl}) is the inverse matrix of (g_{kl}).

6.7.3 Parallel Transport and Geodesics

Definition 6.7.7 (Parallel transport). A vector field V along a curve γ is **parallel** if $\nabla_{\gamma'} V = 0$. **Parallel transport** along γ is the isomorphism $T_{\gamma(a)}M \to T_{\gamma(b)}M$ defined by parallel translation.

Definition 6.7.8 (Geodesic). A **geodesic** is a curve γ whose tangent vector is parallel along itself:

$$\nabla_{\gamma'} \gamma' = 0.$$

In coordinates, $\gamma(t) = (x^1(t), \dots, x^n(t))$ is a geodesic iff:

$$\frac{d^2x^k}{dt^2} + \Gamma^k_{ij}\frac{dx^i}{dt}\frac{dx^j}{dt} = 0.$$

Theorem 6.7.3. *Geodesics locally minimize length. Any two sufficiently close points can be joined by a unique geodesic.*

Definition 6.7.9 (Exponential map)**.** The **exponential map** $\exp_p : T_pM \to M$ sends v to $\gamma_v(1)$, where γ_v is the geodesic with $\gamma_v(0) = p$ and $\gamma_v'(0) = v$.

The exponential map provides **normal coordinates** centered at p, in which $g_{ij}(p) = \delta_{ij}$ and $\Gamma^k_{ij}(p) = 0$.

6.8 Curvature

6.8.1 The Riemann Curvature Tensor

Definition 6.8.1 (Riemann curvature tensor)**.** The **Riemann curvature tensor** is the $(1,3)$-tensor R defined by:

$$R(X,Y)Z := \nabla_X\nabla_Y Z - \nabla_Y\nabla_X Z - \nabla_{[X,Y]}Z.$$

R measures the failure of second covariant derivatives to commute.

Proposition 6.8.1 (Coordinate expression)**.**

$$R^l{}_{kij} = \frac{\partial\Gamma^l_{kj}}{\partial x^i} - \frac{\partial\Gamma^l_{ki}}{\partial x^j} + \Gamma^m_{kj}\Gamma^l_{mi} - \Gamma^m_{ki}\Gamma^l_{mj}.$$

Proposition 6.8.2 (Symmetries of Riemann tensor)**.** *With* $R_{ijkl} := g_{im}R^m{}_{jkl}$*:*

(i) $R_{ijkl} = -R_{ijlk}$ *(antisymmetric in last two indices).*

(ii) $R_{ijkl} = -R_{jikl}$ *(antisymmetric in first two indices).*

(iii) $R_{ijkl} = R_{klij}$ *(symmetric under pair exchange).*

(iv) $R_{ijkl} + R_{iklj} + R_{iljk} = 0$ *(first Bianchi identity).*

Proposition 6.8.3 (Second Bianchi identity)**.**

$$\nabla_m R^l{}_{ijk} + \nabla_i R^l{}_{jmk} + \nabla_j R^l{}_{mik} = 0.$$

6.8.2 Ricci and Scalar Curvature

Definition 6.8.2 (Ricci tensor). The **Ricci tensor** is the contraction:

$$R_{ij} := R^k{}_{ikj} = g^{kl} R_{kilj}.$$

The Ricci tensor is symmetric: $R_{ij} = R_{ji}$.

Definition 6.8.3 (Scalar curvature). The **scalar curvature** is:

$$R := g^{ij} R_{ij}.$$

Definition 6.8.4 (Einstein tensor). The **Einstein tensor** is:

$$G_{ij} := R_{ij} - \frac{1}{2} R g_{ij}.$$

Theorem 6.8.4 (Contracted Bianchi identity). $\nabla^i G_{ij} = 0$, *i.e., the Einstein tensor is divergence-free.*

This identity is crucial for general relativity: it ensures conservation of energy-momentum.

6.8.3 Sectional and Gaussian Curvature

Definition 6.8.5 (Sectional curvature). For linearly independent vectors $X, Y \in T_p M$, the **sectional curvature** of the plane they span is:

$$K(X,Y) := \frac{g(R(X,Y)Y, X)}{g(X,X)g(Y,Y) - g(X,Y)^2}.$$

The sectional curvature measures the Gaussian curvature of the surface swept out by geodesics in the given plane.

Theorem 6.8.5 (Gauss's Theorema Egregium). *The Gaussian curvature of a surface in $\mathbb{R}^3$ depends only on the intrinsic metric, not on how the surface is embedded.*

Example 6.8.1 (Constant curvature spaces).

- $\mathbb{R}^n$: $K = 0$ (flat).
- S^n (radius r): $K = 1/r^2$ (positive curvature).
- Hyperbolic space $\mathbb{H}^n$: $K = -1$ (negative curvature).

6.9 Differential Forms and Physics

6.9.1 Electromagnetism

The electromagnetic field is naturally described by a 2-form on spacetime.

Definition 6.9.1 (Faraday 2-form). In coordinates (t, x, y, z):

$$F = E_x\, dx \wedge dt + E_y\, dy \wedge dt + E_z\, dz \wedge dt + B_x\, dy \wedge dz + B_y\, dz \wedge dx + B_z\, dx \wedge dy.$$

Maxwell's equations become:

$$\begin{aligned} dF &= 0 && (\text{homogeneous: } \nabla \cdot \mathbf{B} = 0,\ \nabla \times \mathbf{E} + \partial_t \mathbf{B} = 0) \\ d{*}F &= {*}J && (\text{inhomogeneous: } \nabla \cdot \mathbf{E} = \rho,\ \nabla \times \mathbf{B} - \partial_t \mathbf{E} = \mathbf{J}) \end{aligned}$$

where $*$ is the Hodge star operator.

Since $dF = 0$ and spacetime is simply connected, $F = dA$ for some 1-form A (the electromagnetic potential).

6.9.2 Symplectic Geometry and Mechanics

Definition 6.9.2 (Symplectic form). A **symplectic form** on a manifold M is a closed, non-degenerate 2-form ω.

Example 6.9.1. On $\mathbb{R}^{2n}$ with coordinates $(q^1, \ldots, q^n, p_1, \ldots, p_n)$:

$$\omega = \sum_{i=1}^{n} dq^i \wedge dp_i.$$

Theorem 6.9.1 (Darboux's theorem). *Every symplectic manifold locally looks like* $(\mathbb{R}^{2n}, \omega_0)$.

Hamilton's equations for a Hamiltonian $H : M \to \mathbb{R}$ are encoded by:

$$\iota_{X_H} \omega = -dH$$

where X_H is the Hamiltonian vector field.

6.10 Summary and Physical Applications

Differential geometry is the natural language for modern physics:

General relativity: Spacetime is a 4-dimensional Lorentzian manifold (metric signature $(-,+,+,+)$). Einstein's equation $G_{\mu\nu} = 8\pi G T_{\mu\nu}$ relates geometry (Einstein tensor) to matter (stress-energy tensor).

Gauge theories: Electromagnetic and Yang-Mills fields are connections on fiber bundles. The curvature of the connection is the field strength.

Classical mechanics: Phase space is a symplectic manifold. Lagrangian mechanics lives on the tangent bundle; Hamiltonian mechanics on the cotangent bundle.

String theory: Worldsheets are 2-dimensional surfaces. Target space geometry and Calabi-Yau manifolds are central.

Condensed matter: Berry phase and topological insulators involve geometric phases and Chern numbers.

Exercises

Exercise 6.10.1. Show that the transition map for stereographic projection on S^2 is smooth.

Exercise 6.10.2. Verify that $v(fg) = f(p)v(g) + g(p)v(f)$ for the derivation $v = \partial/\partial x^i|_p$.

Exercise 6.10.3. Compute the pullback $F^*(dx \wedge dy)$ where $F : \mathbb{R}^2 \to \mathbb{R}^2$ is $F(r, \theta) = (r\cos\theta, r\sin\theta)$.

Exercise 6.10.4. Verify that $d^2 = 0$ directly from the definition.

Exercise 6.10.5. Show that the 1-form $\omega = \frac{-y\,dx + x\,dy}{x^2+y^2}$ on $\mathbb{R}^2 \setminus \{0\}$ is closed but not exact. Compute $\oint_{S^1} \omega$.

Exercise 6.10.6. Compute the Lie bracket $[\partial_r, r\partial_\theta]$ in polar coordinates.

Exercise 6.10.7. Verify Cartan's formula $\mathcal{L}_X\omega = d(\iota_X\omega) + \iota_X(d\omega)$ for a 1-form.

Exercise 6.10.8. Compute the Christoffel symbols for the 2-sphere with metric $ds^2 = d\theta^2 + \sin^2\theta\, d\phi^2$.

Exercise 6.10.9. Show that geodesics on the sphere are great circles.

Exercise 6.10.10. Compute the Riemann tensor for the hyperbolic plane with metric $ds^2 = (dx^2 + dy^2)/y^2$.

Exercise 6.10.11. Verify the second Bianchi identity implies $\nabla^\mu G_{\mu\nu} = 0$.

Exercise 6.10.12. Show that a vector field X generates isometries of (M, g) iff $\mathcal{L}_X g = 0$. Such X is called a **Killing vector field**.

Chapter 7

Topology Essentials

Topology studies properties of spaces that are preserved under continuous deformations. For physics, topological concepts explain phenomena that are robust against perturbations: quantized quantities, topological phases of matter, anomalies, and global aspects of field configurations. This chapter covers the essential topological tools: homotopy, homology, fiber bundles, and characteristic classes.

7.1 Topological Spaces

7.1.1 Basic Definitions

Definition 7.1.1 (Topological space)**.** A **topological space** is a set X together with a collection $\mathcal{T}$ of subsets (called **open sets**) satisfying:

(i) $\varnothing, X \in \mathcal{T}$.

(ii) Arbitrary unions of open sets are open.

(iii) Finite intersections of open sets are open.

The collection $\mathcal{T}$ is called a **topology** on X.

Example 7.1.1 (Standard topologies)**.**

(a) **Discrete topology**: every subset is open.

(b) **Indiscrete topology**: only $\varnothing$ and X are open.

(c) **Metric topology**: open sets are unions of open balls.

(d) **Subspace topology**: $A \subseteq X$ inherits $\{U \cap A : U \in \mathcal{T}\}$.

(e) **Product topology**: on $X \times Y$, generated by $U \times V$ with U, V open.

(f) **Quotient topology**: on $X/\sim$, U open iff $\pi^{-1}(U)$ open in X.

Definition 7.1.2 (Continuous map). A function $f : X \to Y$ between topological spaces is **continuous** if $f^{-1}(V)$ is open in X for every open V in Y.

Definition 7.1.3 (Homeomorphism). A **homeomorphism** is a continuous bijection with continuous inverse. Spaces X and Y are **homeomorphic** ($X \cong Y$) if such a map exists.

Topology studies properties invariant under homeomorphism.

7.1.2 Separation Axioms

Definition 7.1.4 (Hausdorff space). A space is **Hausdorff** (or T_2) if distinct points have disjoint open neighborhoods.

Definition 7.1.5 (Normal space). A space is **normal** (or T_4) if disjoint closed sets have disjoint open neighborhoods.

All metric spaces are Hausdorff. Manifolds are required to be Hausdorff.

7.1.3 Compactness

Definition 7.1.6 (Compact space). A space X is **compact** if every open cover has a finite subcover.

Theorem 7.1.1 (Properties of compact spaces).

(i) Closed subsets of compact spaces are compact.

(ii) Continuous images of compact spaces are compact.

(iii) A compact subset of a Hausdorff space is closed.

(iv) Products of compact spaces are compact (Tychonoff's theorem).

Definition 7.1.7 (Local compactness). X is **locally compact** if every point has a compact neighborhood.

Definition 7.1.8 (Compactification). A **compactification** of X is a compact space $\hat{X}$ containing X as a dense subspace.

Example 7.1.2 (One-point compactification). $\hat{\mathbb{R}}^n = \mathbb{R}^n \cup \{\infty\} \cong S^n$. This is the Alexandroff compactification.

7.1.4 Connectedness

Definition 7.1.9 (Connected space)**.** X is **connected** if it cannot be written as a disjoint union of two nonempty open sets.

Definition 7.1.10 (Path-connected)**.** X is **path-connected** if any two points can be joined by a continuous path.

Proposition 7.1.2. *Path-connected $\Rightarrow$ connected. The converse fails in general but holds for manifolds.*

Definition 7.1.11 (Connected components)**.** The **connected components** of X are the maximal connected subsets. They partition X.

Definition 7.1.12 (Simply connected)**.** X is **simply connected** if it is path-connected and every loop can be continuously contracted to a point.

7.2 Homotopy Theory

7.2.1 Homotopy of Maps

Definition 7.2.1 (Homotopy)**.** Two continuous maps $f, g : X \to Y$ are **homotopic** ($f \simeq g$) if there exists a continuous map $H : X \times [0, 1] \to Y$ with $H(x, 0) = f(x)$ and $H(x, 1) = g(x)$.

H is called a **homotopy** from f to g. Intuitively, f can be continuously deformed into g.

Proposition 7.2.1. *Homotopy is an equivalence relation on continuous maps $X \to Y$.*

Definition 7.2.2 (Homotopy equivalence)**.** Spaces X and Y are **homotopy equivalent** ($X \simeq Y$) if there exist maps $f : X \to Y$ and $g : Y \to X$ such that $g \circ f \simeq \mathrm{id}_X$ and $f \circ g \simeq \mathrm{id}_Y$.

Definition 7.2.3 (Contractible space)**.** X is **contractible** if it is homotopy equivalent to a point, i.e., id_X is homotopic to a constant map.

Example 7.2.1. $\mathbb{R}^n$ is contractible. S^n is not contractible for $n \geq 0$.

Definition 7.2.4 (Retraction and deformation retract)**.** A **retraction** of X onto $A \subseteq X$ is a continuous $r : X \to A$ with $r|_A = \mathrm{id}_A$.

A is a **deformation retract** of X if there exists a homotopy $H : X \times [0, 1] \to X$ with $H(x, 0) = x$, $H(x, 1) \in A$, and $H(a, t) = a$ for all $a \in A$.

Example 7.2.2. S^{n-1} is a deformation retract of $\mathbb{R}^n \setminus \{0\}$.

7.2.2 The Fundamental Group

Definition 7.2.5 (Loop). A **loop** based at $x_0 \in X$ is a continuous map $\gamma : [0,1] \to X$ with $\gamma(0) = \gamma(1) = x_0$.

Definition 7.2.6 (Loop homotopy). Two loops γ_0, γ_1 based at x_0 are **homotopic** (as loops) if there exists a homotopy $H : [0,1] \times [0,1] \to X$ with $H(s,0) = \gamma_0(s)$, $H(s,1) = \gamma_1(s)$, and $H(0,t) = H(1,t) = x_0$.

Definition 7.2.7 (Fundamental group). The **fundamental group** $\pi_1(X, x_0)$ is the set of homotopy classes of loops based at x_0, with group operation given by concatenation:

$$[\gamma_1] \cdot [\gamma_2] := [\gamma_1 * \gamma_2]$$

where $(\gamma_1 * \gamma_2)(s) = \begin{cases} \gamma_1(2s) & s \leq 1/2 \\ \gamma_2(2s-1) & s \geq 1/2 \end{cases}$.

Theorem 7.2.2. *$\pi_1(X, x_0)$ is a group with identity $[c_{x_0}]$ (constant loop) and inverse $[\gamma]^{-1} = [\bar{\gamma}]$ where $\bar{\gamma}(s) = \gamma(1-s)$.*

Proposition 7.2.3. *If X is path-connected, then $\pi_1(X, x_0) \cong \pi_1(X, x_1)$ for any $x_0, x_1 \in X$. We write $\pi_1(X)$.*

Theorem 7.2.4 (Fundamental groups of standard spaces).

$$\begin{aligned} \pi_1(\mathbb{R}^n) &= 0 && \textit{(trivial)} \\ \pi_1(S^1) &\cong \mathbb{Z} && \textit{(winding number)} \\ \pi_1(S^n) &= 0 && n \geq 2 \\ \pi_1(T^n) &\cong \mathbb{Z}^n && \\ \pi_1(\mathbb{R}P^n) &\cong \mathbb{Z}/2\mathbb{Z} && n \geq 2 \\ \pi_1(SO(3)) &\cong \mathbb{Z}/2\mathbb{Z} && \end{aligned}$$

Theorem 7.2.5 (Seifert-van Kampen). *If $X = U \cup V$ with $U, V, U \cap V$ open and path-connected, then $\pi_1(X)$ is determined by $\pi_1(U)$, $\pi_1(V)$, and $\pi_1(U \cap V)$ via a pushout (amalgamated free product).*

Example 7.2.3. For the wedge sum $X \vee Y$ (spaces joined at a point): $\pi_1(X \vee Y) \cong \pi_1(X) * \pi_1(Y)$ (free product).

7.2.3 Higher Homotopy Groups

Definition 7.2.8 (Higher homotopy groups). The ***n*-th homotopy group** $\pi_n(X, x_0)$ is the set of homotopy classes of maps $(S^n, *) \to (X, x_0)$, i.e., maps $S^n \to X$ sending a basepoint to x_0.

Theorem 7.2.6. *$\pi_n(X, x_0)$ is a group for $n \geq 1$, and abelian for $n \geq 2$.*

Theorem 7.2.7 (Homotopy groups of spheres).

$$\begin{aligned} \pi_n(S^n) &\cong \mathbb{Z} && \textit{(degree)} \\ \pi_k(S^n) &= 0 && k < n \\ \pi_3(S^2) &\cong \mathbb{Z} && \textit{(Hopf fibration)} \end{aligned}$$

The groups $\pi_k(S^n)$ for $k > n$ are highly nontrivial and not fully known.

Theorem 7.2.8 (Whitehead's theorem). *A map $f : X \to Y$ between CW complexes is a homotopy equivalence iff it induces isomorphisms on all homotopy groups.*

7.2.4 Covering Spaces

Definition 7.2.9 (Covering space). A **covering space** of X is a space $\tilde{X}$ with a continuous surjection $p : \tilde{X} \to X$ such that every point $x \in X$ has a neighborhood U with $p^{-1}(U)$ a disjoint union of open sets, each mapped homeomorphically onto U by p.

Example 7.2.4. $p : \mathbb{R} \to S^1$, $p(t) = e^{2\pi i t}$, is a covering. $p : S^n \to \mathbb{R}P^n$ (antipodal identification) is a 2-sheeted covering.

Theorem 7.2.9 (Lifting criterion). *A map $f : Y \to X$ lifts to $\tilde{f} : Y \to \tilde{X}$ (with $p \circ \tilde{f} = f$) iff $f_*(\pi_1(Y)) \subseteq p_*(\pi_1(\tilde{X}))$.*

Theorem 7.2.10 (Classification of covering spaces). *For a "nice" space X (path-connected, locally path-connected, semi-locally simply connected):*

(i) Covering spaces of X correspond to subgroups of $\pi_1(X)$.

*(ii) The **universal cover** $\tilde{X}$ (simply connected covering) corresponds to the trivial subgroup.*

(iii) $\pi_1(X) \cong \mathrm{Deck}(\tilde{X})$ (group of deck transformations).

Example 7.2.5. $\mathbb{R}$ is the universal cover of S^1. The deck transformations are translations by integers.

7.3 Homology Theory

Homology provides computable algebraic invariants that detect "holes" of various dimensions.

7.3.1 Simplicial Complexes

Definition 7.3.1 (Simplex). The **standard n-simplex** is:

$$\Delta^n := \{(t_0, \dots, t_n) \in \mathbb{R}^{n+1} : t_i \geq 0, \sum t_i = 1\}.$$

Δ^0 is a point, Δ^1 a segment, Δ^2 a triangle, Δ^3 a tetrahedron.

Definition 7.3.2 (Simplicial complex). A **simplicial complex** K is a collection of simplices such that:

(i) Every face of a simplex in K is in K.

(ii) The intersection of two simplices in K is a face of each.

7.3.2 Simplicial Homology

Definition 7.3.3 (Chain group). The **n-th chain group** $C_n(K)$ is the free abelian group generated by n-simplices of K.

Definition 7.3.4 (Boundary operator). The **boundary operator** $\partial_n : C_n(K) \to C_{n-1}(K)$ is:

$$\partial_n[v_0, \dots, v_n] := \sum_{i=0}^{n} (-1)^i [v_0, \dots, \hat{v}_i, \dots, v_n]$$

where $\hat{v}_i$ means v_i is omitted.

Lemma 7.3.1. $\partial_{n-1} \circ \partial_n = 0$.

Definition 7.3.5 (Cycles, boundaries, homology).

- **n-cycles**: $Z_n := \ker \partial_n$.
- **n-boundaries**: $B_n := \operatorname{im} \partial_{n+1}$.
- **n-th homology group**: $H_n(K) := Z_n / B_n$.

Since $\partial^2 = 0$, we have $B_n \subseteq Z_n$, so the quotient is well-defined.

Example 7.3.1 (Homology of standard spaces).

$$H_k(S^n) \cong \begin{cases} \mathbb{Z} & k = 0, n \\ 0 & \text{otherwise} \end{cases}$$

$$H_k(T^2) \cong \begin{cases} \mathbb{Z} & k = 0, 2 \\ \mathbb{Z}^2 & k = 1 \\ 0 & k > 2 \end{cases}$$

$$H_k(\mathbb{R}P^2) \cong \begin{cases} \mathbb{Z} & k = 0 \\ \mathbb{Z}/2\mathbb{Z} & k = 1 \\ 0 & k \geq 2 \end{cases}$$

7.3.3 Singular Homology

Definition 7.3.6 (Singular simplex). A **singular n-simplex** in X is a continuous map $\sigma : \Delta^n \to X$.

Definition 7.3.7 (Singular homology). The **singular chain group** $C_n(X)$ is the free abelian group on singular n-simplices. With the boundary operator defined analogously, the **singular homology** is $H_n(X) := \ker \partial_n / \operatorname{im} \partial_{n+1}$.

Singular homology is defined for any topological space and is a homotopy invariant.

Theorem 7.3.2 (Properties of homology).

(i) $H_0(X) \cong \mathbb{Z}^{\#(\textit{path components})}$.

(ii) If X *is path-connected,* $H_1(X) \cong \pi_1(X)^{\mathrm{ab}}$ *(abelianization).*

(iii) Homotopy equivalent spaces have isomorphic homology.

(iv) $H_n(X) = 0$ *for* $n > \dim X$ *if* X *is a manifold.*

7.3.4 The Euler Characteristic

Definition 7.3.8 (Euler characteristic). For a space with finitely generated homology:

$$\chi(X) := \sum_{n=0}^{\infty} (-1)^n \operatorname{rank} H_n(X) = \sum_{n=0}^{\infty} (-1)^n b_n$$

where $b_n = \operatorname{rank} H_n(X)$ are the **Betti numbers**.

Theorem 7.3.3 (Euler-Poincaré formula). *For a finite simplicial complex with f_n simplices of dimension n:*

$$\chi(K) = \sum_{n=0}^{\infty} (-1)^n f_n.$$

Example 7.3.2. For a convex polyhedron: $\chi = V - E + F = 2$ where $V =$ vertices, $E =$ edges, $F =$ faces.

For surfaces: $\chi(S^2) = 2$, $\chi(T^2) = 0$, $\chi(\mathbb{R}P^2) = 1$.

Theorem 7.3.4 (Gauss-Bonnet). *For a compact oriented 2-dimensional Riemannian manifold M:*

$$\int_M K\, dA = 2\pi\chi(M)$$

where K is the Gaussian curvature.

7.3.5 Exact Sequences

Definition 7.3.9 (Exact sequence). A sequence of groups and homomorphisms

$$\cdots \to A_{n+1} \xrightarrow{f_{n+1}} A_n \xrightarrow{f_n} A_{n-1} \to \cdots$$

is **exact** if $\ker f_n = \operatorname{im} f_{n+1}$ for all n.

Example 7.3.3 (Short exact sequences). $0 \to A \to B \to C \to 0$ exact means $A \hookrightarrow B$ (injective) and $B \twoheadrightarrow C$ (surjective) with $C \cong B/A$.

Theorem 7.3.5 (Long exact sequence of a pair). *For a pair (X, A) with $A \subseteq X$:*

$$\cdots \to H_n(A) \to H_n(X) \to H_n(X, A) \xrightarrow{\partial_*} H_{n-1}(A) \to \cdots$$

where $H_n(X, A)$ is ***relative homology****.*

Theorem 7.3.6 (Mayer-Vietoris sequence). *If $X = U \cup V$ with U, V open:*

$$\cdots \to H_n(U \cap V) \to H_n(U) \oplus H_n(V) \to H_n(X) \xrightarrow{\partial_*} H_{n-1}(U \cap V) \to \cdots$$

7.4 Cohomology

7.4.1 Singular Cohomology

Definition 7.4.1 (Cochain and cohomology). The **cochain group** is $C^n(X; G) := \operatorname{Hom}(C_n(X), G)$ for an abelian group G.

The **coboundary** $\delta : C^n \to C^{n+1}$ is defined by $(\delta\phi)(c) := \phi(\partial c)$.

The n**-th cohomology group** is $H^n(X; G) := \ker \delta / \operatorname{im} \delta$.

Theorem 7.4.1 (Universal coefficient theorem). *There is a (split) short exact sequence:*

$$0 \to \operatorname{Ext}(H_{n-1}(X), G) \to H^n(X; G) \to \operatorname{Hom}(H_n(X), G) \to 0.$$

For $G = \mathbb{R}$ *or* $\mathbb{Q}$*:* $H^n(X; G) \cong \operatorname{Hom}(H_n(X), G)$.

7.4.2 Cup Product

Definition 7.4.2 (Cup product). The **cup product** $\smile: H^p(X) \times H^q(X) \to H^{p+q}(X)$ makes $H^*(X) := \bigoplus_n H^n(X)$ into a graded ring.

The cup product is associative and graded-commutative: $\alpha \smile \beta = (-1)^{pq} \beta \smile \alpha$.

Example 7.4.1. $H^*(S^n; \mathbb{Z}) \cong \mathbb{Z}[\alpha]/(\alpha^2)$ where α generates H^n.
$H^*(T^n; \mathbb{Z}) \cong \Lambda_{\mathbb{Z}}[\alpha_1, \ldots, \alpha_n]$ (exterior algebra), each $\alpha_i \in H^1$.
$H^*(\mathbb{C}P^n; \mathbb{Z}) \cong \mathbb{Z}[\alpha]/(\alpha^{n+1})$ where $\alpha \in H^2$.

7.4.3 de Rham Cohomology

Theorem 7.4.2 (de Rham's theorem). *For a smooth manifold M:*

$$H^n_{dR}(M) \cong H^n(M; \mathbb{R}).$$

The isomorphism is given by integration: $[\omega] \mapsto \left([\sigma] \mapsto \int_\sigma \omega\right)$.

This connects differential forms to topology.

Theorem 7.4.3 (Poincaré duality). *For a closed oriented n-manifold M:*

$$H^k(M; \mathbb{R}) \cong H_{n-k}(M; \mathbb{R}).$$

7.5 Fiber Bundles

7.5.1 Basic Definitions

Definition 7.5.1 (Fiber bundle). A **fiber bundle** consists of:

- A **total space** E.
- A **base space** B.
- A **projection** $\pi : E \to B$.
- A **fiber** F.

such that B has an open cover $\{U_\alpha\}$ with **local trivializations**:

$$\phi_\alpha : \pi^{-1}(U_\alpha) \xrightarrow{\cong} U_\alpha \times F$$

compatible with projection to U_α.

We write $F \to E \xrightarrow{\pi} B$ or simply $E \to B$.

Definition 7.5.2 (Transition functions). On overlaps $U_\alpha \cap U_\beta$, the **transition functions** are:

$$g_{\alpha\beta} : U_\alpha \cap U_\beta \to \mathrm{Aut}(F)$$

defined by $\phi_\alpha \circ \phi_\beta^{-1}(x, f) = (x, g_{\alpha\beta}(x) \cdot f)$.

Proposition 7.5.1 (Cocycle condition). *$g_{\alpha\beta}g_{\beta\gamma} = g_{\alpha\gamma}$ on triple overlaps.*

Definition 7.5.3 (Section). A **section** of $E \to B$ is a continuous map $s : B \to E$ with $\pi \circ s = \mathrm{id}_B$.

Definition 7.5.4 (Trivial bundle). A bundle is **trivial** if $E \cong B \times F$ (globally).

7.5.2 Vector Bundles

Definition 7.5.5 (Vector bundle). A **vector bundle** of rank k is a fiber bundle with fiber $\mathbb{R}^k$ (or $\mathbb{C}^k$) and transition functions in $GL(k)$.

Example 7.5.1 (Standard vector bundles).

(a) **Trivial bundle**: $B \times \mathbb{R}^k$.

(b) **Tangent bundle**: $TM \to M$, fiber $T_xM \cong \mathbb{R}^n$.

(c) **Cotangent bundle**: $T^*M \to M$.

(d) **Möbius band**: nontrivial line bundle over S^1.

(e) **Tautological bundle**: over $\mathbb{R}P^n$ or $\mathbb{C}P^n$.

Definition 7.5.6 (Operations on vector bundles). Given vector bundles E, E' over B:

- **Direct sum**: $E \oplus E'$ with fiber $E_x \oplus E'_x$.
- **Tensor product**: $E \otimes E'$ with fiber $E_x \otimes E'_x$.
- **Dual bundle**: E^* with fiber E^*_x.
- **Pullback**: for $f : B' \to B$, f^*E over B' with $(f^*E)_y = E_{f(y)}$.

7.5.3 Principal Bundles

Definition 7.5.7 (Principal G-bundle). A **principal G-bundle** is a fiber bundle $P \to B$ with fiber G (a Lie group) and a free right G-action on P such that $B = P/G$ and local trivializations are G-equivariant.

Example 7.5.2.

(a) $S^1 \to S^3 \to S^2$ (Hopf fibration): principal $U(1)$-bundle.

(b) Frame bundle $F(M) \to M$: principal $GL(n)$-bundle.

(c) $G \to G/H$: principal H-bundle for closed subgroup $H \subseteq G$.

Proposition 7.5.2 (Associated bundles). *Given a principal G-bundle $P \to B$ and a G-representation V, the **associated vector bundle** is:*

$$E = P \times_G V := (P \times V)/G$$

where G acts by $g \cdot (p, v) = (pg^{-1}, gv)$.

Every vector bundle arises this way from its frame bundle.

7.5.4 The Hopf Fibration

Definition 7.5.8 (Hopf fibration). The **Hopf fibration** is $S^1 \to S^3 \to S^2$, where:

$$\pi : S^3 \subseteq \mathbb{C}^2 \to S^2 \cong \mathbb{C}P^1, \quad (z_1, z_2) \mapsto [z_1 : z_2].$$

Proposition 7.5.3. *Each fiber is a great circle in S^3. Any two fibers are linked (Hopf link).*

The Hopf fibration generalizes:

- $S^1 \to S^{2n+1} \to \mathbb{C}P^n$ (complex Hopf).
- $S^3 \to S^7 \to S^4$ (quaternionic Hopf).
- $S^7 \to S^{15} \to S^8$ (octonionic Hopf).

The Hopf invariant of $S^3 \to S^2$ generates $\pi_3(S^2) \cong \mathbb{Z}$.

7.6 Connections and Curvature on Bundles

7.6.1 Connections on Principal Bundles

Definition 7.6.1 (Connection (Ehresmann)). A **connection** on a principal G-bundle $P \to B$ is a G-equivariant distribution of horizontal subspaces $H_p \subseteq T_pP$ complementary to the vertical (fiber) directions.

Equivalently, a connection is a $\mathfrak{g}$-valued 1-form $\omega \in \Omega^1(P, \mathfrak{g})$ satisfying:

(i) $\omega(A^*) = A$ for fundamental vector fields $A^* \in \mathfrak{g}$.

(ii) $R_g^*\omega = \mathrm{Ad}_{g^{-1}}\omega$ (G-equivariance).

Definition 7.6.2 (Curvature form). The **curvature** of a connection ω is:

$$\Omega := d\omega + \frac{1}{2}[\omega, \omega] = d\omega + \omega \wedge \omega.$$

Theorem 7.6.1 (Structure equation (Cartan)). $\Omega = d\omega + \omega \wedge \omega$. *Equivalently, for horizontal vector fields X, Y:*

$$\Omega(X, Y) = -\omega([X, Y]).$$

Theorem 7.6.2 (Bianchi identity). $d\Omega = [\Omega, \omega]$, *or in components:* $D\Omega = 0$ *where D is the covariant exterior derivative.*

7.6.2 Gauge Fields in Physics

In physics, connections are **gauge fields**:

Example 7.6.1 (Electromagnetism). Principal $U(1)$-bundle. Connection $A = A_\mu dx^\mu$. Curvature $F = dA$ is the electromagnetic field strength.

Example 7.6.2 (Yang-Mills theory). Principal G-bundle (e.g., $G = SU(2), SU(3)$). Connection A is the gauge potential. Curvature $F = dA + A \wedge A$ is the field strength.

Definition 7.6.3 (Gauge transformation). A **gauge transformation** is an automorphism of the principal bundle. Under gauge transformation $g : B \to G$:

$$A \mapsto g^{-1}Ag + g^{-1}dg, \quad F \mapsto g^{-1}Fg.$$

7.7 Characteristic Classes

Characteristic classes are cohomology classes that measure the "twisting" of bundles.

7.7.1 Chern Classes

Definition 7.7.1 (Chern classes). For a complex vector bundle $E \to B$ of rank k, the **Chern classes** $c_i(E) \in H^{2i}(B;\mathbb{Z})$ for $i = 0, 1, \dots, k$ satisfy:

(i) $c_0(E) = 1$.

(ii) **Naturality**: $c_i(f^*E) = f^*c_i(E)$.

(iii) **Whitney sum**: $c(E \oplus E') = c(E) \smile c(E')$ where $c = 1 + c_1 + c_2 + \cdots$.

(iv) **Normalization**: $c_1(\gamma^1) = -[\omega]$ for the tautological bundle over $\mathbb{C}P^1$.

Theorem 7.7.1 (Chern-Weil theory). *For a connection with curvature F:*

$$c_k(E) = \left[\frac{1}{k!}\left(\frac{i}{2\pi}\right)^k \operatorname{tr}(F^k)\right] \in H^{2k}_{dR}(B).$$

In particular, $c_1(E) = \left[\frac{i}{2\pi}\operatorname{tr}(F)\right]$.

Example 7.7.1. For a line bundle ($k = 1$): c_1 measures the "twist." The first Chern number $\int_M c_1$ is an integer (quantized).

7.7.2 Other Characteristic Classes

Definition 7.7.2 (Pontryagin classes). For a real vector bundle E, the **Pontryagin classes** $p_i(E) \in H^{4i}(B;\mathbb{Z})$ are defined via the Chern classes of the complexification: $p_i(E) = (-1)^i c_{2i}(E \otimes \mathbb{C})$.

Definition 7.7.3 (Euler class). For an oriented real vector bundle of rank n, the **Euler class** $e(E) \in H^n(B;\mathbb{Z})$ satisfies $e(E)^2 = p_{n/2}(E)$ (for n even).

Theorem 7.7.2 (Gauss-Bonnet-Chern). *For a closed oriented $2n$-manifold:*

$$\chi(M) = \int_M e(TM).$$

Definition 7.7.4 (Stiefel-Whitney classes). For a real vector bundle E, the **Stiefel-Whitney classes** $w_i(E) \in H^i(B;\mathbb{Z}/2\mathbb{Z})$ detect:

- $w_1(E) = 0$ iff E is orientable.
- $w_2(E) = 0$ iff E admits a spin structure.

7.7.3 The Chern Number and Monopoles

Example 7.7.2 (Magnetic monopole). A magnetic monopole configuration is a principal $U(1)$-bundle over S^2 (surrounding the monopole). The first Chern number:

$$n = \frac{1}{2\pi} \int_{S^2} F$$

equals the monopole charge (quantized).

Theorem 7.7.3 (Dirac quantization). *If magnetic monopoles exist, electric charge must be quantized: $eg = n\hbar/2$ for integer n.*

7.8 Applications to Physics

7.8.1 Topological Quantum Numbers

Many physical quantities are topological invariants:

Magnetic flux quantization: In a superconductor, magnetic flux through a loop is $\Phi = n\Phi_0$ where $\Phi_0 = h/2e$.

Quantum Hall effect: Hall conductance $\sigma_{xy} = \nu e^2/h$ where ν is an integer (or fraction) related to the Chern number of the band structure.

Skyrmions: Topological solitons in magnetic systems classified by $\pi_2(S^2) \cong \mathbb{Z}$.

Vortices: In superfluids and superconductors, classified by $\pi_1(S^1) \cong \mathbb{Z}$.

7.8.2 Instantons and Anomalies

Definition 7.8.1 (Instanton). An **instanton** is a finite-action solution to the Yang-Mills equations on $\mathbb{R}^4$ (or S^4). Instantons are classified by the second Chern number:

$$k = \frac{1}{8\pi^2} \int \operatorname{tr}(F \wedge F).$$

Theorem 7.8.1 (Atiyah-Singer index theorem). *For a Dirac operator D on a manifold M:*

$$\operatorname{index}(D) = \dim\ker D - \dim\ker D^\dagger = \int_M \hat{A}(M) \wedge \operatorname{ch}(E)$$

where $\hat{A}$ is the A-roof genus and ch *is the Chern character.*

The index theorem relates analytic (zero modes) to topological (characteristic classes) data.

7.8.3 Berry Phase

Definition 7.8.2 (Berry phase). For a quantum system with parameter-dependent Hamiltonian $H(\mathbf{R})$, the **Berry phase** acquired by an eigenstate around a closed loop in parameter space is:

$$\gamma = i \oint \langle n(\mathbf{R}) | \nabla_{\mathbf{R}} | n(\mathbf{R}) \rangle \cdot d\mathbf{R}.$$

Proposition 7.8.2. *The Berry phase is the holonomy of the Berry connection $\mathcal{A} = i\langle n|dn\rangle$ on the line bundle of eigenstates over parameter space.*

7.9 Summary and Outlook

Topology provides the mathematical framework for understanding global, robust properties of physical systems:

Homotopy: Classifies continuous deformations; π_n groups detect "holes" that maps can wind around.

Homology/Cohomology: Computable algebraic invariants; de Rham theory connects to differential forms.

Fiber bundles: Unify gauge theories, parallel transport, and geometric phases.

Characteristic classes: Topological invariants of bundles; quantize physical observables.

These tools are essential for:

- Gauge theories and the Standard Model
- Topological phases of matter
- String theory and M-theory
- Quantum information and topological quantum computing

Exercises

Exercise 7.9.1. Show that $\mathbb{R}^n$ and $\mathbb{R}^m$ are homeomorphic iff $n = m$. (Hint: use invariance of domain or homology.)

Exercise 7.9.2. Compute $\pi_1(T^2)$ using the Seifert-van Kampen theorem.

Exercise 7.9.3. Show that if $\tilde{X} \to X$ is a covering with $\tilde{X}$ simply connected, then $\pi_1(X)$ acts freely and properly discontinuously on $\tilde{X}$.

Exercise 7.9.4. Compute $H_*(S^1 \times S^1)$ using the Künneth formula or Mayer-Vietoris.

Exercise 7.9.5. Verify that $\partial^2 = 0$ for the simplicial boundary operator.

Exercise 7.9.6. Show that the Euler characteristic of a closed orientable surface of genus g is $\chi = 2 - 2g$.

Exercise 7.9.7. Prove that the Möbius band is a nontrivial line bundle over S^1.

Exercise 7.9.8. Show that principal G-bundles over S^2 are classified by $\pi_1(G)$.

Exercise 7.9.9. Verify that the curvature 2-form $\Omega = d\omega + \omega \wedge \omega$ transforms as $\Omega \mapsto g^{-1}\Omega g$ under gauge transformations.

Exercise 7.9.10. For the Hopf fibration $S^1 \to S^3 \to S^2$, compute the first Chern number of the associated line bundle.

Exercise 7.9.11. Show that the total Stiefel-Whitney class of $\mathbb{R}P^n$ is $w(T\mathbb{R}P^n) = (1 + a)^{n+1}$ where $a \in H^1(\mathbb{R}P^n; \mathbb{Z}/2\mathbb{Z})$ is the generator.

Exercise 7.9.12. Derive the Dirac quantization condition $eg = n\hbar/2$ from the requirement that the wavefunction be single-valued on S^2 surrounding a monopole.

Part III

Algebra and Symmetry

Chapter 8

Group Theory

Symmetry is a unifying principle in physics. From the conservation laws of classical mechanics to the gauge symmetries of particle physics, group theory provides the mathematical language for symmetry. This chapter develops the theory of groups, with emphasis on the structures most relevant to physics: Lie groups, their representations, and the classification of symmetries.

8.1 Basic Group Theory

8.1.1 Groups and Subgroups

Definition 8.1.1 (Group). A **group** is a set G with a binary operation $\cdot : G \times G \to G$ satisfying:

(i) **Associativity**: $(ab)c = a(bc)$ for all $a, b, c \in G$.

(ii) **Identity**: There exists $e \in G$ such that $ae = ea = a$ for all $a \in G$.

(iii) **Inverses**: For each $a \in G$, there exists $a^{-1} \in G$ such that $aa^{-1} = a^{-1}a = e$.

Definition 8.1.2 (Abelian group). A group is **abelian** (or **commutative**) if $ab = ba$ for all $a, b \in G$.

Example 8.1.1 (Fundamental examples).

(a) $(\mathbb{Z}, +)$, $(\mathbb{Q}, +)$, $(\mathbb{R}, +)$, $(\mathbb{C}, +)$: additive groups.

(b) $(\mathbb{Q}^*, \cdot)$, $(\mathbb{R}^*, \cdot)$, $(\mathbb{C}^*, \cdot)$: multiplicative groups of nonzero elements.

(c) $(\mathbb{Z}/n\mathbb{Z}, +)$: integers modulo n.

(d) S_n: symmetric group (permutations of n elements).

(e) $GL(n, \mathbb{F})$: general linear group ($n \times n$ invertible matrices).

(f) $O(n)$, $SO(n)$, $U(n)$, $SU(n)$: orthogonal, special orthogonal, unitary, special unitary groups.

Definition 8.1.3 (Order)**.** The **order** of a group G, denoted $|G|$, is its cardinality. The **order** of an element $g \in G$ is the smallest positive integer n such that $g^n = e$, or ∞ if no such n exists.

Definition 8.1.4 (Subgroup)**.** A **subgroup** H of G (written $H \leq G$) is a subset that is itself a group under the same operation.

Proposition 8.1.1 (Subgroup criterion)**.** *$H \subseteq G$ is a subgroup iff $H \neq \varnothing$ and $ab^{-1} \in H$ for all $a, b \in H$.*

Theorem 8.1.2 (Lagrange's theorem)**.** *If $H \leq G$ and G is finite, then $|H|$ divides $|G|$. The **index** $[G : H] := |G|/|H|$ counts the number of cosets.*

Corollary 8.1.3. *The order of any element divides $|G|$. In particular, $g^{|G|} = e$ for all $g \in G$.*

8.1.2 Homomorphisms and Isomorphisms

Definition 8.1.5 (Homomorphism)**.** A **homomorphism** $\phi : G \to H$ is a map satisfying $\phi(ab) = \phi(a)\phi(b)$.

Proposition 8.1.4. *If $\phi : G \to H$ is a homomorphism:*

(i) $\phi(e_G) = e_H$.

(ii) $\phi(a^{-1}) = \phi(a)^{-1}$.

(iii) $\phi(G)$ *is a subgroup of* H.

(iv) $\ker \phi := \{g \in G : \phi(g) = e_H\}$ *is a subgroup of* G.

Definition 8.1.6 (Isomorphism)**.** An **isomorphism** is a bijective homomorphism. Groups G and H are **isomorphic** ($G \cong H$) if an isomorphism exists between them.

Definition 8.1.7 (Automorphism)**.** An **automorphism** is an isomorphism from a group to itself. The set $\mathrm{Aut}(G)$ of all automorphisms forms a group.

Example 8.1.2 (Inner automorphisms)**.** For $g \in G$, conjugation $\phi_g(x) = gxg^{-1}$ is an automorphism. Such automorphisms are **inner**; others are **outer**.

8.1.3 Normal Subgroups and Quotients

Definition 8.1.8 (Coset). For $H \leq G$ and $g \in G$, the **left coset** is $gH := \{gh : h \in H\}$ and the **right coset** is $Hg := \{hg : h \in H\}$.

Definition 8.1.9 (Normal subgroup). $N \leq G$ is **normal** (written $N \trianglelefteq G$) if $gNg^{-1} = N$ for all $g \in G$, equivalently $gN = Ng$ for all g.

Proposition 8.1.5. *N is normal iff it is the kernel of some homomorphism.*

Definition 8.1.10 (Quotient group). For $N \trianglelefteq G$, the **quotient group** G/N is the set of cosets $\{gN : g \in G\}$ with operation $(gN)(hN) := (gh)N$.

Theorem 8.1.6 (First isomorphism theorem). *If $\phi : G \to H$ is a homomorphism, then $G/\ker\phi \cong \phi(G)$.*

Theorem 8.1.7 (Second isomorphism theorem). *If $H \leq G$ and $N \trianglelefteq G$, then $HN \leq G$, $H \cap N \trianglelefteq H$, and $H/(H \cap N) \cong HN/N$.*

Theorem 8.1.8 (Third isomorphism theorem). *If $N \trianglelefteq M \trianglelefteq G$, then $M/N \trianglelefteq G/N$ and $(G/N)/(M/N) \cong G/M$.*

8.1.4 Group Actions

Definition 8.1.11 (Group action). A **(left) action** of G on a set X is a map $G \times X \to X$, $(g, x) \mapsto g \cdot x$, satisfying:

(i) $e \cdot x = x$ for all $x \in X$.

(ii) $(gh) \cdot x = g \cdot (h \cdot x)$ for all $g, h \in G$, $x \in X$.

Definition 8.1.12 (Orbit and stabilizer). For $x \in X$:

- The **orbit** of x is $G \cdot x := \{g \cdot x : g \in G\}$.
- The **stabilizer** (or **isotropy group**) is $G_x := \{g \in G : g \cdot x = x\}$.

Theorem 8.1.9 (Orbit-stabilizer theorem). $|G \cdot x| = [G : G_x] = |G|/|G_x|$.

Definition 8.1.13 (Types of actions). An action is:

- **Transitive** if there is only one orbit ($G \cdot x = X$).
- **Free** if $G_x = \{e\}$ for all x.
- **Faithful** if $g \cdot x = x$ for all x implies $g = e$.

Example 8.1.3. $SO(3)$ acts on S^2 transitively. The stabilizer of the north pole is $SO(2)$. Thus $S^2 \cong SO(3)/SO(2)$.

Theorem 8.1.10 (Burnside's lemma). *The number of orbits of G acting on X is:*

$$|orbits| = \frac{1}{|G|}\sum_{g\in G}|X^g|$$

where $X^g := \{x \in X : g \cdot x = x\}$.

8.2 Important Classes of Groups

8.2.1 Cyclic and Abelian Groups

Definition 8.2.1 (Cyclic group). A group is **cyclic** if it is generated by a single element: $G = \langle g \rangle = \{g^n : n \in \mathbb{Z}\}$.

Theorem 8.2.1. *Every cyclic group is abelian. Cyclic groups are isomorphic to* $\mathbb{Z}$ *(infinite) or* $\mathbb{Z}/n\mathbb{Z}$ *(finite of order* n*).*

Theorem 8.2.2 (Structure theorem for finitely generated abelian groups). *Every finitely generated abelian group is isomorphic to:*

$$\mathbb{Z}^r \oplus \mathbb{Z}/n_1\mathbb{Z} \oplus \mathbb{Z}/n_2\mathbb{Z} \oplus \cdots \oplus \mathbb{Z}/n_k\mathbb{Z}$$

where $n_1|n_2|\cdots|n_k$ *(invariant factor form) or equivalently to* $\mathbb{Z}^r \oplus \bigoplus_i \mathbb{Z}/p_i^{a_i}\mathbb{Z}$ *(primary decomposition).*

8.2.2 Symmetric and Alternating Groups

Definition 8.2.2 (Symmetric group). The **symmetric group** S_n is the group of all permutations of $\{1, 2, \ldots, n\}$. It has order $n!$.

Definition 8.2.3 (Cycle notation). A k**-cycle** $(a_1\, a_2\, \cdots\, a_k)$ maps $a_i \mapsto a_{i+1}$ (indices mod k) and fixes other elements.

Theorem 8.2.3. *Every permutation is a product of disjoint cycles. Disjoint cycles commute.*

Definition 8.2.4 (Sign of permutation). The **sign** (or **parity**) of $\sigma \in S_n$ is $\text{sgn}(\sigma) = (-1)^k$ where k is the number of transpositions in any decomposition of σ.

Definition 8.2.5 (Alternating group). The **alternating group** A_n consists of even permutations: $A_n = \ker(\text{sgn}) \trianglelefteq S_n$. It has order $n!/2$.

Theorem 8.2.4. A_n *is simple (has no nontrivial normal subgroups) for* $n \geq 5$.

Corollary 8.2.5 (Insolubility of quintic). *The general polynomial of degree* ≥ 5 *is not solvable by radicals.*

8.2.3 Matrix Groups

Definition 8.2.6 (General linear group). $GL(n, \mathbb{F}) := \{A \in M_n(\mathbb{F}) : \det A \neq 0\}$ is the group of $n \times n$ invertible matrices over field $\mathbb{F}$.

Definition 8.2.7 (Special linear group). $SL(n, \mathbb{F}) := \{A \in GL(n, \mathbb{F}) : \det A = 1\}$.

Definition 8.2.8 (Orthogonal and special orthogonal groups).

$$O(n) := \{A \in GL(n, \mathbb{R}) : A^T A = I\}$$
$$SO(n) := \{A \in O(n) : \det A = 1\}$$

$O(n)$ preserves the Euclidean inner product. $SO(n)$ is the connected component of the identity (rotations).

Definition 8.2.9 (Unitary and special unitary groups).

$$U(n) := \{A \in GL(n, \mathbb{C}) : A^\dagger A = I\}$$
$$SU(n) := \{A \in U(n) : \det A = 1\}$$

Definition 8.2.10 (Symplectic group). $Sp(2n, \mathbb{F}) := \{A \in GL(2n, \mathbb{F}) : A^T J A = J\}$ where $J = \begin{pmatrix} 0 & I_n \\ -I_n & 0 \end{pmatrix}$.

Proposition 8.2.6 (Dimensions).

$$\begin{aligned} \dim GL(n, \mathbb{R}) &= n^2, & \dim GL(n, \mathbb{C}) &= 2n^2 \\ \dim O(n) &= \frac{n(n-1)}{2}, & \dim SO(n) &= \frac{n(n-1)}{2} \\ \dim U(n) &= n^2, & \dim SU(n) &= n^2 - 1 \\ \dim Sp(2n) &= n(2n+1) \end{aligned}$$

8.2.4 The Lorentz and Poincaré Groups

Definition 8.2.11 (Lorentz group). The **Lorentz group** $O(1, 3)$ preserves the Minkowski metric $\eta = \operatorname{diag}(-1, 1, 1, 1)$:

$$O(1, 3) := \{\Lambda \in GL(4, \mathbb{R}) : \Lambda^T \eta \Lambda = \eta\}.$$

Proposition 8.2.7 (Components of Lorentz group). *$O(1, 3)$ has four connected components:*

- *$SO^+(1, 3)$: proper orthochronous (identity component).*
- *Space inversion P:* $\det \Lambda = -1$, $\Lambda^0{}_0 > 0$.

- *Time reversal T:* $\det\Lambda = 1$, $\Lambda^0{}_0 < 0$.
- *PT:* $\det\Lambda = -1$, $\Lambda^0{}_0 < 0$.

Definition 8.2.12 (Poincaré group). The **Poincaré group** (or **inhomogeneous Lorentz group**) is the semidirect product:

$$\mathrm{ISO}(1,3) = \mathbb{R}^{1,3} \rtimes O(1,3)$$

consisting of Lorentz transformations and spacetime translations.

The Poincaré group is the symmetry group of special relativity.

8.3 Lie Groups

8.3.1 Definition and Examples

Definition 8.3.1 (Lie group). A **Lie group** is a smooth manifold G that is also a group, such that multiplication $G \times G \to G$ and inversion $G \to G$ are smooth maps.

Example 8.3.1 (Classical Lie groups). $GL(n,\mathbb{R})$, $GL(n,\mathbb{C})$, $SL(n,\mathbb{R})$, $SL(n,\mathbb{C})$, $O(n)$, $SO(n)$, $U(n)$, $SU(n)$, $Sp(2n)$ are all Lie groups.

Definition 8.3.2 (Lie subgroup). A **Lie subgroup** is a subgroup that is also an immersed submanifold.

Theorem 8.3.1 (Closed subgroup theorem). *A closed subgroup of a Lie group is a Lie subgroup (embedded).*

Definition 8.3.3 (Connected and simply connected). A Lie group G is:

- **Connected** if it has one connected component.
- **Simply connected** if $\pi_1(G) = 0$.

Example 8.3.2. $SO(3)$ is connected but not simply connected ($\pi_1(SO(3)) = \mathbb{Z}/2\mathbb{Z}$). $SU(2)$ is simply connected and is the universal cover of $SO(3)$.

8.3.2 The Lie Algebra

Definition 8.3.4 (Lie algebra of a Lie group). The **Lie algebra** $\mathfrak{g}$ of a Lie group G is the tangent space at the identity T_eG, equipped with the Lie bracket $[X,Y]$ induced by the commutator of left-invariant vector fields.

For matrix groups, the Lie algebra consists of matrices with the commutator bracket $[X,Y] = XY - YX$.

Example 8.3.3 (Lie algebras of classical groups).

$$\begin{aligned}
\mathfrak{gl}(n, \mathbb{R}) &= M_n(\mathbb{R}) && \text{(all matrices)} \\
\mathfrak{sl}(n, \mathbb{R}) &= \{X \in M_n(\mathbb{R}) : \mathrm{tr}(X) = 0\} \\
\mathfrak{o}(n) = \mathfrak{so}(n) &= \{X \in M_n(\mathbb{R}) : X^T = -X\} && \text{(antisymmetric)} \\
\mathfrak{u}(n) &= \{X \in M_n(\mathbb{C}) : X^\dagger = -X\} && \text{(anti-Hermitian)} \\
\mathfrak{su}(n) &= \{X \in \mathfrak{u}(n) : \mathrm{tr}(X) = 0\}
\end{aligned}$$

Definition 8.3.5 (Abstract Lie algebra). A **Lie algebra** over $\mathbb{F}$ is a vector space $\mathfrak{g}$ with a bilinear bracket $[\cdot, \cdot] : \mathfrak{g} \times \mathfrak{g} \to \mathfrak{g}$ satisfying:

(i) **Antisymmetry**: $[X, Y] = -[Y, X]$.

(ii) **Jacobi identity**: $[[X, Y], Z] + [[Y, Z], X] + [[Z, X], Y] = 0$.

8.3.3 The Exponential Map

Definition 8.3.6 (Exponential map). The **exponential map** $\exp : \mathfrak{g} \to G$ is defined by $\exp(X) := \gamma_X(1)$, where $\gamma_X(t)$ is the unique one-parameter subgroup with $\gamma'_X(0) = X$.

For matrix groups:

$$\exp(X) = e^X = \sum_{n=0}^{\infty} \frac{X^n}{n!} = I + X + \frac{X^2}{2!} + \cdots$$

Theorem 8.3.2 (Properties of exponential).

(i) $\exp(0) = e$.

(ii) $\exp((s + t)X) = \exp(sX) \exp(tX)$.

(iii) $\exp(X)^{-1} = \exp(-X)$.

(iv) $\exp$ *is a local diffeomorphism near* $0 \in \mathfrak{g}$.

(v) $\exp(gXg^{-1}) = g \exp(X) g^{-1}$ *(equivariance).*

Theorem 8.3.3 (Baker-Campbell-Hausdorff formula). *For* $X, Y \in \mathfrak{g}$ *sufficiently small:*

$$\exp(X) \exp(Y) = \exp\left(X + Y + \frac{1}{2}[X, Y] + \frac{1}{12}([X, [X, Y]] + [Y, [Y, X]]) + \cdots\right).$$

This shows that the group structure near the identity is determined by the Lie bracket.

Theorem 8.3.4 (Lie's theorems).

(i) Every Lie group has a Lie algebra.

(ii) Every finite-dimensional Lie algebra is the Lie algebra of some Lie group.

(iii) Simply connected Lie groups are determined (up to isomorphism) by their Lie algebras.

8.3.4 One-Parameter Subgroups

Definition 8.3.7 (One-parameter subgroup). A **one-parameter subgroup** of G is a smooth homomorphism $\gamma : (\mathbb{R}, +) \to G$.

Theorem 8.3.5. *One-parameter subgroups are exactly the maps $t \mapsto \exp(tX)$ for $X \in \mathfrak{g}$.*

Example 8.3.4 (Rotations in $SO(3)$). Rotation by angle θ about axis $\hat{n}$ is $R(\theta, \hat{n}) = \exp(\theta\hat{n} \cdot \mathbf{J})$ where J_i are generators of $\mathfrak{so}(3)$:

$$(J_i)_{jk} = -\epsilon_{ijk}.$$

8.4 Structure of Lie Algebras

8.4.1 Subalgebras and Ideals

Definition 8.4.1 (Subalgebra and ideal). A **subalgebra** $\mathfrak{h} \subseteq \mathfrak{g}$ is a subspace closed under brackets.

An **ideal** $\mathfrak{i} \trianglelefteq \mathfrak{g}$ is a subalgebra with $[\mathfrak{g}, \mathfrak{i}] \subseteq \mathfrak{i}$.

Definition 8.4.2 (Simple and semisimple). A Lie algebra is:

- **Simple** if it is non-abelian and has no nontrivial ideals.
- **Semisimple** if it is a direct sum of simple Lie algebras.

Definition 8.4.3 (Derived series and solvability). The **derived series** is $\mathfrak{g}^{(0)} = \mathfrak{g}$, $\mathfrak{g}^{(k+1)} = [\mathfrak{g}^{(k)}, \mathfrak{g}^{(k)}]$.

$\mathfrak{g}$ is **solvable** if $\mathfrak{g}^{(n)} = 0$ for some n.

Theorem 8.4.1 (Levi decomposition). *Every finite-dimensional Lie algebra $\mathfrak{g}$ over $\mathbb{C}$ or $\mathbb{R}$ decomposes as $\mathfrak{g} = \mathfrak{r} \rtimes \mathfrak{s}$ where $\mathfrak{r}$ is the maximal solvable ideal (radical) and $\mathfrak{s}$ is semisimple.*

8.4.2 The Killing Form

Definition 8.4.4 (Killing form). The **Killing form** on $\mathfrak{g}$ is the symmetric bilinear form:

$$\kappa(X, Y) := \operatorname{tr}(\operatorname{ad}_X \circ \operatorname{ad}_Y)$$

where $\operatorname{ad}_X(Y) := [X, Y]$ is the adjoint representation.

Theorem 8.4.2 (Cartan's criterion). *$\mathfrak{g}$ is semisimple iff the Killing form is non-degenerate.*

Example 8.4.1. For $\mathfrak{su}(n)$: $\kappa(X, Y) = 2n \operatorname{tr}(XY)$.
For $\mathfrak{so}(n)$: $\kappa(X, Y) = (n-2) \operatorname{tr}(XY)$.

8.4.3 Cartan Subalgebras and Root Systems

Definition 8.4.5 (Cartan subalgebra). A **Cartan subalgebra** $\mathfrak{h}$ of a semisimple $\mathfrak{g}$ is a maximal abelian subalgebra consisting of semisimple (diagonalizable) elements.

Definition 8.4.6 (Root). A **root** is a nonzero $\alpha \in \mathfrak{h}^*$ such that:

$$\mathfrak{g}_\alpha := \{X \in \mathfrak{g} : [H, X] = \alpha(H)X \text{ for all } H \in \mathfrak{h}\} \neq 0.$$

Theorem 8.4.3 (Root space decomposition). *For semisimple $\mathfrak{g}$ over $\mathbb{C}$:*

$$\mathfrak{g} = \mathfrak{h} \oplus \bigoplus_{\alpha \in \Phi} \mathfrak{g}_\alpha$$

where Φ is the set of roots and each $\mathfrak{g}_\alpha$ is one-dimensional.

Definition 8.4.7 (Root system). The set of roots $\Phi \subseteq \mathfrak{h}^*$ forms a **root system**: a finite set satisfying certain axioms involving reflections.

Theorem 8.4.4 (Classification of simple Lie algebras). *Simple Lie algebras over $\mathbb{C}$ are classified by Dynkin diagrams:*

- *Classical: A_n ($\mathfrak{sl}_{n+1}$), B_n ($\mathfrak{so}_{2n+1}$), C_n ($\mathfrak{sp}_{2n}$), D_n ($\mathfrak{so}_{2n}$).*
- *Exceptional: E_6, E_7, E_8, F_4, G_2.*

8.5 Representation Theory

8.5.1 Basic Definitions

Definition 8.5.1 (Representation of a group). A **representation** of G on a vector space V is a homomorphism $\rho : G \to GL(V)$. The **dimension** of the representation is $\dim V$.

Definition 8.5.2 (Representation of a Lie algebra). A **representation** of $\mathfrak{g}$ on V is a Lie algebra homomorphism $\rho : \mathfrak{g} \to \mathfrak{gl}(V)$, i.e., $\rho([X,Y]) = [\rho(X), \rho(Y)]$.

Example 8.5.1 (Standard representations).

- **Trivial**: $\rho(g) = I$ for all g.
- **Defining/fundamental**: $GL(n)$ acting on $\mathbb{C}^n$.
- **Adjoint**: $\mathrm{Ad} : G \to GL(\mathfrak{g})$, $\mathrm{Ad}_g(X) = gXg^{-1}$.
- **Regular**: G acting on $L^2(G)$ by left translation.

Definition 8.5.3 (Invariant subspace and irreducibility). A subspace $W \subseteq V$ is **invariant** if $\rho(g)W \subseteq W$ for all g.

A representation is **irreducible** (or **simple**) if the only invariant subspaces are $\{0\}$ and V.

Definition 8.5.4 (Completely reducible). A representation is **completely reducible** (or **semisimple**) if it is a direct sum of irreducible representations.

Theorem 8.5.1 (Maschke's theorem). *Every representation of a finite group over $\mathbb{C}$ is completely reducible.*

Theorem 8.5.2 (Weyl's theorem). *Every finite-dimensional representation of a semisimple Lie algebra (or compact Lie group) is completely reducible.*

8.5.2 Schur's Lemma

Theorem 8.5.3 (Schur's lemma). *Let $\rho : G \to GL(V)$ and $\sigma : G \to GL(W)$ be irreducible representations.*

(i) If $T : V \to W$ is an intertwiner ($T\rho(g) = \sigma(g)T$ for all g), then $T = 0$ or T is an isomorphism.

(ii) If $V = W$ and $\rho = \sigma$, then $T = \lambda I$ for some $\lambda \in \mathbb{C}$ (over algebraically closed fields).

Corollary 8.5.4. *Irreducible representations of abelian groups are one-dimensional.*

8.5.3 Characters

Definition 8.5.5 (Character). The **character** of a representation ρ is the function $\chi_\rho : G \to \mathbb{C}$ defined by:

$$\chi_\rho(g) := \mathrm{tr}(\rho(g)).$$

Proposition 8.5.5 (Properties of characters).

(i) $\chi_\rho(e) = \dim V$.

(ii) $\chi_\rho(hgh^{-1}) = \chi_\rho(g)$ *(class function).*

(iii) $\chi_{\rho\oplus\sigma} = \chi_\rho + \chi_\sigma$.

(iv) $\chi_{\rho\otimes\sigma} = \chi_\rho \cdot \chi_\sigma$.

(v) $\chi_{\rho^*}(g) = \overline{\chi_\rho(g)}$.

Theorem 8.5.6 (Orthogonality of characters). *For a finite group G, irreducible characters satisfy:*

$$\langle \chi_\rho, \chi_\sigma \rangle := \frac{1}{|G|} \sum_{g\in G} \overline{\chi_\rho(g)}\chi_\sigma(g) = \delta_{\rho\sigma}.$$

Theorem 8.5.7. *The number of irreducible representations of a finite group equals the number of conjugacy classes.*

8.5.4 Representations of $SU(2)$ and $SO(3)$

Theorem 8.5.8 (Irreducible representations of $SU(2)$). *For each $j \in \{0, \frac{1}{2}, 1, \frac{3}{2}, 2, \ldots\}$, there is a unique (up to equivalence) irreducible representation $D^{(j)}$ of dimension $2j+1$.*

The Lie algebra $\mathfrak{su}(2)$ has basis $\{J_1, J_2, J_3\}$ with:

$$[J_i, J_j] = i\epsilon_{ijk} J_k.$$

In representation $D^{(j)}$, the Casimir $\mathbf{J}^2 = J_1^2 + J_2^2 + J_3^2$ has eigenvalue $j(j+1)$.

Proposition 8.5.9. *Integer j representations correspond to $SO(3)$. Half-integer j are "spinor" representations of $SU(2)$ only (double-valued on $SO(3)$).*

Example 8.5.2. $j = 0$: trivial (scalar).
$j = 1/2$: fundamental spinor (dimension 2).
$j = 1$: vector representation (dimension 3, isomorphic to $\mathfrak{so}(3)$).

8.5.5 Tensor Products and Clebsch-Gordan

Definition 8.5.6 (Tensor product of representations). For representations $\rho : G \to GL(V)$ and $\sigma : G \to GL(W)$:

$$(\rho \otimes \sigma)(g) := \rho(g) \otimes \sigma(g).$$

Theorem 8.5.10 (Clebsch-Gordan decomposition). *For $SU(2)$:*

$$D^{(j_1)} \otimes D^{(j_2)} = \bigoplus_{j=|j_1-j_2|}^{j_1+j_2} D^{(j)}.$$

Example 8.5.3. $D^{(1/2)} \otimes D^{(1/2)} = D^{(0)} \oplus D^{(1)}$ (two spin-1/2 particles combine to spin 0 or 1).

8.6 Physical Applications

8.6.1 Symmetry and Conservation Laws

Theorem 8.6.1 (Noether's theorem). *Every continuous symmetry of a physical system corresponds to a conserved quantity.*

Example 8.6.1 (Symmetries and conserved quantities).

Symmetry	Conserved Quantity
Time translation	Energy
Space translation	Momentum
Rotation $SO(3)$	Angular momentum
Lorentz boosts	Center of mass motion
$U(1)$ gauge	Electric charge
$SU(3)$ color	Color charge

8.6.2 Quantum Mechanics

In quantum mechanics:

- States transform under representations of symmetry groups.
- Observables are Hermitian operators in representation spaces.
- Selection rules follow from representation theory.

Example 8.6.2 (Angular momentum). Angular momentum operators J_i form the Lie algebra $\mathfrak{su}(2)$. States $|j, m\rangle$ form bases for irreducible representations with $J^2|j, m\rangle = j(j+1)\hbar^2|j, m\rangle$ and $J_z|j, m\rangle = m\hbar|j, m\rangle$.

8.6.3 Particle Physics

Example 8.6.3 (Standard Model gauge group). The Standard Model has gauge group $SU(3)_C \times SU(2)_L \times U(1)_Y$:

- $SU(3)_C$: color (strong force), quarks in fundamental rep.
- $SU(2)_L$: weak isospin, left-handed fermions in doublets.
- $U(1)_Y$: hypercharge.

Example 8.6.4 (Quark model). Hadrons are classified by representations of flavor $SU(3)$:

- Mesons: $\mathbf{3} \otimes \bar{\mathbf{3}} = \mathbf{1} \oplus \mathbf{8}$ (singlet + octet).
- Baryons: $\mathbf{3} \otimes \mathbf{3} \otimes \mathbf{3} = \mathbf{1} \oplus \mathbf{8} \oplus \mathbf{8} \oplus \mathbf{10}$.

8.7 Summary

Group theory provides the mathematical framework for symmetry in physics:

Abstract groups: Capture symmetry operations, with subgroups, quotients, and actions.

Lie groups: Continuous symmetries, locally described by Lie algebras.

Lie algebras: Infinitesimal symmetries, classified by root systems and Dynkin diagrams.

Representations: How symmetries act on physical states and fields.

Physical applications: Conservation laws (Noether), angular momentum, particle classification, gauge theories.

Exercises

Exercise 8.7.1. Prove that any group of prime order is cyclic.

Exercise 8.7.2. Show that A_4 has no subgroup of order 6 (despite $6|12$).

Exercise 8.7.3. Prove the orbit-stabilizer theorem.

Exercise 8.7.4. Show that $SU(2)$ is diffeomorphic to S^3.

Exercise 8.7.5. Compute the Lie algebra of $SO(3)$ and verify the commutation relations $[J_i, J_j] = \epsilon_{ijk} J_k$.

Exercise 8.7.6. Show that $\exp : \mathfrak{so}(3) \to SO(3)$ is surjective but not injective.

Exercise 8.7.7. Verify the Jacobi identity for the commutator bracket $[X, Y] = XY - YX$.

Exercise 8.7.8. Show that the Killing form of $\mathfrak{sl}(2, \mathbb{C})$ is $\kappa(X, Y) = 4\,\mathrm{tr}(XY)$.

Exercise 8.7.9. Find all irreducible representations of S_3 and construct the character table.

Exercise 8.7.10. Decompose $D^{(1)} \otimes D^{(1)}$ into irreducible representations of $SU(2)$.

Exercise 8.7.11. Show that the center of $SU(n)$ is $\mathbb{Z}/n\mathbb{Z}$ and that $SU(n)/Z(SU(n)) \cong PSU(n)$.

Exercise 8.7.12. Derive the dimension formula $\dim D^{(j)} = 2j + 1$ using the highest weight method.

Chapter 9

Representation Theory

Building on the foundations of Chapter 8, we now develop representation theory in greater depth. This chapter focuses on the representation theory of Lie groups and algebras most important for physics: the classification by highest weights, representations of the Lorentz and Poincaré groups, and the spinor representations essential for describing fermions.

9.1 Representations of Lie Algebras

9.1.1 Weight Space Decomposition

Definition 9.1.1 (Weight). Let $\mathfrak{g}$ be a semisimple Lie algebra with Cartan subalgebra $\mathfrak{h}$. For a representation (ρ, V), a **weight** is a linear functional $\lambda \in \mathfrak{h}^*$ such that the **weight space**

$$V_\lambda := \{v \in V : \rho(H)v = \lambda(H)v \text{ for all } H \in \mathfrak{h}\}$$

is nonzero. The elements of V_λ are **weight vectors**.

Theorem 9.1.1 (Weight space decomposition). *Every finite-dimensional representation V of a semisimple Lie algebra decomposes as:*

$$V = \bigoplus_\lambda V_\lambda$$

where the sum is over all weights of the representation.

Definition 9.1.2 (Weight lattice). The **weight lattice** $\Lambda_W \subseteq \mathfrak{h}^*$ is the set of all weights that appear in some finite-dimensional representation.

Definition 9.1.3 (Root lattice). The **root lattice** Λ_R is the $\mathbb{Z}$-span of the roots. We have $\Lambda_R \subseteq \Lambda_W$.

9.1.2 Highest Weight Theory

Definition 9.1.4 (Positive roots and ordering)**.** A choice of **positive roots** Φ^+ induces a partial order on $\mathfrak{h}^*$: $\lambda \geq \mu$ if $\lambda - \mu$ is a non-negative integer combination of positive roots.

Definition 9.1.5 (Highest weight)**.** A weight λ in a representation V is a **highest weight** if $\lambda + \alpha$ is not a weight for any positive root α. A vector $v \in V_\lambda$ is a **highest weight vector**.

Theorem 9.1.2 (Classification by highest weight)**.**

(i) Every finite-dimensional irreducible representation has a unique highest weight.

(ii) The highest weight space is one-dimensional.

(iii) Two irreducible representations are equivalent iff they have the same highest weight.

*(iv) A weight λ is the highest weight of an irreducible representation iff λ is **dominant integral**: $\langle \lambda, \alpha^\vee \rangle \in \mathbb{Z}_{\geq 0}$ for all simple roots α.*

Definition 9.1.6 (Simple roots and fundamental weights)**.** The **simple roots** $\{\alpha_1, \ldots, \alpha_r\}$ are a basis for Φ such that every positive root is a non-negative combination.

The **fundamental weights** $\{\omega_1, \ldots, \omega_r\}$ form the dual basis: $\langle \omega_i, \alpha_j^\vee \rangle = \delta_{ij}$.

Proposition 9.1.3. *Dominant integral weights have the form $\lambda = \sum_{i=1}^r n_i \omega_i$ with $n_i \in \mathbb{Z}_{\geq 0}$. The **Dynkin labels** $(n_1, \ldots, n_r)$ specify the representation.*

9.1.3 The Weyl Group and Weyl Character Formula

Definition 9.1.7 (Weyl group)**.** The **Weyl group** W is generated by reflections s_α through hyperplanes perpendicular to roots:

$$s_\alpha(\lambda) = \lambda - \langle \lambda, \alpha^\vee \rangle \alpha.$$

Theorem 9.1.4 (Properties of Weyl group)**.**

(i) W is finite.

(ii) W permutes the weights of any representation.

(iii) W acts simply transitively on the set of Weyl chambers.

Theorem 9.1.5 (Weyl dimension formula). *For the irreducible representation $V(\lambda)$ with highest weight λ:*

$$\dim V(\lambda) = \prod_{\alpha \in \Phi^+} \frac{\langle \lambda + \rho, \alpha \rangle}{\langle \rho, \alpha \rangle}$$

where $\rho = \frac{1}{2}\sum_{\alpha \in \Phi^+} \alpha$ is the Weyl vector.

Theorem 9.1.6 (Weyl character formula). *The character of $V(\lambda)$ is:*

$$\chi_\lambda = \frac{\sum_{w \in W} \det(w) e^{w(\lambda+\rho)}}{\sum_{w \in W} \det(w) e^{w(\rho)}} = \frac{\sum_{w \in W} \det(w) e^{w(\lambda+\rho)}}{\prod_{\alpha \in \Phi^+}(e^{\alpha/2} - e^{-\alpha/2})}.$$

9.2 Representations of Classical Lie Algebras

9.2.1 $\mathfrak{sl}(n, \mathbb{C})$ (Type A_{n-1})

The Lie algebra $\mathfrak{sl}(n, \mathbb{C})$ consists of traceless $n \times n$ complex matrices.

Proposition 9.2.1 (Root system of A_{n-1}). *Cartan subalgebra: diagonal traceless matrices. Roots: $\alpha_{ij} = e_i - e_j$ for $i \neq j$. Simple roots: $\alpha_i = e_i - e_{i+1}$.*

Theorem 9.2.2 (Representations of $\mathfrak{sl}(n)$). *Irreducible representations are labeled by Young diagrams with at most $n-1$ rows, or equivalently by dominant weights $(n_1, \ldots, n_{n-1})$.*

Example 9.2.1 (Representations of $\mathfrak{sl}(3)$).

- $(1, 0)$: fundamental, dimension 3 (quarks).
- $(0, 1)$: anti-fundamental $\bar{\mathbf{3}}$, dimension 3 (antiquarks).
- $(1, 1)$: adjoint, dimension 8 (gluons).
- $(2, 0)$: symmetric tensor, dimension 6.
- $(0, 2)$: dimension 6.
- $(3, 0)$: dimension 10 (decuplet).

9.2.2 $\mathfrak{so}(n, \mathbb{C})$ (Types B_r, D_r)

Definition 9.2.1. $\mathfrak{so}(n, \mathbb{C}) = \{X \in \mathfrak{gl}(n, \mathbb{C}) : X^T = -X\}$ (antisymmetric matrices).

Proposition 9.2.3 (Types). *$\mathfrak{so}(2r+1)$ is type B_r with* $\dim = r(2r+1)$. *$\mathfrak{so}(2r)$ is type D_r with* $\dim = r(2r-1)$.

Theorem 9.2.4 (Spinor representations).

- $\mathfrak{so}(2r+1)$: *one spinor representation of dimension* 2^r.
- $\mathfrak{so}(2r)$: *two half-spinor representations of dimension* 2^{r-1} *each (Weyl spinors).*

These are not representations of $SO(n)$ *but of its double cover* $\mathrm{Spin}(n)$.

Example 9.2.2 (Low dimensions).

- $\mathfrak{so}(3) \cong \mathfrak{su}(2)$: spinor is the fundamental of $SU(2)$.
- $\mathfrak{so}(4) \cong \mathfrak{su}(2) \oplus \mathfrak{su}(2)$: two Weyl spinors.
- $\mathfrak{so}(6) \cong \mathfrak{su}(4)$: Weyl spinors are **4** and $\bar{\mathbf{4}}$.

9.2.3 $\mathfrak{sp}(2n, \mathbb{C})$ (Type C_n)

Definition 9.2.2. $\mathfrak{sp}(2n, \mathbb{C}) = \{X \in \mathfrak{gl}(2n, \mathbb{C}) : X^T J + JX = 0\}$ where $J = \begin{pmatrix} 0 & I \\ -I & 0 \end{pmatrix}$.

Proposition 9.2.5. $\dim \mathfrak{sp}(2n) = n(2n+1)$. *The fundamental representation has dimension* $2n$.

9.3 Tensor Methods

9.3.1 Tensor Products of Representations

Theorem 9.3.1 (Complete reducibility of tensor products). *For semisimple Lie algebras, any tensor product of finite-dimensional representations decomposes into irreducibles:*

$$V(\lambda) \otimes V(\mu) = \bigoplus_{\nu} N_{\lambda\mu}^{\nu} V(\nu)$$

where $N_{\lambda\mu}^{\nu} \in \mathbb{Z}_{\geq 0}$ *are* ***Littlewood-Richardson coefficients****.*

Proposition 9.3.2 (Weight addition). *The weights of* $V \otimes W$ *are* $\{\lambda + \mu : \lambda \in \mathrm{wt}(V), \mu \in \mathrm{wt}(W)\}$.

9.3.2 Young Tableaux

Definition 9.3.1 (Young diagram). A **Young diagram** is a collection of boxes arranged in left-justified rows with weakly decreasing row lengths.

Definition 9.3.2 (Young tableau). A **Young tableau** is a Young diagram filled with numbers according to certain rules:

- **Standard**: entries $1, \ldots, n$ each once, increasing along rows and columns.
- **Semistandard**: entries from $\{1, \ldots, N\}$, weakly increasing along rows, strictly increasing down columns.

Theorem 9.3.3 (Young tableaux and $GL(N)$). *Irreducible polynomial representations of $GL(N, \mathbb{C})$ correspond to Young diagrams with at most N rows. The dimension equals the number of semistandard tableaux with entries in $\{1, \ldots, N\}$.*

Example 9.3.1. The diagram $(2, 1)$ (two boxes in first row, one in second) corresponds to the mixed symmetry representation. For $GL(3)$: dimension $= 8$.

9.3.3 Symmetry and Antisymmetry

Definition 9.3.3 (Symmetric and exterior powers). The **k-th symmetric power S^kV** consists of symmetric tensors.

The **k-th exterior power Λ^kV** consists of antisymmetric tensors.

Proposition 9.3.4 (Correspondence with Young diagrams).

- *S^kV corresponds to a single row of k boxes.*
- *Λ^kV corresponds to a single column of k boxes.*

Example 9.3.2 (Decomposition of $V \otimes V$). For any representation V:

$$V \otimes V = S^2V \oplus \Lambda^2V.$$

For the fundamental of $SU(N)$: $\mathbf{N} \otimes \mathbf{N} = \frac{\mathbf{N(N+1)}}{\mathbf{2}} \oplus \frac{\mathbf{N(N-1)}}{\mathbf{2}}$.

9.4 The Lorentz Group and Its Representations

9.4.1 Structure of the Lorentz Group

Proposition 9.4.1 (Lie algebra of Lorentz group). *The Lie algebra $\mathfrak{so}(1,3)$ has generators J_i (rotations) and K_i (boosts) with:*

$$[J_i, J_j] = i\epsilon_{ijk}J_k$$
$$[J_i, K_j] = i\epsilon_{ijk}K_k$$
$$[K_i, K_j] = -i\epsilon_{ijk}J_k$$

Theorem 9.4.2 (Complexification and decomposition). *Define $A_i := \frac{1}{2}(J_i + iK_i)$ and $B_i := \frac{1}{2}(J_i - iK_i)$. Then:*

$$[A_i, A_j] = i\epsilon_{ijk}A_k, \quad [B_i, B_j] = i\epsilon_{ijk}B_k, \quad [A_i, B_j] = 0.$$

Thus $\mathfrak{so}(1,3)_\mathbb{C} \cong \mathfrak{su}(2)_\mathbb{C} \oplus \mathfrak{su}(2)_\mathbb{C} \cong \mathfrak{sl}(2,\mathbb{C}) \oplus \mathfrak{sl}(2,\mathbb{C})$.

Corollary 9.4.3. *Finite-dimensional irreducible representations of $\mathfrak{so}(1,3)_\mathbb{C}$ are labeled by pairs (j_1, j_2) with $j_1, j_2 \in \{0, \frac{1}{2}, 1, \ldots\}$. The dimension is $(2j_1 + 1)(2j_2 + 1)$.*

9.4.2 Important Lorentz Representations

Example 9.4.1 (Standard representations).

- $(0,0)$: **Scalar**, dimension 1.
- $(\frac{1}{2}, 0)$: **Left-handed Weyl spinor**, dimension 2.
- $(0, \frac{1}{2})$: **Right-handed Weyl spinor**, dimension 2.
- $(\frac{1}{2}, 0) \oplus (0, \frac{1}{2})$: **Dirac spinor**, dimension 4.
- $(\frac{1}{2}, \frac{1}{2})$: **Vector**, dimension 4.
- $(1,0) \oplus (0,1)$: **Antisymmetric tensor** (electromagnetic field), dimension 6.
- $(1,1)$: **Traceless symmetric tensor**, dimension 9.

Definition 9.4.1 (Chirality). Under parity P: $(j_1, j_2) \leftrightarrow (j_2, j_1)$. Representations with $j_1 \neq j_2$ have definite **chirality** (handedness).

9.4.3 Spinors

Definition 9.4.2 (Weyl spinors)**.** A **left-handed Weyl spinor** ψ_L transforms in $(\frac{1}{2}, 0)$:

$$\psi_L \mapsto \exp\left(\frac{i}{2}(\boldsymbol{\theta} - i\boldsymbol{\phi}) \cdot \boldsymbol{\sigma}\right) \psi_L$$

where $\boldsymbol{\theta}$ parametrizes rotations and $\boldsymbol{\phi}$ boosts.

A **right-handed Weyl spinor** ψ_R transforms in $(0, \frac{1}{2})$:

$$\psi_R \mapsto \exp\left(\frac{i}{2}(\boldsymbol{\theta} + i\boldsymbol{\phi}) \cdot \boldsymbol{\sigma}\right) \psi_R.$$

Definition 9.4.3 (Dirac spinor)**.** A **Dirac spinor** combines both chiralities:

$$\psi = \begin{pmatrix} \psi_L \\ \psi_R \end{pmatrix}.$$

Definition 9.4.4 (Gamma matrices)**.** The **Dirac gamma matrices** γ^μ satisfy the Clifford algebra:

$$\{\gamma^\mu, \gamma^\nu\} = 2\eta^{\mu\nu} I.$$

The Lorentz generators on Dirac spinors are $S^{\mu\nu} = \frac{i}{4}[\gamma^\mu, \gamma^\nu]$.

Definition 9.4.5 (Chirality matrix)**.** $\gamma^5 := i\gamma^0\gamma^1\gamma^2\gamma^3$ satisfies $(\gamma^5)^2 = I$ and $\{\gamma^5, \gamma^\mu\} = 0$. Projection operators: $P_{L,R} = \frac{1}{2}(1 \mp \gamma^5)$.

9.5 The Poincaré Group and Its Representations

9.5.1 Structure of the Poincaré Group

Definition 9.5.1 (Poincaré algebra)**.** The Poincaré algebra has generators P_μ (translations) and $M_{\mu\nu}$ (Lorentz) with:

$$\begin{aligned} [P_\mu, P_\nu] &= 0 \\ [M_{\mu\nu}, P_\rho] &= i(\eta_{\mu\rho} P_\nu - \eta_{\nu\rho} P_\mu) \\ [M_{\mu\nu}, M_{\rho\sigma}] &= i(\eta_{\mu\rho} M_{\nu\sigma} - \eta_{\mu\sigma} M_{\nu\rho} - \eta_{\nu\rho} M_{\mu\sigma} + \eta_{\nu\sigma} M_{\mu\rho}) \end{aligned}$$

Definition 9.5.2 (Casimir operators)**.** The Poincaré algebra has two Casimir operators:

- $P^2 = P_\mu P^\mu$ (mass squared).
- $W^2 = W_\mu W^\mu$ where $W_\mu = \frac{1}{2}\epsilon_{\mu\nu\rho\sigma} P^\nu M^{\rho\sigma}$ is the **Pauli-Lubanski pseudovector**.

9.5.2 Wigner's Classification

Theorem 9.5.1 (Wigner's classification). *Unitary irreducible representations of the Poincaré group are classified by:*

(i) The value of $P^2 = m^2$ (mass).

(ii) The value of W^2 (related to spin).

(iii) The sign of P^0 (for $m^2 > 0$) or additional discrete labels.

Definition 9.5.3 (Little group). The **little group** (or **stabilizer**) of a standard momentum k^μ is the subgroup of Lorentz transformations leaving k^μ invariant.

Theorem 9.5.2 (Little groups by mass).

- ***Massive*** *($m^2 > 0$): standard $k = (m, 0, 0, 0)$, little group $SO(3)$.*
- ***Massless*** *($m^2 = 0$, $P^\mu \neq 0$): standard $k = (E, 0, 0, E)$, little group $ISO(2) = \mathbb{R}^2 \rtimes SO(2)$.*
- ***Tachyonic*** *($m^2 < 0$): little group $SO(2, 1)$.*
- ***Vacuum*** *($P^\mu = 0$): little group is full Lorentz group.*

9.5.3 Massive Particles

Theorem 9.5.3 (Massive representations). *For $m > 0$, irreducible representations are labeled by mass m and spin $s \in \{0, \frac{1}{2}, 1, \frac{3}{2}, \ldots\}$. The representation space is:*

$$\mathcal{H} = L^2(\mathcal{O}_m^+) \otimes V_s$$

where $\mathcal{O}_m^+$ is the positive mass hyperboloid and V_s is the spin-s representation of $SO(3)$.

States $|p, \sigma\rangle$ with $p^2 = m^2$, $p^0 > 0$, and $\sigma \in \{-s, \ldots, s\}$.

$W^2 = -m^2 s(s + 1)$.

Example 9.5.1 (Massive particles in nature).

- Electron: $m \approx 0.511$ MeV, $s = 1/2$.
- $W^\pm$, Z bosons: $m \approx 80, 91$ GeV, $s = 1$.
- Higgs: $m \approx 125$ GeV, $s = 0$.

9.5.4 Massless Particles

Theorem 9.5.4 (Massless representations). *For $m = 0$, the little group $ISO(2)$ has representations labeled by* ***helicity*** $\lambda \in \{0, \pm\frac{1}{2}, \pm 1, \ldots\}$.

Physical massless particles have $|\lambda|$ (spin) with two helicity states $\pm\lambda$, except:

- *Scalars ($\lambda = 0$): one state.*
- *Particles with only one helicity (neutrinos in SM, if massless).*

$W^2 = 0$ for all massless representations.

Example 9.5.2 (Massless particles).

- Photon: $\lambda = \pm 1$ (two polarizations).
- Graviton: $\lambda = \pm 2$ (two polarizations).
- Neutrino (if massless): $\lambda = -1/2$ (left-handed only in SM).

Remark 9.5.1 (CPT and spin-statistics). CPT symmetry relates particles to antiparticles. The spin-statistics theorem requires:

- Integer spin → bosons (symmetric wavefunctions).
- Half-integer spin → fermions (antisymmetric wavefunctions).

9.6 Induced Representations

9.6.1 The Method of Induced Representations

Definition 9.6.1 (Induced representation). Let $H \leq G$ and $\sigma : H \to GL(W)$ be a representation. The **induced representation** $\mathrm{Ind}_H^G \sigma$ acts on:

$$\mathrm{Ind}_H^G W := \{f : G \to W : f(gh) = \sigma(h)^{-1} f(g) \text{ for all } h \in H\}$$

by $(g \cdot f)(x) = f(g^{-1}x)$.

Theorem 9.6.1 (Frobenius reciprocity). *For $H \leq G$, representation σ of H, and representation ρ of G:*

$$\mathrm{Hom}_G(\rho, \mathrm{Ind}_H^G \sigma) \cong \mathrm{Hom}_H(\mathrm{Res}_H^G \rho, \sigma).$$

9.6.2 Application to Poincaré Group

Wigner's method constructs Poincaré representations as induced representations:

1. Choose standard momentum k (e.g., $(m, \mathbf{0})$ for massive).
2. Identify little group H_k (e.g., $SO(3)$ for massive).
3. Choose representation σ of H_k (e.g., spin s).
4. Induce to full Poincaré group.

The resulting representation space consists of wavefunctions on the mass shell satisfying covariance conditions.

9.7 Unitary Representations

9.7.1 Unitarity and Physics

Definition 9.7.1 (Unitary representation). A representation $\rho : G \to GL(V)$ is **unitary** if V is a Hilbert space and $\rho(g)$ is unitary for all g.

In quantum mechanics, symmetries must act unitarily to preserve probability.

Theorem 9.7.1 (Unitarity constraints). *For a connected Lie group G:*

- *If G is compact, all finite-dimensional representations are equivalent to unitary ones.*
- *If G is non-compact (e.g., Lorentz), finite-dimensional representations cannot be unitary (except trivial).*

Corollary 9.7.2. *The finite-dimensional Lorentz representations (j_1, j_2) are not unitary for $(j_1, j_2) \neq (0, 0)$. Physical (unitary) representations are infinite-dimensional.*

9.7.2 Unitary Representations of Compact Groups

Theorem 9.7.3 (Peter-Weyl theorem). *For compact Lie group G:*

$$L^2(G) \cong \bigoplus_{\lambda} V_\lambda \otimes V_\lambda^*$$

where the sum is over all irreducible representations, each appearing with multiplicity $\dim V_\lambda$.

Corollary 9.7.4. *The matrix coefficients $\rho_{ij}^\lambda(g) = \langle e_i, \rho^\lambda(g) e_j \rangle$ form an orthonormal basis of $L^2(G)$ (after normalization).*

9.7.3 Unitary Representations of Non-Compact Groups

Definition 9.7.2 (Principal and discrete series). Non-compact semisimple groups have:

- **Principal series**: induced from characters of minimal parabolic.
- **Discrete series**: square-integrable representations (exist iff $\text{rank}(G) = \text{rank}(K)$ for maximal compact K).
- **Complementary series**: other unitary representations.

Example 9.7.1 ($SL(2,\mathbb{R})$). $SL(2,\mathbb{R})$ has:

- Principal series: $\lambda \in i\mathbb{R}$.
- Discrete series: $\mathcal{D}_n^{\pm}$ for $n \geq 2$.
- Complementary series: $0 < \lambda < 1$.

9.8 Applications to Quantum Field Theory

9.8.1 Fields and Representations

Definition 9.8.1 (Quantum field). A **quantum field** $\phi(x)$ transforms under Poincaré as:

$$U(\Lambda, a)\phi(x)U(\Lambda, a)^{-1} = D(\Lambda^{-1})\phi(\Lambda x + a)$$

where D is a finite-dimensional Lorentz representation.

Example 9.8.1 (Standard fields).

- **Scalar field** ϕ: $(0, 0)$, spin 0.
- **Dirac field** ψ: $(\frac{1}{2}, 0) \oplus (0, \frac{1}{2})$, spin $\frac{1}{2}$.
- **Vector field** A_μ: $(\frac{1}{2}, \frac{1}{2})$, spin 1.
- **Weyl field** ψ_L: $(\frac{1}{2}, 0)$, spin $\frac{1}{2}$.
- **Graviton field** $h_{\mu\nu}$: $(1, 1) \oplus \cdots$, spin 2.

9.8.2 The Spin-Statistics Connection

Theorem 9.8.1 (Spin-statistics theorem). *In a local relativistic quantum field theory:*

- *Fields with integer spin satisfy Bose-Einstein statistics.*
- *Fields with half-integer spin satisfy Fermi-Dirac statistics.*

This deep result follows from Lorentz invariance, locality, and positive energy.

9.8.3 Selection Rules

Proposition 9.8.2 (Selection rules from symmetry). *If a system has symmetry group G and initial state transforms as ρ_i, final state as ρ_f, and interaction as ρ_{int}, then the transition amplitude vanishes unless ρ_f appears in $\rho_i \otimes \rho_{\text{int}}$.*

Example 9.8.2 (Angular momentum selection rules). Electric dipole transitions ($L = 1$): $\Delta j \in \{0, \pm 1\}$ (but not $j = 0 \to j' = 0$).

9.9 Summary

Representation theory connects abstract symmetry to observable physics:

Highest weight theory: Classifies representations of semisimple Lie algebras by dominant integral weights.

Tensor methods: Young tableaux systematize the construction and decomposition of representations.

Lorentz representations: Classified by (j_1, j_2), includes spinors essential for fermions.

Poincaré representations: Wigner's classification by mass and spin determines particle types.

Unitarity: Physical representations must be unitary; this constrains allowed theories.

Applications: Fields transform in Lorentz representations; particles are Poincaré representations; selection rules follow from representation decomposition.

Exercises

Exercise 9.9.1. Compute the weights of the adjoint representation of $\mathfrak{sl}(3, \mathbb{C})$ and verify they are the roots plus zero (with multiplicity 2).

Exercise 9.9.2. Use the Weyl dimension formula to verify $\dim V(n\omega_1) = n+1$ for $\mathfrak{sl}(2)$.

Exercise 9.9.3. Decompose $\mathbf{3} \otimes \mathbf{3}$ for $SU(3)$ using Young tableaux. Identify the resulting representations.

Exercise 9.9.4. Show that the Weyl group of $\mathfrak{sl}(n)$ is the symmetric group S_n.

Exercise 9.9.5. Verify that $[A_i, B_j] = 0$ for the Lorentz algebra decomposition into $\mathfrak{su}(2) \oplus \mathfrak{su}(2)$.

Exercise 9.9.6. Show that under parity, left-handed and right-handed Weyl spinors are exchanged.

Exercise 9.9.7. Compute the helicity content of the electromagnetic field strength tensor $F_{\mu\nu}$.

Exercise 9.9.8. Verify that the Pauli-Lubanski vector satisfies $W_\mu P^\mu = 0$.

Exercise 9.9.9. For a massive spin-1 particle, show that $W^2 = -2m^2$ and identify the three polarization states.

Exercise 9.9.10. Prove that the little group of a lightlike vector is $ISO(2)$.

Exercise 9.9.11. Show that the gamma matrices satisfy $\mathrm{tr}(\gamma^\mu \gamma^\nu) = 4\eta^{\mu\nu}$.

Exercise 9.9.12. Using Frobenius reciprocity, show that the multiplicity of the trivial representation in $\mathrm{Ind}_H^G \mathbf{1}$ equals $[G : H]$ for finite groups.

Chapter 10

Tensor Analysis

Tensors are the natural language for expressing physical laws in a coordinate-independent way. From the stress tensor in mechanics to the metric tensor in general relativity, tensors appear throughout physics. This chapter develops tensor algebra and calculus systematically, with emphasis on the index notation used extensively in physics.

10.1 Tensors on Vector Spaces

10.1.1 Multilinear Maps

Definition 10.1.1 (Multilinear map). A map $T : V_1 \times \cdots \times V_k \to W$ is **multilinear** (or **k-linear**) if it is linear in each argument:

$$T(v_1, \ldots, av_i + bv_i', \ldots, v_k) = aT(v_1, \ldots, v_i, \ldots, v_k) + bT(v_1, \ldots, v_i', \ldots, v_k).$$

Definition 10.1.2 (Tensor). A **tensor of type** (r, s) on a vector space V is a multilinear map:

$$T : \underbrace{V^* \times \cdots \times V^*}_{r} \times \underbrace{V \times \cdots \times V}_{s} \to \mathbb{F}.$$

We say T is **r-contravariant** and **s-covariant**, with **rank** $r + s$.

Example 10.1.1 (Basic tensors).

- Type $(0, 0)$: scalars.
- Type $(1, 0)$: vectors $v \in V$ (via $v(\omega) = \omega(v)$ for $\omega \in V^*$).
- Type $(0, 1)$: covectors (linear forms) $\omega \in V^*$.

- Type $(1, 1)$: linear maps $V \to V$.
- Type $(0, 2)$: bilinear forms on V.
- Type $(2, 0)$: bilinear forms on V^*.

10.1.2 Tensor Product

Definition 10.1.3 (Tensor product of tensors). Given tensors S of type (r, s) and T of type (p, q), their **tensor product** $S \otimes T$ is the tensor of type $(r + p, s + q)$ defined by:

$$(S\otimes T)(\omega^1, \ldots, \omega^{r+p}, v_1, \ldots, v_{s+q}) = S(\omega^1, \ldots, \omega^r, v_1, \ldots, v_s)\cdot T(\omega^{r+1}, \ldots, \omega^{r+p}, v_{s+1}, \ldots,$$

Definition 10.1.4 (Tensor space). The space of all (r, s)-tensors is:

$$T^r_s(V) := \underbrace{V \otimes \cdots \otimes V}_{r} \otimes \underbrace{V^* \otimes \cdots \otimes V^*}_{s}.$$

Proposition 10.1.1. *If* $\dim V = n$, *then* $\dim T^r_s(V) = n^{r+s}$.

10.1.3 Components and Index Notation

Let $\{e_1, \ldots, e_n\}$ be a basis for V and $\{e^1, \ldots, e^n\}$ the dual basis for V^*.

Definition 10.1.5 (Components). The **components** of a tensor T of type (r, s) are:

$$T^{i_1\cdots i_r}{}_{j_1\cdots j_s} := T(e^{i_1}, \ldots, e^{i_r}, e_{j_1}, \ldots, e_{j_s}).$$

Proposition 10.1.2 (Tensor expansion). *Any tensor can be written:*

$$T = T^{i_1\cdots i_r}{}_{j_1\cdots j_s}\, e_{i_1} \otimes \cdots \otimes e_{i_r} \otimes e^{j_1} \otimes \cdots \otimes e^{j_s}$$

where summation over repeated indices is implied (Einstein convention).

Definition 10.1.6 (Einstein summation convention). Repeated indices (one upper, one lower) are summed:

$$A^i B_i := \sum_{i=1}^{n} A^i B_i.$$

10.1.4 Change of Basis

Proposition 10.1.3 (Transformation law). *Under change of basis $\tilde{e}_i = A^j{}_i e_j$ (with inverse $e_i = (A^{-1})^j{}_i \tilde{e}_j$):*

$$\tilde{T}^{i_1 \cdots i_r}{}_{j_1 \cdots j_s} = (A^{-1})^{i_1}{}_{k_1} \cdots (A^{-1})^{i_r}{}_{k_r} A^{l_1}{}_{j_1} \cdots A^{l_s}{}_{j_s} T^{k_1 \cdots k_r}{}_{l_1 \cdots l_s}.$$

This transformation law is often taken as the definition of a tensor: a tensor is an object whose components transform in this specific way.

Remark 10.1.1 (Contravariant vs covariant).

- Upper indices transform with (A^{-1}) (contravariant, opposite to basis).
- Lower indices transform with A (covariant, same as basis).

10.2 Tensor Operations

10.2.1 Contraction

Definition 10.2.1 (Contraction). The **contraction** of a tensor T of type (r, s) over the i-th contravariant and j-th covariant index produces a tensor of type $(r-1, s-1)$:

$$(C^i_j T)^{i_1 \cdots \hat{i}_p \cdots i_r}{}_{j_1 \cdots \hat{j}_q \cdots j_s} := T^{i_1 \cdots k \cdots i_r}{}_{j_1 \cdots k \cdots j_s}$$

where k appears in position p (upper) and q (lower), and the hat denotes omission.

Example 10.2.1. The trace of a $(1, 1)$ tensor: $\mathrm{tr}(T) = T^i{}_i$.

For vectors v and covectors ω: $\omega(v) = \omega_i v^i$ is the contraction of $\omega \otimes v$.

10.2.2 Raising and Lowering Indices

Definition 10.2.2 (Metric tensor). A **metric tensor** g is a non-degenerate symmetric $(0, 2)$-tensor. In components: $g_{ij} = g_{ji}$, $\det(g_{ij}) \neq 0$.

Definition 10.2.3 (Inverse metric). The **inverse metric** g^{ij} satisfies $g^{ik} g_{kj} = \delta^i_j$.

Definition 10.2.4 (Index raising and lowering). Given a metric:

$$\begin{aligned} \text{Lowering:} \quad & T_{i\cdots} := g_{ij} T^{j\cdots} \\ \text{Raising:} \quad & T^{i\cdots} := g^{ij} T_{j\cdots} \end{aligned}$$

Example 10.2.2. For a vector v^i: $v_i = g_{ij} v^j$ is the associated covector.

In Euclidean space with $g_{ij} = \delta_{ij}$: $v_i = v^i$ (no distinction).

In Minkowski space with $\eta_{\mu\nu} = \mathrm{diag}(-1, 1, 1, 1)$: $v_0 = -v^0$, $v_i = v^i$ for $i = 1, 2, 3$.

10.2.3 Symmetry and Antisymmetry

Definition 10.2.5 (Symmetric and antisymmetric tensors). A $(0, 2)$-tensor T is:

- **Symmetric** if $T_{ij} = T_{ji}$.
- **Antisymmetric** (or **skew-symmetric**) if $T_{ij} = -T_{ji}$.

Similarly for higher-rank tensors in specified index pairs.

Definition 10.2.6 (Symmetrization and antisymmetrization).

$$T_{(ij)} := \frac{1}{2}(T_{ij} + T_{ji}) \qquad \text{(symmetrization)}$$
$$T_{[ij]} := \frac{1}{2}(T_{ij} - T_{ji}) \qquad \text{(antisymmetrization)}$$

For k indices:

$$T_{(i_1 \cdots i_k)} := \frac{1}{k!} \sum_{\sigma \in S_k} T_{i_{\sigma(1)} \cdots i_{\sigma(k)}}$$
$$T_{[i_1 \cdots i_k]} := \frac{1}{k!} \sum_{\sigma \in S_k} \operatorname{sgn}(\sigma) T_{i_{\sigma(1)} \cdots i_{\sigma(k)}}$$

Proposition 10.2.1. *Any tensor can be decomposed into symmetric and antisymmetric parts:*

$$T_{ij} = T_{(ij)} + T_{[ij]}.$$

10.2.4 The Levi-Civita Symbol

Definition 10.2.7 (Levi-Civita symbol). The **Levi-Civita symbol** (or **permutation symbol**) in n dimensions:

$$\epsilon_{i_1 \cdots i_n} = \epsilon^{i_1 \cdots i_n} := \begin{cases} +1 & \text{if } (i_1, \ldots, i_n) \text{ is an even permutation of } (1, \ldots, n) \\ -1 & \text{if odd permutation} \\ 0 & \text{if any indices repeated} \end{cases}$$

Remark 10.2.1. In a curved space with metric g, the **Levi-Civita tensor** is:

$$\varepsilon_{i_1 \cdots i_n} = \sqrt{|g|}\epsilon_{i_1 \cdots i_n}, \quad \varepsilon^{i_1 \cdots i_n} = \frac{\operatorname{sgn}(g)}{\sqrt{|g|}}\epsilon^{i_1 \cdots i_n}$$

where $g = \det(g_{ij})$.

Proposition 10.2.2 (Useful identities in 3D).

$$\epsilon_{ijk}\epsilon^{lmn} = \delta_i^l\delta_j^m\delta_k^n + \delta_j^l\delta_k^m\delta_i^n + \delta_k^l\delta_i^m\delta_j^n - \delta_i^l\delta_k^m\delta_j^n - \delta_j^l\delta_i^m\delta_k^n - \delta_k^l\delta_j^m\delta_i^n$$
$$\epsilon_{ijk}\epsilon^{ljk} = 2\delta_i^l$$
$$\epsilon_{ijk}\epsilon^{ijk} = 6$$

Proposition 10.2.3 (Cross product). *In 3D:* $(A \times B)^i = \epsilon^{ijk}A_jB_k$.

10.3 Tensors on Manifolds

10.3.1 Tensor Fields

Definition 10.3.1 (Tensor field). A **tensor field** of type (r, s) on a manifold M assigns to each point $p \in M$ a tensor $T_p \in T_s^r(T_pM)$, smoothly varying with p.

Proposition 10.3.1 (Local expression). *In local coordinates* $(x^1, \ldots, x^n)$, *a tensor field has components:*

$$T = T^{i_1\cdots i_r}{}_{j_1\cdots j_s}(x)\,\frac{\partial}{\partial x^{i_1}} \otimes \cdots \otimes \frac{\partial}{\partial x^{i_r}} \otimes dx^{j_1} \otimes \cdots \otimes dx^{j_s}.$$

Proposition 10.3.2 (Coordinate transformation). *Under coordinate change* $x \to \tilde{x}$*:*

$$\tilde{T}^{i_1\cdots i_r}{}_{j_1\cdots j_s} = \frac{\partial \tilde{x}^{i_1}}{\partial x^{k_1}} \cdots \frac{\partial \tilde{x}^{i_r}}{\partial x^{k_r}}\frac{\partial x^{l_1}}{\partial \tilde{x}^{j_1}} \cdots \frac{\partial x^{l_s}}{\partial \tilde{x}^{j_s}} T^{k_1\cdots k_r}{}_{l_1\cdots l_s}.$$

10.3.2 The Metric Tensor on Manifolds

Definition 10.3.2 (Riemannian metric). A **Riemannian metric** on M is a smooth, symmetric, positive-definite $(0, 2)$-tensor field g.

Definition 10.3.3 (Pseudo-Riemannian metric). A **pseudo-Riemannian metric** allows indefinite signature. The **Lorentzian** case has signature $(-, +, +, +)$ or $(+, -, -, -)$.

Definition 10.3.4 (Line element). The **line element** is:

$$ds^2 = g_{ij}\,dx^i dx^j.$$

For a curve $\gamma(t)$: $ds^2 = g_{ij}\dot{x}^i\dot{x}^j\,dt^2$.

Example 10.3.1 (Standard metrics).

- Euclidean $\mathbb{R}^3$: $ds^2 = dx^2 + dy^2 + dz^2$.
- Spherical: $ds^2 = dr^2 + r^2(d\theta^2 + \sin^2\theta\,d\phi^2)$.
- Minkowski: $ds^2 = -c^2dt^2 + dx^2 + dy^2 + dz^2$.
- Schwarzschild: $ds^2 = -(1 - \frac{2GM}{c^2r})c^2dt^2 + (1 - \frac{2GM}{c^2r})^{-1}dr^2 + r^2d\Omega^2$.

10.3.3 Volume Element

Definition 10.3.5 (Volume element). The **volume element** on a (pseudo-)Riemannian manifold:

$$dV = \sqrt{|g|}\, dx^1 \wedge \cdots \wedge dx^n$$

where $g = \det(g_{ij})$.

Proposition 10.3.3. *The volume element is coordinate-independent (transforms as a density).*

10.4 Covariant Differentiation

10.4.1 The Problem with Partial Derivatives

Proposition 10.4.1. *Partial derivatives of tensor components do not form tensor components:*

$$\frac{\partial T^i}{\partial x^j} \quad \textit{does not transform as a } (1,1)\textit{-tensor.}$$

Proof sketch. Under $x \to \tilde{x}$:

$$\frac{\partial \tilde{T}^i}{\partial \tilde{x}^j} = \frac{\partial \tilde{x}^i}{\partial x^k} \frac{\partial x^l}{\partial \tilde{x}^j} \frac{\partial T^k}{\partial x^l} + \frac{\partial^2 \tilde{x}^i}{\partial x^k \partial x^m} \frac{\partial x^m}{\partial \tilde{x}^j} T^k.$$

The second term spoils tensorial transformation. □

10.4.2 Covariant Derivative

Definition 10.4.1 (Covariant derivative). The **covariant derivative** ∇ is a connection that maps (r, s)-tensors to $(r, s+1)$-tensors, satisfying:

(i) Linearity: $\nabla(aS + bT) = a\nabla S + b\nabla T$.

(ii) Leibniz rule: $\nabla(S \otimes T) = (\nabla S) \otimes T + S \otimes (\nabla T)$.

(iii) Reduces to partial derivative on scalars: $\nabla_i f = \partial_i f$.

(iv) Commutes with contraction.

Proposition 10.4.2 (Covariant derivative of tensors). *In components with Christoffel symbols* Γ^k_{ij}*:*

$$\begin{aligned}
\nabla_j v^i &= \partial_j v^i + \Gamma^i_{jk} v^k \\
\nabla_j \omega_i &= \partial_j \omega_i - \Gamma^k_{ji} \omega_k \\
\nabla_k T^{ij} &= \partial_k T^{ij} + \Gamma^i_{kl} T^{lj} + \Gamma^j_{kl} T^{il} \\
\nabla_k T_{ij} &= \partial_k T_{ij} - \Gamma^l_{ki} T_{lj} - \Gamma^l_{kj} T_{il}
\end{aligned}$$

General pattern: $+\Gamma$ *for each upper index,* $-\Gamma$ *for each lower index.*

10.4.3 Levi-Civita Connection

Theorem 10.4.3 (Levi-Civita connection). *On a (pseudo-)Riemannian manifold, there is a unique connection that is:*

(i) ***Torsion-free:*** $\Gamma^k_{ij} = \Gamma^k_{ji}$.

(ii) ***Metric-compatible:*** $\nabla_k g_{ij} = 0$.

Proposition 10.4.4 (Christoffel symbols). *The Christoffel symbols of the Levi-Civita connection:*

$$\Gamma^k_{ij} = \frac{1}{2} g^{kl} \left(\frac{\partial g_{il}}{\partial x^j} + \frac{\partial g_{jl}}{\partial x^i} - \frac{\partial g_{ij}}{\partial x^l} \right).$$

Corollary 10.4.5. *Metric compatibility implies:*

- $\nabla_k g_{ij} = 0$ *and* $\nabla_k g^{ij} = 0$.
- *Index raising/lowering commutes with covariant differentiation.*
- $\nabla_k \varepsilon_{i_1 \cdots i_n} = 0$ *(covariant constancy of volume form).*

10.4.4 Geodesics Revisited

Definition 10.4.2 (Geodesic equation). A geodesic satisfies:

$$\frac{d^2 x^k}{d\tau^2} + \Gamma^k_{ij} \frac{dx^i}{d\tau} \frac{dx^j}{d\tau} = 0$$

or equivalently $\nabla_{\dot\gamma} \dot\gamma = 0$ (parallel transport of velocity).

Proposition 10.4.6 (Geodesic as extremal). *Geodesics extremize the action:*

$$S = \int g_{ij} \dot{x}^i \dot{x}^j \, d\tau$$

(length for spacelike, proper time for timelike in Lorentzian signature).

10.5 Curvature Tensors

10.5.1 The Riemann Tensor

Definition 10.5.1 (Riemann curvature tensor). The **Riemann tensor** measures non-commutativity of covariant derivatives:

$$(\nabla_i \nabla_j - \nabla_j \nabla_i) v^k = R^k{}_{lij} v^l.$$

Proposition 10.5.1 (Component expression).

$$R^k{}_{lij} = \partial_i \Gamma^k_{jl} - \partial_j \Gamma^k_{il} + \Gamma^k_{im} \Gamma^m_{jl} - \Gamma^k_{jm} \Gamma^m_{il}.$$

Proposition 10.5.2 (Symmetries). *With $R_{klij} := g_{km} R^m{}_{lij}$:*

$$\begin{aligned} R_{klij} &= -R_{klji} && \textit{(antisymmetric in last pair)} \\ R_{klij} &= -R_{lkij} && \textit{(antisymmetric in first pair)} \\ R_{klij} &= R_{ijkl} && \textit{(symmetric under pair exchange)} \\ R_{klij} + R_{kijl} + R_{kjli} &= 0 && \textit{(first Bianchi identity)} \end{aligned}$$

Proposition 10.5.3 (Independent components). *In n dimensions, the Riemann tensor has $\frac{n^2(n^2-1)}{12}$ independent components.*

n	*2*	*3*	*4*	*n*
components	*1*	*6*	*20*	$\frac{n^2(n^2-1)}{12}$

10.5.2 Ricci Tensor and Scalar

Definition 10.5.2 (Ricci tensor). The **Ricci tensor** is the contraction:

$$R_{ij} := R^k{}_{ikj} = g^{kl} R_{kilj}.$$

Proposition 10.5.4. *R_{ij} is symmetric: $R_{ij} = R_{ji}$.*

Definition 10.5.3 (Ricci scalar). The **Ricci scalar** (or **scalar curvature**):

$$R := g^{ij} R_{ij}.$$

Definition 10.5.4 (Einstein tensor). The **Einstein tensor**:

$$G_{ij} := R_{ij} - \frac{1}{2} R g_{ij}.$$

Theorem 10.5.5 (Contracted Bianchi identity).

$$\nabla^i G_{ij} = 0.$$

This is the mathematical foundation for conservation of energy-momentum in general relativity.

10.5.3 Weyl Tensor

Definition 10.5.5 (Weyl tensor). The **Weyl tensor** (conformal curvature tensor) is the trace-free part of Riemann:

$$C_{ijkl} = R_{ijkl} - \frac{2}{n-2}(g_{i[k}R_{l]j} - g_{j[k}R_{l]i}) + \frac{2}{(n-1)(n-2)}Rg_{i[k}g_{l]j}.$$

Proposition 10.5.6. *C_{ijkl} has all symmetries of R_{ijkl} plus is trace-free: $C^i{}_{jik} = 0$.*

Theorem 10.5.7 (Decomposition of Riemann). *In $n \geq 3$ dimensions:*

$$R_{ijkl} = C_{ijkl} + \frac{2}{n-2}(g_{i[k}R_{l]j} - g_{j[k}R_{l]i}) - \frac{2}{(n-1)(n-2)}Rg_{i[k}g_{l]j}.$$

The three pieces encode: conformal curvature (Weyl), Ricci curvature, and scalar curvature.

Proposition 10.5.8. *In 3D: $C_{ijkl} = 0$ (Weyl vanishes). Riemann is determined by Ricci.*

In 4D: Weyl has 10 independent components.

10.6 Tensor Densities

10.6.1 Definition and Transformation

Definition 10.6.1 (Tensor density). A **tensor density of weight** w transforms as:

$$\tilde{\mathcal{T}}^{i_1\cdots}{}_{j_1\cdots} = \left|\frac{\partial x}{\partial \tilde{x}}\right|^w \frac{\partial \tilde{x}^{i_1}}{\partial x^{k_1}} \cdots \frac{\partial x^{l_1}}{\partial \tilde{x}^{j_1}} \cdots \mathcal{T}^{k_1\cdots}{}_{l_1\cdots}$$

where $|\partial x/\partial \tilde{x}|$ is the absolute Jacobian determinant.

Example 10.6.1.

- $\sqrt{|g|}$ is a scalar density of weight $+1$.
- $1/\sqrt{|g|}$ is a scalar density of weight -1.
- The Levi-Civita symbol $\epsilon_{i_1\cdots i_n}$ is a tensor density of weight -1.
- $\varepsilon_{i_1\cdots i_n} = \sqrt{|g|}\epsilon_{i_1\cdots i_n}$ is a true tensor.

10.6.2 Covariant Derivative of Densities

Proposition 10.6.1. *For a tensor density $\mathcal{T}$ of weight w:*

$$\nabla_k \mathcal{T}^{i\cdots}{}_{j\cdots} = \partial_k \mathcal{T}^{i\cdots}{}_{j\cdots} + \Gamma^i_{kl}\mathcal{T}^{l\cdots}{}_{j\cdots} + \cdots - \Gamma^l_{kj}\mathcal{T}^{i\cdots}{}_{l\cdots} - \cdots - w\Gamma^l_{kl}\mathcal{T}^{i\cdots}{}_{j\cdots}.$$

Note the extra term $-w\Gamma^l_{kl}$ involving the trace.

Corollary 10.6.2. *For the metric determinant: $\partial_k\sqrt{|g|} = \sqrt{|g|}\Gamma^l_{kl}$.*
Thus: $\nabla_k\sqrt{|g|} = 0$ (automatically, from metric compatibility).

10.7 Differential Forms Revisited

10.7.1 Forms as Antisymmetric Tensors

Proposition 10.7.1. *A differential k-form ω is a totally antisymmetric $(0,k)$-tensor:*

$$\omega_{i_1\cdots i_k} = \omega_{[i_1\cdots i_k]}.$$

Proposition 10.7.2 (Wedge product in indices)**.** $(\omega\wedge\eta)_{i_1\cdots i_{k+l}} = \frac{(k+l)!}{k!l!}\omega_{[i_1\cdots i_k}\eta_{i_{k+1}\cdots i_{k+l}]}$.

Proposition 10.7.3 (Exterior derivative in indices)**.** $(d\omega)_{i_0 i_1\cdots i_k} = (k+1)\partial_{[i_0}\omega_{i_1\cdots i_k]}$.

10.7.2 Hodge Dual

Definition 10.7.1 (Hodge star operator)**.** On an n-dimensional (pseudo-)Riemannian manifold, the **Hodge dual** of a k-form ω is the $(n-k)$-form:

$$(*\omega)_{i_{k+1}\cdots i_n} = \frac{1}{k!}\omega^{i_1\cdots i_k}\varepsilon_{i_1\cdots i_k i_{k+1}\cdots i_n}.$$

Proposition 10.7.4 (Properties)**.**

(i) $*1 = \varepsilon_{i_1\cdots i_n}\, dx^{i_1}\wedge\cdots\wedge dx^{i_n}/n! = \sqrt{|g|}\, dx^1\wedge\cdots\wedge dx^n$.

(ii) $**\omega = (-1)^{k(n-k)}\text{sgn}(g)\omega$ *where* $\text{sgn}(g) = \pm 1$ *is the signature.*

(iii) For α, β both k-forms: $\alpha\wedge *\beta = \langle\alpha,\beta\rangle * 1$ where $\langle\cdot,\cdot\rangle$ is the natural inner product on forms.

Example 10.7.1 (3D Euclidean space)**.** $*dx = dy\wedge dz$, $*dy = dz\wedge dx$, $*dz = dx\wedge dy$.
$*(dx\wedge dy) = dz$, etc.

Example 10.7.2 (4D Minkowski space)**.** $*(dt\wedge dx) = dy\wedge dz$, etc.
For electromagnetic field $F = E_i\, dx^i\wedge dt + \frac{1}{2}\epsilon_{ijk}B^k\, dx^i\wedge dx^j$:
$*F$ exchanges electric and magnetic fields (up to signs).

10.7.3 Codifferential

Definition 10.7.2 (Codifferential). The **codifferential** $\delta : \Omega^k \to \Omega^{k-1}$ is:

$$\delta := (-1)^{nk+n+1}\mathrm{sgn}(g) * d* = (-1)^k *^{-1} d * .$$

Proposition 10.7.5 (Relation to divergence). *For a vector field X with associated 1-form $X^\flat$:*

$$\delta X^\flat = -\nabla_i X^i = -\mathrm{div}(X).$$

Definition 10.7.3 (Laplace-Beltrami operator). The **Laplace-Beltrami operator** (or Hodge Laplacian):

$$\Delta := d\delta + \delta d.$$

On functions: $\Delta f = -\nabla^i \nabla_i f = -\frac{1}{\sqrt{|g|}}\partial_i(\sqrt{|g|}g^{ij}\partial_j f)$.

10.8 Applications to Physics

10.8.1 Electromagnetism

Proposition 10.8.1 (Maxwell's equations in tensor form). *With $F_{\mu\nu} = \partial_\mu A_\nu - \partial_\nu A_\mu$ (field strength) and J^μ (4-current):*

$\partial_{[\rho} F_{\mu\nu]} = 0$	*(or $dF = 0$)*	*Bianchi identity*
$\nabla_\mu F^{\mu\nu} = \mu_0 J^\nu$	*(or $d{*}F = \mu_0 {*}J$)*	*inhomogeneous Maxwell*

Proposition 10.8.2 (Stress-energy tensor of electromagnetic field).

$$T_{\mu\nu} = \frac{1}{\mu_0}\left(F_{\mu\rho}F_\nu{}^\rho - \frac{1}{4}g_{\mu\nu}F_{\rho\sigma}F^{\rho\sigma}\right).$$

10.8.2 General Relativity

Theorem 10.8.3 (Einstein field equations).

$$G_{\mu\nu} + \Lambda g_{\mu\nu} = \frac{8\pi G}{c^4} T_{\mu\nu}$$

where $G_{\mu\nu} = R_{\mu\nu} - \frac{1}{2}R g_{\mu\nu}$ is the Einstein tensor, Λ is the cosmological constant, and $T_{\mu\nu}$ is the stress-energy tensor.

Proposition 10.8.4 (Conservation law). *The Bianchi identity $\nabla^\mu G_{\mu\nu} = 0$ implies:*

$$\nabla^\mu T_{\mu\nu} = 0$$

(covariant conservation of energy-momentum).

10.8.3 Continuum Mechanics

Definition 10.8.1 (Stress tensor). The **Cauchy stress tensor** σ_{ij} gives the force per unit area:

$$dF_i = \sigma_{ij} n_j \, dA$$

where n_j is the unit normal to the surface.

Proposition 10.8.5 (Symmetry). *Angular momentum conservation implies* $\sigma_{ij} = \sigma_{ji}$.

Definition 10.8.2 (Strain tensor). The **strain tensor** for small deformations:

$$\epsilon_{ij} = \frac{1}{2}\left(\frac{\partial u_i}{\partial x_j} + \frac{\partial u_j}{\partial x_i}\right)$$

where u_i is the displacement field.

Proposition 10.8.6 (Hooke's law). *For isotropic linear elastic materials:*

$$\sigma_{ij} = \lambda \epsilon_{kk} \delta_{ij} + 2\mu \epsilon_{ij}$$

where λ, μ *are Lamé coefficients.*

10.9 Summary

Tensor analysis provides coordinate-independent methods for physics:

Tensors as multilinear maps: Type (r, s) tensors transform predictably under coordinate changes.

Index notation: Einstein convention streamlines calculations; upper/lower indices distinguish transformation properties.

Metric operations: Raising/lowering indices, inner products, volumes all depend on the metric.

Covariant differentiation: The Levi-Civita connection provides a tensorial derivative respecting the metric.

Curvature: Riemann, Ricci, and Weyl tensors encode geometric properties; Einstein tensor governs gravity.

Forms and Hodge theory: Exterior calculus with metric structure; essential for electromagnetism and gauge theories.

Exercises

Exercise 10.9.1. Show that $\delta^i_j := e^i(e_j)$ transforms as a $(1,1)$-tensor and that $\delta^i_j = g^{ik} g_{kj}$.

Exercise 10.9.2. Prove that contraction of a (r, s)-tensor gives a $(r-1, s-1)$-tensor.

Exercise 10.9.3. Verify the identity $\epsilon_{ijk}\epsilon^{lmk} = \delta^l_i\delta^m_j - \delta^l_j\delta^m_i$.

Exercise 10.9.4. Compute the Christoffel symbols for the 2-sphere with metric $ds^2 = d\theta^2 + \sin^2\theta\, d\phi^2$.

Exercise 10.9.5. Show that $\nabla_k g_{ij} = 0$ for the Levi-Civita connection.

Exercise 10.9.6. Verify that the geodesic equation follows from extremizing $\int g_{ij}\dot{x}^i\dot{x}^j\, d\tau$.

Exercise 10.9.7. Prove the first Bianchi identity $R_{ijkl} + R_{iklj} + R_{iljk} = 0$.

Exercise 10.9.8. Show that in 2D, $R_{ijkl} = K(g_{ik}g_{jl} - g_{il}g_{jk})$ where K is the Gaussian curvature.

Exercise 10.9.9. Derive the contracted Bianchi identity $\nabla^i G_{ij} = 0$ from the second Bianchi identity.

Exercise 10.9.10. Verify that $**\omega = (-1)^{k(n-k)}\omega$ for a k-form in n-dimensional Euclidean space.

Exercise 10.9.11. Show that Maxwell's equations $dF = 0$ and $d{*F} = 0$ (in vacuum) imply the wave equation for F.

Exercise 10.9.12. For the Schwarzschild metric, compute Γ^r_{tt} and verify it gives Newtonian gravity in the weak-field limit.

Part IV

Functional Analysis

Chapter 11

Hilbert Spaces

Hilbert spaces provide the mathematical foundation for quantum mechanics. The state of a quantum system is a vector in a Hilbert space, observables are self-adjoint operators, and measurements are described by the spectral theorem. This chapter develops the theory of Hilbert spaces and bounded operators, with emphasis on the structures most relevant to physics.

11.1 Inner Product Spaces

11.1.1 Definitions

Definition 11.1.1 (Inner product). An **inner product** on a complex vector space V is a map $\langle\cdot,\cdot\rangle : V \times V \to \mathbb{C}$ satisfying:

(i) **Conjugate symmetry**: $\langle y, x\rangle = \overline{\langle x, y\rangle}$.

(ii) **Linearity in second argument**: $\langle x, \alpha y + \beta z\rangle = \alpha\langle x, y\rangle + \beta\langle x, z\rangle$.

(iii) **Positive definiteness**: $\langle x, x\rangle \geq 0$, with equality iff $x = 0$.

A space with an inner product is an **inner product space** (or **pre-Hilbert space**).

Remark 11.1.1 (Convention). Physicists typically use linearity in the second argument (as above). Mathematicians often use linearity in the first. We follow the physics convention, consistent with Dirac notation.

Definition 11.1.2 (Induced norm). The inner product induces a norm:

$$\|x\| := \sqrt{\langle x, x\rangle}.$$

Theorem 11.1.1 (Cauchy-Schwarz inequality). *For all x, y in an inner product space:*

$$|\langle x, y\rangle| \leq \|x\|\|y\|$$

with equality iff x and y are linearly dependent.

Corollary 11.1.2 (Triangle inequality). $\|x+y\| \leq \|x\| + \|y\|$.

Proposition 11.1.3 (Parallelogram law).

$$\|x+y\|^2 + \|x-y\|^2 = 2\|x\|^2 + 2\|y\|^2.$$

This characterizes norms coming from inner products.

Proposition 11.1.4 (Polarization identity). *The inner product can be recovered from the norm:*

$$\langle x, y\rangle = \frac{1}{4}\left(\|x+y\|^2 - \|x-y\|^2 + i\|x+iy\|^2 - i\|x-iy\|^2\right).$$

11.1.2 Examples of Inner Product Spaces

Example 11.1.1 ($\mathbb{C}^n$). The standard inner product on $\mathbb{C}^n$:

$$\langle x, y\rangle = \sum_{i=1}^{n} \bar{x}_i y_i = x^\dagger y.$$

Example 11.1.2 (L^2 spaces). For a measure space (X, μ):

$$L^2(X,\mu) := \left\{f : X \to \mathbb{C} : \int_X |f|^2 d\mu < \infty\right\}$$

with inner product:

$$\langle f, g\rangle = \int_X \bar{f}(x) g(x)\, d\mu(x).$$

Example 11.1.3 (ℓ^2). The space of square-summable sequences:

$$\ell^2 := \left\{(x_n)_{n=1}^{\infty} : \sum_{n=1}^{\infty} |x_n|^2 < \infty\right\}$$

with $\langle x, y\rangle = \sum_{n=1}^{\infty} \bar{x}_n y_n$.

Example 11.1.4 (Function spaces on $[a, b)$.]$\mathrm{L}^2([a, b])$ with Lebesgue measure, or $L^2([a,b], w)$ with weight function $w(x) > 0$:

$$\langle f, g\rangle_w = \int_a^b \bar{f}(x) g(x) w(x)\, dx.$$

11.2 Hilbert Spaces

11.2.1 Completeness

Definition 11.2.1 (Cauchy sequence). A sequence (x_n) in a normed space is **Cauchy** if for every $\epsilon > 0$, there exists N such that $\|x_m - x_n\| < \epsilon$ for all $m, n > N$.

Definition 11.2.2 (Completeness). A normed space is **complete** if every Cauchy sequence converges (to a limit in the space).

Definition 11.2.3 (Hilbert space). A **Hilbert space** is a complete inner product space.

Example 11.2.1 (Hilbert spaces).

- $\mathbb{C}^n$ with standard inner product.
- ℓ^2 (complete by construction).
- $L^2(X, \mu)$ for any measure space (Riesz-Fischer theorem).

Example 11.2.2 (Non-example). The space of continuous functions $C([0,1])$ with L^2 inner product is not complete. Its completion is $L^2([0,1])$.

11.2.2 Separability

Definition 11.2.4 (Separable space). A Hilbert space is **separable** if it has a countable dense subset.

Proposition 11.2.1. *A Hilbert space is separable iff it has a countable orthonormal basis.*

Example 11.2.3. $L^2([0, 2\pi])$ is separable with orthonormal basis $\{e^{inx}/\sqrt{2\pi}\}_{n\in\mathbb{Z}}$. ℓ^2 is separable with basis $\{e_n\}$ where $(e_n)_k = \delta_{nk}$.

Most Hilbert spaces in physics are separable. We assume separability unless stated otherwise.

11.2.3 Orthogonality

Definition 11.2.5 (Orthogonality). Vectors x, y are **orthogonal** ($x \perp y$) if $\langle x, y\rangle = 0$.

A set S is **orthogonal** if distinct elements are orthogonal.

A set is **orthonormal** if orthogonal and $\|x\| = 1$ for all $x \in S$.

Definition 11.2.6 (Orthogonal complement). For a subset $S \subseteq \mathcal{H}$:

$$S^\perp := \{x \in \mathcal{H} : \langle x, y \rangle = 0 \text{ for all } y \in S\}.$$

Proposition 11.2.2. *$S^\perp$ is always a closed subspace.*

Theorem 11.2.3 (Pythagorean theorem). *If $x_1, \dots, x_n$ are pairwise orthogonal:*

$$\|x_1 + \cdots + x_n\|^2 = \|x_1\|^2 + \cdots + \|x_n\|^2.$$

11.3 Orthonormal Bases

11.3.1 Gram-Schmidt Process

Theorem 11.3.1 (Gram-Schmidt orthogonalization). *Given linearly independent vectors $\{v_1, v_2, \dots\}$, we can construct orthonormal vectors $\{e_1, e_2, \dots\}$ spanning the same space:*

$$\begin{aligned} u_1 &= v_1, & e_1 &= u_1/\|u_1\| \\ u_k &= v_k - \sum_{j=1}^{k-1} \langle e_j, v_k \rangle e_j, & e_k &= u_k/\|u_k\|. \end{aligned}$$

11.3.2 Orthonormal Bases and Expansions

Definition 11.3.1 (Orthonormal basis). An **orthonormal basis** (ONB) for a Hilbert space $\mathcal{H}$ is a maximal orthonormal set, equivalently, an orthonormal set whose closed linear span is $\mathcal{H}$.

Theorem 11.3.2 (Existence). *Every Hilbert space has an orthonormal basis. For separable spaces, it is countable.*

Theorem 11.3.3 (Fourier expansion). *Let $\{e_n\}$ be an orthonormal basis. For any $x \in \mathcal{H}$:*

$$x = \sum_n \langle e_n, x \rangle e_n$$

*with convergence in norm. The coefficients $c_n = \langle e_n, x \rangle$ are called **Fourier coefficients**.*

Theorem 11.3.4 (Parseval's identity).

$$\|x\|^2 = \sum_n |\langle e_n, x \rangle|^2.$$

Theorem 11.3.5 (Bessel's inequality). *For any orthonormal set* $\{e_n\}$ *(not necessarily complete):*

$$\sum_n |\langle e_n, x\rangle|^2 \leq \|x\|^2.$$

Equality holds iff $\{e_n\}$ *is complete (an ONB).*

Corollary 11.3.6. *All separable infinite-dimensional Hilbert spaces are isomorphic to* ℓ^2.

11.3.3 Classical Orthonormal Systems

Example 11.3.1 (Fourier basis). On $L^2([0, 2\pi])$: $\{e^{inx}/\sqrt{2\pi}\}_{n\in\mathbb{Z}}$.

Real form on $L^2([0, L])$: $\{1/\sqrt{L}\}\cup\{\sqrt{2/L}\cos(2\pi nx/L)\}_{n\geq 1}\cup\{\sqrt{2/L}\sin(2\pi nx/L)\}_{n\geq}$

Example 11.3.2 (Legendre polynomials). On $L^2([-1, 1])$, the normalized Legendre polynomials:

$$\tilde{P}_n(x) = \sqrt{\frac{2n+1}{2}}P_n(x), \quad P_n(x) = \frac{1}{2^n n!}\frac{d^n}{dx^n}(x^2 - 1)^n.$$

Example 11.3.3 (Hermite functions). On $L^2(\mathbb{R})$, the normalized Hermite functions:

$$\psi_n(x) = \frac{1}{\sqrt{2^n n!\sqrt{\pi}}}H_n(x)e^{-x^2/2}$$

where $H_n(x) = (-1)^n e^{x^2}\frac{d^n}{dx^n}e^{-x^2}$ are Hermite polynomials.

These are eigenfunctions of the quantum harmonic oscillator.

Example 11.3.4 (Spherical harmonics). On $L^2(S^2)$: $\{Y_l^m(\theta, \phi)\}$ with $l = 0, 1, 2, \ldots$ and $m = -l, \ldots, l$:

$$Y_l^m(\theta, \phi) = \sqrt{\frac{2l+1}{4\pi}\frac{(l-m)!}{(l+m)!}}P_l^m(\cos\theta)e^{im\phi}.$$

11.4 Projection and Decomposition

11.4.1 Orthogonal Projection

Theorem 11.4.1 (Projection theorem). *Let* M *be a closed subspace of a Hilbert space* $\mathcal{H}$. *For any* $x \in \mathcal{H}$, *there exists a unique* $y \in M$ *such that:*

$$\|x - y\| = \inf_{z\in M}\|x - z\|.$$

This y *is characterized by:* $y \in M$ *and* $(x - y) \perp M$.

Definition 11.4.1 (Orthogonal projection). The **orthogonal projection** onto M is the map $P_M : \mathcal{H} \to M$ sending x to its best approximation y.

Proposition 11.4.2 (Properties of projections).

(i) P_M *is linear and bounded with* $\|P_M\| = 1$ *(if* $M \neq \{0\}$*).*

(ii) $P_M^2 = P_M$ *(idempotent).*

(iii) $P_M^* = P_M$ *(self-adjoint).*

(iv) $\ker P_M = M^\perp$, $\operatorname{ran} P_M = M$.

Theorem 11.4.3 (Orthogonal decomposition). *For any closed subspace* M*:*

$$\mathcal{H} = M \oplus M^\perp$$

meaning every x *has a unique decomposition* $x = y + z$ *with* $y \in M$, $z \in M^\perp$.

Corollary 11.4.4. $(M^\perp)^\perp = M$ *for closed subspaces.*

11.4.2 Direct Sums

Definition 11.4.2 (Internal direct sum). If $\mathcal{H} = M \oplus N$ with $M \perp N$, we call this an **orthogonal direct sum**.

Definition 11.4.3 (External direct sum). For Hilbert spaces $\mathcal{H}_1, \mathcal{H}_2$, their direct sum:

$$\mathcal{H}_1 \oplus \mathcal{H}_2 := \{(x_1, x_2) : x_1 \in \mathcal{H}_1, x_2 \in \mathcal{H}_2\}$$

with $\langle (x_1, x_2), (y_1, y_2) \rangle = \langle x_1, y_1 \rangle_1 + \langle x_2, y_2 \rangle_2$.

Proposition 11.4.5. *The direct sum of Hilbert spaces is a Hilbert space.*

11.5 The Riesz Representation Theorem

Definition 11.5.1 (Bounded linear functional). A **linear functional** on $\mathcal{H}$ is a linear map $\phi : \mathcal{H} \to \mathbb{C}$. It is **bounded** if $|\phi(x)| \leq C\|x\|$ for some C.

Theorem 11.5.1 (Riesz representation theorem). *For every bounded linear functional* ϕ *on a Hilbert space* $\mathcal{H}$*, there exists a unique* $y \in \mathcal{H}$ *such that:*

$$\phi(x) = \langle y, x \rangle \quad \text{for all } x \in \mathcal{H}.$$

Moreover, $\|\phi\| = \|y\|$.

Corollary 11.5.2. *The dual space* $\mathcal{H}^*$ *is isometrically anti-isomorphic to* $\mathcal{H}$ *via* $y \mapsto \langle y, \cdot \rangle$.

This theorem is fundamental: it allows us to identify a Hilbert space with its dual, justifying Dirac's bra-ket notation.

11.6 Dirac Notation

11.6.1 Bras and Kets

Definition 11.6.1 (Dirac notation). In **Dirac notation**:

- A **ket** $|\psi\rangle$ represents a vector in $\mathcal{H}$.
- A **bra** $\langle\phi|$ represents a linear functional on $\mathcal{H}$.
- The **bracket** $\langle\phi|\psi\rangle := \langle\phi, \psi\rangle$ is the inner product.

By the Riesz theorem, every bra corresponds to a unique ket: $\langle\phi| \leftrightarrow |\phi\rangle$.

Proposition 11.6.1 (Properties).

$$\langle\psi|\phi\rangle = \overline{\langle\phi|\psi\rangle}$$
$$\langle\phi|(\alpha|\psi_1\rangle + \beta|\psi_2\rangle) = \alpha\langle\phi|\psi_1\rangle + \beta\langle\phi|\psi_2\rangle$$
$$(\alpha\langle\phi_1| + \beta\langle\phi_2|)|\psi\rangle = \bar{\alpha}\langle\phi_1|\psi\rangle + \bar{\beta}\langle\phi_2|\psi\rangle$$

11.6.2 Operators in Dirac Notation

Definition 11.6.2 (Outer product). For $|\phi\rangle, |\psi\rangle \in \mathcal{H}$, the **outer product** $|\phi\rangle\langle\psi|$ is the operator:

$$(|\phi\rangle\langle\psi|)|\chi\rangle := \langle\psi|\chi\rangle|\phi\rangle.$$

Proposition 11.6.2. *$|\phi\rangle\langle\psi|$ is a rank-one operator. Its adjoint is $|\psi\rangle\langle\phi|$.*

Proposition 11.6.3 (Projection operators). *$P_\psi := |\psi\rangle\langle\psi|$ is the orthogonal projection onto* $\text{span}\{|\psi\rangle\}$ *(assuming $\|\psi\| = 1$).*

Theorem 11.6.4 (Resolution of identity). *For an orthonormal basis $\{|e_n\rangle\}$:*

$$\sum_n |e_n\rangle\langle e_n| = I$$

(the identity operator, convergence in strong operator topology).

Corollary 11.6.5 (Basis expansion). $|\psi\rangle = \sum_n |e_n\rangle\langle e_n|\psi\rangle = \sum_n c_n|e_n\rangle$ *where $c_n = \langle e_n|\psi\rangle$.*

11.6.3 Matrix Elements

Definition 11.6.3 (Matrix elements). For an operator A and orthonormal basis $\{|e_n\rangle\}$:

$$A_{mn} := \langle e_m|A|e_n\rangle.$$

Proposition 11.6.6. *Using resolution of identity:*

$$A = \sum_{m,n} A_{mn}|e_m\rangle\langle e_n| = \sum_{m,n}\langle e_m|A|e_n\rangle|e_m\rangle\langle e_n|.$$

11.7 Bounded Operators

11.7.1 Basic Definitions

Definition 11.7.1 (Bounded operator). A linear operator $A : \mathcal{H} \to \mathcal{K}$ between Hilbert spaces is **bounded** if:

$$\|A\| := \sup_{\|x\|=1} \|Ax\| < \infty.$$

Equivalently, $\|Ax\| \leq \|A\|\|x\|$ for all x.

Proposition 11.7.1. *For linear operators, bounded $\Leftrightarrow$ continuous.*

Definition 11.7.2 (Operator space $\mathcal{B}(\mathcal{H})$). $\mathcal{B}(\mathcal{H})$ denotes the space of all bounded linear operators $\mathcal{H} \to \mathcal{H}$. It is a Banach algebra under the operator norm.

11.7.2 The Adjoint Operator

Theorem 11.7.2 (Adjoint existence). *For every $A \in \mathcal{B}(\mathcal{H})$, there exists a unique $A^* \in \mathcal{B}(\mathcal{H})$ such that:*

$$\langle A^*x, y\rangle = \langle x, Ay\rangle \quad \text{for all } x, y \in \mathcal{H}.$$

In Dirac notation: $\langle\phi|A|\psi\rangle = \langle A^\dagger\phi|\psi\rangle = \overline{\langle\psi|A^\dagger|\phi\rangle}$.

Proposition 11.7.3 (Properties of adjoint).

(i) $(A^*)^* = A$.

(ii) $(A + B)^* = A^* + B^*$.

(iii) $(\alpha A)^* = \bar{\alpha}A^*$.

(iv) $(AB)^* = B^*A^*$.

(v) $\|A^*\| = \|A\|$.

(vi) $\|A^*A\| = \|A\|^2$.

11.7.3 Special Classes of Operators

Definition 11.7.3 (Self-adjoint operator). A is **self-adjoint** (or **Hermitian**) if $A^* = A$.

Proposition 11.7.4. *A is self-adjoint iff $\langle x, Ax\rangle \in \mathbb{R}$ for all x.*

Definition 11.7.4 (Normal operator). A is **normal** if $A^*A = AA^*$.

Self-adjoint operators are normal.

Definition 11.7.5 (Unitary operator). U is **unitary** if $U^*U = UU^* = I$, equivalently $U^* = U^{-1}$.

Proposition 11.7.5. *U is unitary iff $\langle Ux, Uy\rangle = \langle x, y\rangle$ for all x, y (preserves inner product).*

Definition 11.7.6 (Positive operator). A is **positive** (written $A \geq 0$) if $\langle x, Ax\rangle \geq 0$ for all x.

Proposition 11.7.6. *Positive operators are self-adjoint.*

Definition 11.7.7 (Projection). P is a **(orthogonal) projection** if $P^2 = P$ and $P^* = P$.

Proposition 11.7.7. *P is a projection iff P is the orthogonal projection onto a closed subspace.*

11.7.4 Compact Operators

Definition 11.7.8 (Compact operator). $A \in \mathcal{B}(\mathcal{H})$ is **compact** if the image of the unit ball has compact closure.

Equivalently: every bounded sequence (x_n) has a subsequence such that (Ax_{n_k}) converges.

Proposition 11.7.8 (Properties).

(i) Finite-rank operators are compact.

(ii) Compact operators form a closed two-sided ideal in $\mathcal{B}(\mathcal{H})$.

(iii) The limit (in operator norm) of compact operators is compact.

(iv) A compact $\Rightarrow$ A^* compact.

Example 11.7.1. Integral operators $Af(x) = \int K(x, y)f(y)\,dy$ with square-integrable kernel K are compact (Hilbert-Schmidt operators).

11.8 Spectral Theory for Bounded Operators

11.8.1 Spectrum and Resolvent

Definition 11.8.1 (Spectrum). The **spectrum** of $A \in \mathcal{B}(\mathcal{H})$ is:

$$\sigma(A) := \{\lambda \in \mathbb{C} : A - \lambda I \text{ is not invertible}\}.$$

The **resolvent set** is $\rho(A) := \mathbb{C} \setminus \sigma(A)$.

Definition 11.8.2 (Types of spectrum).

- **Point spectrum** $\sigma_p(A)$: eigenvalues ($\ker(A - \lambda I) \neq \{0\}$).
- **Continuous spectrum** $\sigma_c(A)$: $\ker(A - \lambda I) = \{0\}$, $\text{ran}(A - \lambda I)$ dense but $\neq \mathcal{H}$.
- **Residual spectrum** $\sigma_r(A)$: $\ker(A - \lambda I) = \{0\}$, $\text{ran}(A - \lambda I)$ not dense.

Theorem 11.8.1 (Properties of spectrum).

(i) $\sigma(A)$ is nonempty and compact.

(ii) $\sigma(A) \subseteq \{\lambda : |\lambda| \leq \|A\|\}$ (closed disk).

(iii) $\sigma(A^*) = \overline{\sigma(A)}$ (complex conjugate).

(iv) For normal A: $\sigma_r(A) = \varnothing$.

(v) For self-adjoint A: $\sigma(A) \subseteq \mathbb{R}$.

(vi) For unitary U: $\sigma(U) \subseteq \{z : |z| = 1\}$ (unit circle).

Definition 11.8.3 (Spectral radius). $r(A) := \sup\{|\lambda| : \lambda \in \sigma(A)\} = \lim_{n\to\infty} \|A^n\|^{1/n}$.

Proposition 11.8.2. *For normal operators: $r(A) = \|A\|$.*

11.8.2 Spectral Theorem for Compact Self-Adjoint Operators

Theorem 11.8.3 (Spectral theorem - compact self-adjoint). *Let A be a compact self-adjoint operator on a Hilbert space $\mathcal{H}$. Then:*

(i) The nonzero spectrum consists of eigenvalues only: $\sigma(A) \setminus \{0\} = \sigma_p(A) \setminus \{0\}$.

(ii) Each nonzero eigenvalue has finite multiplicity.

(iii) Nonzero eigenvalues form a sequence converging to 0 (if infinitely many).

(iv) Eigenvectors for distinct eigenvalues are orthogonal.

(v) There exists an orthonormal basis $\{e_n\}$ of eigenvectors with $Ae_n = \lambda_n e_n$.

Corollary 11.8.4 (Spectral decomposition). *For compact self-adjoint A:*

$$A = \sum_n \lambda_n |e_n\rangle\langle e_n| = \sum_n \lambda_n P_n$$

where P_n is the projection onto the eigenspace for λ_n.

11.8.3 Spectral Theorem for Bounded Self-Adjoint Operators

Theorem 11.8.5 (Spectral theorem - bounded self-adjoint). *For any bounded self-adjoint operator A, there exists a unique projection-valued measure E on $\mathbb{R}$ such that:*

$$A = \int_{\sigma(A)} \lambda \, dE(\lambda).$$

Definition 11.8.4 (Projection-valued measure). A **projection-valued measure** (PVM) assigns to each Borel set $B \subseteq \mathbb{R}$ a projection $E(B)$ such that:

(i) $E(\mathbb{R}) = I$, $E(\varnothing) = 0$.

(ii) $E(B_1 \cap B_2) = E(B_1)E(B_2)$.

(iii) For disjoint (B_n): $E(\bigcup_n B_n) = \sum_n E(B_n)$ (strong convergence).

Corollary 11.8.6 (Functional calculus). *For bounded Borel function f:*

$$f(A) := \int_{\sigma(A)} f(\lambda) \, dE(\lambda)$$

defines a bounded operator with $\|f(A)\| \leq \sup_{\lambda \in \sigma(A)} |f(\lambda)|$.

Example 11.8.1. $A^n = \int \lambda^n \, dE(\lambda)$, $e^{itA} = \int e^{it\lambda} \, dE(\lambda)$.

11.8.4 Spectral Theorem for Normal Operators

Theorem 11.8.7 (Spectral theorem - normal). *For any bounded normal operator A, there exists a unique PVM E on $\mathbb{C}$ supported on $\sigma(A)$ such that:*

$$A = \int_{\sigma(A)} z \, dE(z).$$

11.9 Tensor Products of Hilbert Spaces

11.9.1 Definition

Definition 11.9.1 (Tensor product of Hilbert spaces). For Hilbert spaces $\mathcal{H}_1, \mathcal{H}_2$, the **tensor product** $\mathcal{H}_1 \otimes \mathcal{H}_2$ is the completion of the algebraic tensor product with inner product:

$$\langle x_1 \otimes x_2, y_1 \otimes y_2 \rangle = \langle x_1, y_1 \rangle_1 \langle x_2, y_2 \rangle_2.$$

Proposition 11.9.1. *If $\{e_i\}$ is an ONB for $\mathcal{H}_1$ and $\{f_j\}$ for $\mathcal{H}_2$, then $\{e_i \otimes f_j\}$ is an ONB for $\mathcal{H}_1 \otimes \mathcal{H}_2$.*

Example 11.9.1. $L^2(\mathbb{R}^n) \cong L^2(\mathbb{R}) \otimes \cdots \otimes L^2(\mathbb{R})$ (n factors).

$L^2(\mathbb{R}^3) \otimes \mathbb{C}^2$ describes a spin-1/2 particle in 3D.

11.9.2 Operators on Tensor Products

Definition 11.9.2. For $A \in \mathcal{B}(\mathcal{H}_1)$, $B \in \mathcal{B}(\mathcal{H}_2)$:

$$(A \otimes B)(x \otimes y) := (Ax) \otimes (By).$$

Proposition 11.9.2. $\|A \otimes B\| = \|A\|\|B\|$, $(A \otimes B)^* = A^* \otimes B^*$.

11.9.3 Entanglement

Definition 11.9.3 (Product and entangled states)**.** A state $|\psi\rangle \in \mathcal{H}_1 \otimes \mathcal{H}_2$ is a **product state** if $|\psi\rangle = |\phi_1\rangle \otimes |\phi_2\rangle$.

Otherwise, it is **entangled**.

Example 11.9.2 (Bell states)**.** The maximally entangled Bell states in $\mathbb{C}^2 \otimes \mathbb{C}^2$:

$$\begin{aligned}
|\Phi^+\rangle &= \frac{1}{\sqrt{2}}(|00\rangle + |11\rangle) \\
|\Phi^-\rangle &= \frac{1}{\sqrt{2}}(|00\rangle - |11\rangle) \\
|\Psi^+\rangle &= \frac{1}{\sqrt{2}}(|01\rangle + |10\rangle) \\
|\Psi^-\rangle &= \frac{1}{\sqrt{2}}(|01\rangle - |10\rangle)
\end{aligned}$$

These cannot be written as product states.

11.10 Applications to Quantum Mechanics

11.10.1 Postulates of Quantum Mechanics

The mathematical framework of Hilbert spaces underlies quantum mechanics:

1. **States**: Pure states are rays in a Hilbert space $\mathcal{H}$ (equivalently, rank-1 projections or normalized vectors up to phase).
2. **Observables**: Physical observables are self-adjoint operators on $\mathcal{H}$.
3. **Measurement**: Measuring observable A in state $|\psi\rangle$ yields eigenvalue λ with probability $|\langle\lambda|\psi\rangle|^2$ (for discrete spectrum) or $\langle\psi|E(\Delta)|\psi\rangle$ for the value in Δ.
4. **Dynamics**: Time evolution is given by $|\psi(t)\rangle = e^{-iHt/\hbar}|\psi(0)\rangle$ where H is the Hamiltonian.
5. **Composition**: The state space of a composite system is the tensor product of component state spaces.

11.10.2 Position and Momentum

Example 11.10.1 (Position representation). For a particle in 1D: $\mathcal{H} = L^2(\mathbb{R})$.

Position operator: $(Qf)(x) = xf(x)$.

Momentum operator: $(Pf)(x) = -i\hbar\frac{df}{dx}$.

Canonical commutation relation: $[Q, P] = i\hbar I$.

Remark 11.10.1. Q and P are unbounded operators (discussed in Chapter 12), but the formalism extends appropriately.

11.10.3 Angular Momentum

Example 11.10.2 (Angular momentum operators). J_x, J_y, J_z satisfy $[J_i, J_j] = i\hbar\epsilon_{ijk}J_k$.

$J^2 = J_x^2 + J_y^2 + J_z^2$ commutes with all J_i.

Simultaneous eigenstates $|j, m\rangle$: $J^2|j, m\rangle = \hbar^2 j(j+1)|j, m\rangle$, $J_z|j, m\rangle = \hbar m|j, m\rangle$.

11.10.4 Density Operators

Definition 11.10.1 (Density operator). A **density operator** (or **density matrix**) ρ is a positive trace-class operator with $\mathrm{tr}(\rho) = 1$.

Proposition 11.10.1. *ρ represents a* ***pure state*** *if $\rho = |\psi\rangle\langle\psi|$ (equivalently $\rho^2 = \rho$).*

Otherwise ρ represents a ***mixed state****: $\rho = \sum_i p_i|\psi_i\rangle\langle\psi_i|$ with $p_i > 0$, $\sum p_i = 1$.*

Proposition 11.10.2 (Expectation value). $\langle A\rangle_\rho = \mathrm{tr}(\rho A)$.

11.11 Summary

Hilbert spaces provide the mathematical foundation for quantum mechanics:

Structure: Complete inner product spaces with orthonormal bases allowing Fourier-type expansions.

Riesz theorem: Identifies $\mathcal{H}$ with $\mathcal{H}^*$, enabling Dirac notation.

Operators: Self-adjoint (observables), unitary (symmetries), projections (measurements).

Spectral theory: Self-adjoint operators decompose via projection-valued measures; eigenvalues correspond to measurement outcomes.

Tensor products: Describe composite quantum systems; enable entanglement.

Quantum mechanics: States are vectors, observables are operators, probabilities come from inner products.

Exercises

Exercise 11.11.1. Prove the Cauchy-Schwarz inequality using $\|x+\lambda y\|^2 \geq 0$ for appropriate λ.

Exercise 11.11.2. Show that ℓ^2 is complete.

Exercise 11.11.3. Apply Gram-Schmidt to $\{1, x, x^2\}$ in $L^2([-1, 1])$ to obtain the first three Legendre polynomials.

Exercise 11.11.4. Prove that $\{e_n\}$ is an ONB iff $\sum_n |\langle e_n, x\rangle|^2 = \|x\|^2$ for all x.

Exercise 11.11.5. Show that the orthogonal projection onto a closed subspace M is unique.

Exercise 11.11.6. Prove the Riesz representation theorem.

Exercise 11.11.7. Verify that $(AB)^* = B^* A^*$ for bounded operators.

Exercise 11.11.8. Show that eigenvalues of self-adjoint operators are real and eigenvectors for distinct eigenvalues are orthogonal.

Exercise 11.11.9. Prove that compact operators map weakly convergent sequences to strongly convergent sequences.

Exercise 11.11.10. For self-adjoint A, show that $\|A\| = \sup_{\|x\|=1} |\langle x, Ax\rangle|$.

Exercise 11.11.11. Verify that $|\Phi^+\rangle = \frac{1}{\sqrt{2}}(|00\rangle + |11\rangle)$ cannot be written as a product state.

Exercise 11.11.12. Show that $[Q, P] = i\hbar I$ implies the Heisenberg uncertainty relation $\Delta Q \cdot \Delta P \geq \hbar/2$.

Chapter 12

Unbounded Operators and Spectral Theory

The most important operators in quantum mechanics—position, momentum, and the Hamiltonian—are unbounded. This chapter develops the theory of unbounded operators, their domains, self-adjoint extensions, and the spectral theorem in its full generality. These tools are essential for rigorous quantum mechanics.

12.1 Unbounded Operators

12.1.1 Motivation and Definition

Example 12.1.1 (Position operator). On $L^2(\mathbb{R})$, define $(Qf)(x) = xf(x)$. For $f_n(x) = n^{1/2}\chi_{[0,1/n]}(x)$:

$$\|f_n\| = 1, \quad \|Qf_n\| \leq n^{-1/2} \to 0.$$

But for $g_n(x) = n^{1/2}\chi_{[n,n+1/n]}(x)$:

$$\|g_n\| = 1, \quad \|Qg_n\| \geq n.$$

So Q cannot be bounded on all of $L^2(\mathbb{R})$.

Definition 12.1.1 (Unbounded operator). An **unbounded operator** on a Hilbert space $\mathcal{H}$ is a linear map $A : \mathcal{D}(A) \to \mathcal{H}$ where the **domain** $\mathcal{D}(A)$ is a linear subspace of $\mathcal{H}$.

Remark 12.1.1. "Unbounded" means "not necessarily bounded." The operator may or may not actually be unbounded.

Definition 12.1.2 (Dense domain). A is **densely defined** if $\overline{\mathcal{D}(A)} = \mathcal{H}$.

Most operators in quantum mechanics are densely defined.

12.1.2 Graph and Closedness

Definition 12.1.3 (Graph)**.** The **graph** of A is:

$$\Gamma(A) := \{(x, Ax) : x \in \mathcal{D}(A)\} \subseteq \mathcal{H} \times \mathcal{H}.$$

Definition 12.1.4 (Closed operator)**.** A is **closed** if $\Gamma(A)$ is closed in $\mathcal{H} \times \mathcal{H}$.

Equivalently: if $x_n \in \mathcal{D}(A)$, $x_n \to x$, and $Ax_n \to y$, then $x \in \mathcal{D}(A)$ and $Ax = y$.

Definition 12.1.5 (Closable operator)**.** A is **closable** if $\overline{\Gamma(A)}$ is the graph of an operator (called the **closure** $\bar{A}$).

Equivalently: if $x_n \to 0$ and $Ax_n \to y$, then $y = 0$.

Theorem 12.1.1 (Closed graph theorem)**.** *A closed operator defined on all of $\mathcal{H}$ is bounded.*

Corollary 12.1.2. *If A is unbounded and closed, then $\mathcal{D}(A) \neq \mathcal{H}$.*

12.1.3 Extension and Restriction

Definition 12.1.6 (Extension)**.** B is an **extension** of A (written $A \subseteq B$) if $\mathcal{D}(A) \subseteq \mathcal{D}(B)$ and $Bx = Ax$ for all $x \in \mathcal{D}(A)$.

Proposition 12.1.3. *$A \subseteq B$ iff $\Gamma(A) \subseteq \Gamma(B)$.*

Definition 12.1.7 (Core)**.** A subspace $D \subseteq \mathcal{D}(A)$ is a **core** for A if $\overline{A|_D} = A$ (where $A|_D$ is the restriction to D).

12.2 Symmetric and Self-Adjoint Operators

12.2.1 The Adjoint of an Unbounded Operator

Definition 12.2.1 (Adjoint)**.** Let A be densely defined. The **adjoint** A^* has domain:

$$\mathcal{D}(A^*) := \{y \in \mathcal{H} : \exists z \in \mathcal{H} \text{ with } \langle y, Ax\rangle = \langle z, x\rangle \text{ for all } x \in \mathcal{D}(A)\}.$$

For $y \in \mathcal{D}(A^*)$, define $A^*y := z$ (unique by density of $\mathcal{D}(A)$).

Proposition 12.2.1 (Properties of adjoint)**.**

(i) *A^* is always closed.*

(ii) *$A \subseteq B$ implies $B^* \subseteq A^*$.*

(iii) *A is closable iff A^* is densely defined, and then $\bar{A} = A^{**}$.*

(iv) $\ker(A^*) = (\operatorname{ran} A)^{\perp}$.

12.2.2 Symmetric Operators

Definition 12.2.2 (Symmetric operator). A densely defined operator A is **symmetric** if:

$$\langle Ax, y\rangle = \langle x, Ay\rangle \quad \text{for all } x, y \in \mathcal{D}(A).$$

Equivalently: $A \subseteq A^*$.

Proposition 12.2.2. *A is symmetric iff $\langle x, Ax\rangle \in \mathbb{R}$ for all $x \in \mathcal{D}(A)$.*

Proposition 12.2.3. *Symmetric operators are closable. The closure $\bar{A}$ is also symmetric.*

Example 12.2.1 (Momentum operator). On $\mathcal{D}(P) = C_c^\infty(\mathbb{R}) \subset L^2(\mathbb{R})$, define $P = -i\hbar\frac{d}{dx}$.

Integration by parts shows P is symmetric.

12.2.3 Self-Adjoint Operators

Definition 12.2.3 (Self-adjoint operator). A densely defined operator A is **self-adjoint** if $A = A^*$.

This means: (1) A is symmetric, and (2) $\mathcal{D}(A) = \mathcal{D}(A^*)$.

Remark 12.2.1 (Symmetric $\neq$ self-adjoint). For unbounded operators, symmetric does not imply self-adjoint! The domain condition $\mathcal{D}(A) = \mathcal{D}(A^*)$ is crucial.

Example 12.2.2 (Momentum on half-line). On $L^2([0,\infty))$, consider $P = -i\frac{d}{dx}$ with:

- $\mathcal{D}_1 = \{f \in H^1 : f(0) = 0\}$: symmetric but not self-adjoint.
- $\mathcal{D}_2 = H^1([0,\infty))$: P^* has domain $\mathcal{D}_1$, so $(P|_{\mathcal{D}_2})^* = P|_{\mathcal{D}_1}$.

Neither choice gives a self-adjoint operator on the half-line!

Definition 12.2.4 (Essentially self-adjoint). A is **essentially self-adjoint** if $\bar{A}$ is self-adjoint.

Equivalently: A has a unique self-adjoint extension.

12.2.4 Criteria for Self-Adjointness

Theorem 12.2.4 (Basic criterion). *A symmetric operator A is self-adjoint iff $\operatorname{ran}(A \pm iI) = \mathcal{H}$.*

Theorem 12.2.5 (Essential self-adjointness criterion). *A symmetric operator A is essentially self-adjoint iff $\ker(A^* \pm iI) = \{0\}$.*

Definition 12.2.5 (Deficiency indices). For symmetric A, the **deficiency indices** are:

$$n_\pm := \dim\ker(A^* \mp iI).$$

Theorem 12.2.6 (von Neumann's theorem). *A symmetric operator A has:*

- *Self-adjoint extensions iff* $n_+ = n_-$.
- *A unique self-adjoint extension (essentially self-adjoint) iff* $n_+ = n_- = 0$.
- *If* $n_+ = n_- = n < \infty$*, there is a* $U(n)$*-family of self-adjoint extensions.*

Example 12.2.3 (Momentum on interval). $P = -i\frac{d}{dx}$ on $C_c^\infty((0,1)) \subset L^2([0,1])$.

Solutions to $P^*f = \pm if$: $f(x) = ce^{\pm x}$.

Both are in $L^2([0,1])$, so $n_+ = n_- = 1$.

Self-adjoint extensions parametrized by $U(1)$: boundary conditions $f(1) = e^{i\theta}f(0)$.

12.3 Important Examples

12.3.1 Position Operator

Definition 12.3.1 (Position operator on $L^2(\mathbb{R})$).

$$(Qf)(x) := xf(x), \quad \mathcal{D}(Q) := \{f \in L^2(\mathbb{R}) : xf \in L^2(\mathbb{R})\}.$$

Theorem 12.3.1. *Q is self-adjoint with purely continuous spectrum $\sigma(Q) = \sigma_c(Q) = \mathbb{R}$.*

Proof sketch. Symmetry is clear. For self-adjointness, show $\operatorname{ran}(Q \pm iI) = L^2(\mathbb{R})$ using $(Q \pm iI)^{-1}f(x) = f(x)/(x \pm i)$.

No eigenvalues: $xf(x) = \lambda f(x)$ implies $f = 0$ a.e. □

12.3.2 Momentum Operator

Definition 12.3.2 (Momentum operator on $L^2(\mathbb{R})$).

$$(Pf)(x) := -i\hbar\frac{df}{dx}, \quad \mathcal{D}(P) := H^1(\mathbb{R}) = \{f \in L^2(\mathbb{R}) : f' \in L^2(\mathbb{R})\}.$$

Theorem 12.3.2. *P is self-adjoint with purely continuous spectrum $\sigma(P) = \sigma_c(P) = \mathbb{R}$.*

Proposition 12.3.3 (Fourier relationship). *P and Q are related by Fourier transform $\mathcal{F}$:*

$$\mathcal{F}Q\mathcal{F}^{-1} = \hbar P, \quad \mathcal{F}P\mathcal{F}^{-1} = -\hbar Q.$$

12.3.3 Laplacian

Definition 12.3.3 (Laplacian on $L^2(\mathbb{R}^n)$).

$$(-\Delta f)(x) := -\sum_{j=1}^{n} \frac{\partial^2 f}{\partial x_j^2}, \quad \mathcal{D}(-\Delta) := H^2(\mathbb{R}^n).$$

Theorem 12.3.4. *$-\Delta$ is self-adjoint, positive ($\langle f, -\Delta f\rangle \geq 0$), with spectrum $\sigma(-\Delta) = [0, \infty)$.*

12.3.4 Harmonic Oscillator

Definition 12.3.4 (Harmonic oscillator Hamiltonian).

$$H := -\frac{\hbar^2}{2m}\frac{d^2}{dx^2} + \frac{1}{2}m\omega^2 x^2 = \frac{P^2}{2m} + \frac{1}{2}m\omega^2 Q^2.$$

Theorem 12.3.5. *H is self-adjoint on $\mathcal{D}(H) = \{f \in L^2(\mathbb{R}) : Hf \in L^2(\mathbb{R})\}$ with purely discrete spectrum:*

$$\sigma(H) = \left\{\hbar\omega\left(n + \frac{1}{2}\right) : n = 0, 1, 2, \ldots\right\}.$$

Eigenfunctions are Hermite functions $\psi_n(x)$.

12.3.5 Hydrogen Atom

Definition 12.3.5 (Hydrogen Hamiltonian). On $L^2(\mathbb{R}^3)$:

$$H := -\frac{\hbar^2}{2m}\Delta - \frac{e^2}{4\pi\epsilon_0 r}.$$

Theorem 12.3.6 (Kato's theorem). *H is self-adjoint on $\mathcal{D}(H) = H^2(\mathbb{R}^3)$. The spectrum consists of:*

- *Discrete spectrum: $E_n = -\frac{me^4}{32\pi^2\epsilon_0^2\hbar^2 n^2}$ for $n = 1, 2, \ldots$ (bound states).*
- *Continuous spectrum: $[0, \infty)$ (scattering states).*

12.4 The Spectral Theorem

12.4.1 Statement

Theorem 12.4.1 (Spectral theorem for unbounded self-adjoint operators). *Let A be a self-adjoint operator on $\mathcal{H}$. There exists a unique projection-valued measure E on $\mathbb{R}$ such that:*

(i) $\mathcal{D}(A) = \{x \in \mathcal{H} : \int_{\mathbb{R}} \lambda^2 \, d\langle x, E(\lambda)x\rangle < \infty\}$.

(ii) For $x \in \mathcal{D}(A)$ *and* $y \in \mathcal{H}$: $\langle y, Ax\rangle = \int_{\mathbb{R}} \lambda \, d\langle y, E(\lambda)x\rangle$.

We write $A = \int_{\mathbb{R}} \lambda \, dE(\lambda)$.

Definition 12.4.1 (Spectral measure). For $x \in \mathcal{H}$, the **spectral measure** μ_x is defined by $\mu_x(B) = \langle x, E(B)x\rangle = \|E(B)x\|^2$.

Proposition 12.4.2. $\|Ax\|^2 = \int_{\mathbb{R}} \lambda^2 \, d\mu_x(\lambda)$ *for* $x \in \mathcal{D}(A)$.

12.4.2 Functional Calculus

Definition 12.4.2 (Functional calculus). For a Borel function $f : \mathbb{R} \to \mathbb{C}$:

$$f(A) := \int_{\mathbb{R}} f(\lambda) \, dE(\lambda)$$

with domain $\mathcal{D}(f(A)) = \{x \in \mathcal{H} : \int |f(\lambda)|^2 \, d\mu_x(\lambda) < \infty\}$.

Theorem 12.4.3 (Properties).

(i) $(f + g)(A) = f(A) + g(A)$ *on appropriate domains.*

(ii) $(fg)(A) = f(A)g(A)$.

(iii) $\overline{f}(A) = f(A)^*$.

(iv) $\|f(A)x\|^2 = \int |f(\lambda)|^2 \, d\mu_x(\lambda)$.

(v) If f *is bounded,* $f(A)$ *is bounded with* $\|f(A)\| \leq \sup_{\lambda \in \sigma(A)} |f(\lambda)|$.

Example 12.4.1 (Important functions).

- $f(\lambda) = e^{it\lambda}$: e^{itA} is unitary (Stone's theorem).
- $f(\lambda) = (\lambda - z)^{-1}$ for $z \notin \sigma(A)$: resolvent $(A - zI)^{-1}$.
- $f(\lambda) = \chi_B(\lambda)$: spectral projection $E(B)$.
- $f(\lambda) = \sqrt{\lambda}$ for $A \geq 0$: positive square root $\sqrt{A}$.

12.4.3 Spectral Types

Definition 12.4.3 (Spectral decomposition of Hilbert space). $\mathcal{H}$ decomposes as:

$$\mathcal{H} = \mathcal{H}_{pp} \oplus \mathcal{H}_{ac} \oplus \mathcal{H}_{sc}$$

where:

- $\mathcal{H}_{pp}$: **pure point** (spectral measures are pure point).
- $\mathcal{H}_{ac}$: **absolutely continuous** (spectral measures absolutely continuous w.r.t. Lebesgue).
- $\mathcal{H}_{sc}$: **singular continuous** (spectral measures singular but no atoms).

Definition 12.4.4 (Spectral types).

$$\begin{aligned}
&\sigma_{pp}(A) := \overline{\{\text{eigenvalues}\}} && \text{(pure point spectrum)}\\
&\sigma_{ac}(A) := \sigma(A|_{\mathcal{H}_{ac}}) && \text{(absolutely continuous spectrum)}\\
&\sigma_{sc}(A) := \sigma(A|_{\mathcal{H}_{sc}}) && \text{(singular continuous spectrum)}
\end{aligned}$$

Example 12.4.2.

- Harmonic oscillator: purely pure point spectrum.
- Free particle ($-\Delta$): purely absolutely continuous spectrum.
- Hydrogen atom: pure point (bound) + absolutely continuous (scattering).
- Almost Mathieu operator: can have singular continuous spectrum (at critical coupling).

12.4.4 RAGE Theorem

Theorem 12.4.4 (RAGE theorem). *Let A be self-adjoint and C compact. Then:*

(i) For $\psi \in \mathcal{H}_c := \mathcal{H}_{ac} \oplus \mathcal{H}_{sc}$ (continuous spectral subspace):

$$\lim_{T\to\infty} \frac{1}{T} \int_0^T \|Ce^{-itA}\psi\|^2 \, dt = 0.$$

(ii) For $\psi \in \mathcal{H}_{pp}$: $\|Ce^{-itA}\psi\|$ does not decay on average.

Physical interpretation: Continuous spectrum corresponds to scattering states that escape to infinity; pure point spectrum corresponds to bound states that remain localized.

12.5 Stone's Theorem and One-Parameter Groups

12.5.1 Strongly Continuous Groups

Definition 12.5.1 (Strongly continuous unitary group)**.** A **strongly continuous one-parameter unitary group** is a family $\{U(t)\}_{t\in\mathbb{R}}$ of unitary operators satisfying:

(i) $U(0) = I$.

(ii) $U(t+s) = U(t)U(s)$ for all $t, s \in \mathbb{R}$.

(iii) $t \mapsto U(t)x$ is continuous for each $x \in \mathcal{H}$.

Definition 12.5.2 (Generator)**.** The **generator** of $\{U(t)\}$ is:

$$Ax := \lim_{t\to 0} \frac{U(t)x - x}{it}$$

with domain $\mathcal{D}(A) = \{x : \text{limit exists}\}$.

12.5.2 Stone's Theorem

Theorem 12.5.1 (Stone's theorem)**.**

(i) If A is self-adjoint, then $U(t) := e^{itA}$ (defined by functional calculus) is a strongly continuous unitary group with generator A.

(ii) Conversely, every strongly continuous unitary group has a self-adjoint generator.

Thus there is a one-to-one correspondence:

$$\{\textit{self-adjoint operators}\} \longleftrightarrow \{\textit{strongly continuous unitary groups}\}.$$

Corollary 12.5.2. *The Schrödinger equation $i\hbar\frac{\partial}{\partial t}|\psi(t)\rangle = H|\psi(t)\rangle$ has solution $|\psi(t)\rangle = e^{-iHt/\hbar}|\psi(0)\rangle$ iff H is self-adjoint.*

12.5.3 Applications

Example 12.5.1 (Translation group)**.** $U(t)f(x) = f(x-t)$ on $L^2(\mathbb{R})$. Generator: $A = -i\frac{d}{dx} = P/\hbar$.

Momentum generates translations!

Example 12.5.2 (Multiplication group)**.** $U(t)f(x) = e^{itx}f(x)$ on $L^2(\mathbb{R})$. Generator: $A = Q$ (position).

Position generates multiplication by phase.

Example 12.5.3 (Rotation group)**.** $SO(3)$ acts unitarily on $L^2(\mathbb{R}^3)$. Generators are angular momentum operators L_x, L_y, L_z.

12.6 Commutation Relations

12.6.1 Canonical Commutation Relations

Theorem 12.6.1 (Canonical commutation relations). *On appropriate domains:*

$$[Q, P] := QP - PQ = i\hbar I.$$

More precisely, for $f \in \mathcal{S}(\mathbb{R})$ (Schwartz space):

$$(QP - PQ)f = i\hbar f.$$

Theorem 12.6.2 (Weyl relations). *The unitary groups satisfy:*

$$e^{isQ}e^{itP} = e^{-ist\hbar}e^{itP}e^{isQ}.$$

Theorem 12.6.3 (Stone-von Neumann theorem). *Any irreducible representation of the Weyl relations on a separable Hilbert space is unitarily equivalent to the Schrödinger representation on $L^2(\mathbb{R})$.*

12.6.2 Uncertainty Relations

Theorem 12.6.4 (Robertson uncertainty relation). *For any self-adjoint A, B and state $|\psi\rangle$:*

$$\Delta A \cdot \Delta B \geq \frac{1}{2}|\langle [A, B] \rangle|$$

where $\Delta A = \sqrt{\langle A^2 \rangle - \langle A \rangle^2}$ is the standard deviation.

Corollary 12.6.5 (Heisenberg uncertainty principle).

$$\Delta Q \cdot \Delta P \geq \frac{\hbar}{2}.$$

Proof. $[Q, P] = i\hbar I$, so $|\langle [Q, P] \rangle| = \hbar$. □

Theorem 12.6.6 (Energy-time uncertainty). *For time-independent Hamiltonian H and observable A:*

$$\Delta E \cdot \Delta t_A \geq \frac{\hbar}{2}$$

where $\Delta t_A := \Delta A / |d\langle A \rangle / dt|$ is the characteristic time for A to change.

12.7 Self-Adjoint Extensions

12.7.1 Boundary Conditions

Example 12.7.1 (Particle in a box). $H = -\frac{\hbar^2}{2m}\frac{d^2}{dx^2}$ on $L^2([0, L])$.
Starting from $C_c^\infty((0, L))$: deficiency indices $(2, 2)$.
Self-adjoint extensions parametrized by boundary conditions:

- Dirichlet: $f(0) = f(L) = 0$.
- Neumann: $f'(0) = f'(L) = 0$.
- Periodic: $f(0) = f(L)$, $f'(0) = f'(L)$.
- General: $U(1) \times U(1)$ family.

Different extensions give different spectra!

12.7.2 Friedrichs Extension

Theorem 12.7.1 (Friedrichs extension). *Let A be a positive symmetric operator ($\langle x, Ax\rangle \geq 0$). Then A has a distinguished self-adjoint extension A_F (the Friedrichs extension), which is also positive.*

Proposition 12.7.2. *A_F is the largest positive self-adjoint extension: if B is a positive self-adjoint extension of A, then $A_F \leq B$ (in the sense of quadratic forms).*

Example 12.7.2. For $-\Delta$ on $C_c^\infty(\Omega)$ where $\Omega \subset \mathbb{R}^n$ is bounded, the Friedrichs extension corresponds to Dirichlet boundary conditions.

12.7.3 Kato-Rellich Theorem

Theorem 12.7.3 (Kato-Rellich theorem). *Let A be self-adjoint and B symmetric with $\mathcal{D}(A) \subseteq \mathcal{D}(B)$. If there exist $a < 1$ and $b \geq 0$ such that:*

$$\|Bx\| \leq a\|Ax\| + b\|x\| \quad \text{for all } x \in \mathcal{D}(A),$$

then $A + B$ is self-adjoint on $\mathcal{D}(A)$.

Example 12.7.3. Hydrogen atom: $V(r) = -e^2/r$ is relatively bounded with respect to $-\Delta$ with relative bound 0 in 3D. Hence $H = -\Delta + V$ is self-adjoint on $\mathcal{D}(-\Delta) = H^2(\mathbb{R}^3)$.

12.8 Spectral Theory and Scattering

12.8.1 Resolvent and Spectrum

Definition 12.8.1 (Resolvent). For $z \in \rho(A)$, the **resolvent** is $R_z(A) := (A - zI)^{-1}$.

Proposition 12.8.1 (Resolvent identity). $R_z - R_w = (z - w)R_z R_w$.

Proposition 12.8.2 (Spectral characterization). *$\lambda \in \sigma(A)$ iff there exists a **Weyl sequence**: (x_n) with $\|x_n\| = 1$ and $\|(A - \lambda)x_n\| \to 0$.*

12.8.2 Scattering Theory

Definition 12.8.2 (Wave operators). For self-adjoint H_0 (free) and H (full):

$$\Omega_\pm :=_{t\to\mp\infty} e^{iHt}e^{-iH_0t}P_{ac}(H_0)$$

where P_{ac} projects onto absolutely continuous subspace.

Definition 12.8.3 (Scattering operator). If wave operators exist:

$$S := \Omega_+^*\Omega_-.$$

Theorem 12.8.3 (Asymptotic completeness). *For short-range potentials (e.g., $|V(x)| \leq C(1 + |x|)^{-1-\epsilon}$):*

(i) Wave operators exist.

(ii) $\operatorname{ran}\Omega_\pm = \mathcal{H}_{ac}(H)$.

(iii) S is unitary on $\mathcal{H}_{ac}(H_0)$.

12.9 Applications to Quantum Mechanics

12.9.1 Observables and Measurement

Proposition 12.9.1 (Measurement axiom - precise version). *For self-adjoint A with spectral measure E:*

- *Probability that measurement yields value in $B \subseteq \mathbb{R}$:* $\operatorname{Prob}(B) = \|E(B)|\psi\rangle\|^2$.
- *State after measurement with result in B: $|\psi'\rangle = E(B)|\psi\rangle/\|E(B)|\psi\rangle\|$.*
- *Expected value: $\langle A\rangle = \int \lambda\, d\mu_\psi(\lambda)$.*

12.9.2 Time Evolution

Proposition 12.9.2 (Schrödinger vs Heisenberg picture).

- ***Schrödinger***: *states evolve,* $|\psi(t)\rangle = e^{-iHt/\hbar}|\psi(0)\rangle$.
- ***Heisenberg***: *operators evolve,* $A(t) = e^{iHt/\hbar}Ae^{-iHt/\hbar}$.

Expectation values agree: $\langle\psi(t)|A|\psi(t)\rangle = \langle\psi(0)|A(t)|\psi(0)\rangle$.

Proposition 12.9.3 (Heisenberg equation).

$$\frac{dA}{dt} = \frac{i}{\hbar}[H, A] + \frac{\partial A}{\partial t}.$$

Corollary 12.9.4 (Constants of motion). *A is conserved ($dA/dt = 0$) iff $[H, A] = 0$ (for time-independent A).*

12.9.3 Symmetries

Theorem 12.9.5 (Wigner's theorem). *Every symmetry transformation (bijection preserving transition probabilities) is implemented by a unitary or antiunitary operator.*

Proposition 12.9.6 (Continuous symmetries). *A strongly continuous one-parameter group $U(t)$ is a symmetry ($[U(t), H] = 0$) iff its generator G is conserved: $[H, G] = 0$.*

Example 12.9.1 (Symmetries and conservation).

Symmetry	Generator	Conserved
Translation	P	Momentum
Rotation	L	Angular momentum
Time translation	H	Energy
Gauge $U(1)$	Q (charge)	Electric charge

12.10 Summary

Unbounded operators require careful treatment of domains:

Symmetric vs self-adjoint: Symmetric means $A \subseteq A^*$; self-adjoint means $A = A^*$. The distinction is crucial for physical observables.

Self-adjoint extensions: Not all symmetric operators have unique extensions. Deficiency indices determine possibilities; physics chooses via boundary conditions.

Spectral theorem: Self-adjoint operators have real spectrum and spectral decomposition $A = \int \lambda\, dE(\lambda)$.

Stone's theorem: Self-adjoint operators generate unitary time evolution; this is why observables must be self-adjoint.

Commutation relations: $[Q, P] = i\hbar$ leads to uncertainty relations and is essentially unique (Stone-von Neumann).

Scattering theory: Continuous spectrum corresponds to scattering states; asymptotic completeness connects past and future asymptotics.

Exercises

Exercise 12.10.1. Show that the position operator Q on $L^2(\mathbb{R})$ is unbounded.

Exercise 12.10.2. Prove that the adjoint A^* is always closed.

Exercise 12.10.3. Show that if A is symmetric and $\mathrm{ran}(A + iI) = \mathcal{H}$, then A is self-adjoint.

Exercise 12.10.4. Compute the deficiency indices of $P = -i\frac{d}{dx}$ on $C_c^\infty((0,\infty)) \subset L^2([0,\infty))$.

Exercise 12.10.5. Verify that $e^{isQ}e^{itP} = e^{-ist\hbar}e^{itP}e^{isQ}$ using Baker-Campbell-Hausdorff.

Exercise 12.10.6. Prove the Robertson uncertainty relation.

Exercise 12.10.7. Find all self-adjoint extensions of $-\frac{d^2}{dx^2}$ on $C_c^\infty((0,1))$ and their spectra.

Exercise 12.10.8. Show that the harmonic oscillator Hamiltonian is essentially self-adjoint on $C_c^\infty(\mathbb{R})$.

Exercise 12.10.9. Verify that $[H, L^2] = 0$ and $[H, L_z] = 0$ for the hydrogen atom Hamiltonian.

Exercise 12.10.10. Prove the Heisenberg equation of motion $\frac{dA}{dt} = \frac{i}{\hbar}[H, A]$.

Exercise 12.10.11. Show that the spectrum of the momentum operator P on $L^2([0, L])$ with periodic boundary conditions is $\{2\pi\hbar n/L : n \in \mathbb{Z}\}$.

Exercise 12.10.12. For the free particle, show that $\langle x^2\rangle(t) = \langle x^2\rangle(0) + \frac{2t}{m}\langle xp + px\rangle(0) + \frac{t^2}{m^2}\langle p^2\rangle(0)$ and interpret physically.

Chapter 13

Distributions and Green's Functions

Physical systems often involve singular objects: point charges, impulsive forces, sharp boundaries. The theory of distributions provides a rigorous framework for such singularities. Green's functions, intimately connected to distributions, are powerful tools for solving differential equations. This chapter develops distribution theory and its applications to physics.

13.1 Motivation: Beyond Functions

13.1.1 The Need for Generalized Functions

Classical functions are insufficient for many physical situations:

Example 13.1.1 (Point charge). The charge density of a point charge q at the origin "should be" zero everywhere except at the origin, where it is infinite, yet integrate to q. No classical function has this property.

Example 13.1.2 (Derivative of step function). The Heaviside step function $\theta(x)$ has derivative 0 for $x \neq 0$, but its "derivative" should somehow capture the jump at $x = 0$.

Example 13.1.3 (Plane wave normalization). In quantum mechanics, $\psi_k(x) = e^{ikx}$ is not in $L^2(\mathbb{R})$, yet we want $\langle \psi_k | \psi_{k'} \rangle = \delta(k - k')$.

Distributions (generalized functions) resolve all these issues.

13.1.2 The Key Idea

Instead of asking "what is $f(x)$?" we ask "what is $\int f(x)\phi(x)\, dx$?" for nice test functions ϕ.

A distribution is defined by how it "pairs" with test functions, not by pointwise values.

13.2 Test Functions and Distributions

13.2.1 Test Function Spaces

Definition 13.2.1 (Space $\mathcal{D}(\mathbb{R}^n)$)**.** The space $\mathcal{D}(\mathbb{R}^n) = C_c^\infty(\mathbb{R}^n)$ consists of smooth functions with compact support:

$$\mathcal{D}(\mathbb{R}^n) := \{\phi \in C^\infty(\mathbb{R}^n) : \text{supp}(\phi) \text{ is compact}\}.$$

Definition 13.2.2 (Convergence in $\mathcal{D}$)**.** $\phi_n \to \phi$ in $\mathcal{D}$ if:

(i) All ϕ_n have support in a common compact set K.

(ii) $D^\alpha \phi_n \to D^\alpha \phi$ uniformly for all multi-indices α.

Definition 13.2.3 (Space $\mathcal{S}(\mathbb{R}^n)$)**.** The **Schwartz space** consists of smooth functions that decay rapidly with all derivatives:

$$\mathcal{S}(\mathbb{R}^n) := \{\phi \in C^\infty(\mathbb{R}^n) : \sup_x |x^\alpha D^\beta \phi(x)| < \infty \text{ for all } \alpha, \beta\}.$$

Proposition 13.2.1. *$\mathcal{D}(\mathbb{R}^n) \subset \mathcal{S}(\mathbb{R}^n) \subset L^p(\mathbb{R}^n)$ for all $p \geq 1$.*

Example 13.2.1 (Standard test functions)**.**

- Bump function: $\phi(x) = \begin{cases} e^{-1/(1-|x|^2)} & |x| < 1 \\ 0 & |x| \geq 1 \end{cases}$ is in $\mathcal{D}(\mathbb{R}^n)$.
- Gaussian: $\phi(x) = e^{-|x|^2}$ is in $\mathcal{S}(\mathbb{R}^n)$ but not $\mathcal{D}(\mathbb{R}^n)$.

13.2.2 Distributions

Definition 13.2.4 (Distribution)**.** A **distribution** on $\mathbb{R}^n$ is a continuous linear functional on $\mathcal{D}(\mathbb{R}^n)$:

$$T : \mathcal{D}(\mathbb{R}^n) \to \mathbb{C}, \quad \phi \mapsto \langle T, \phi \rangle$$

where continuity means: if $\phi_n \to \phi$ in $\mathcal{D}$, then $\langle T, \phi_n \rangle \to \langle T, \phi \rangle$.

The space of distributions is denoted $\mathcal{D}'(\mathbb{R}^n)$.

Definition 13.2.5 (Tempered distribution)**.** A **tempered distribution** is a continuous linear functional on $\mathcal{S}(\mathbb{R}^n)$.

The space is denoted $\mathcal{S}'(\mathbb{R}^n)$.

Proposition 13.2.2. *$\mathcal{S}'(\mathbb{R}^n) \subset \mathcal{D}'(\mathbb{R}^n)$ (tempered distributions are distributions).*

13.2.3 Regular Distributions

Definition 13.2.6 (Regular distribution). A **regular distribution** is defined by a locally integrable function $f \in L^1_{\text{loc}}(\mathbb{R}^n)$:

$$\langle T_f, \phi \rangle := \int_{\mathbb{R}^n} f(x)\phi(x)\,dx.$$

Proposition 13.2.3. *The map $f \mapsto T_f$ is injective: if $\langle T_f, \phi \rangle = \langle T_g, \phi \rangle$ for all $\phi \in \mathcal{D}$, then $f = g$ a.e.*

We often identify f with T_f and write $\langle f, \phi \rangle = \int f\phi$.

Example 13.2.2 (Regular distributions).

- Any $f \in L^p(\mathbb{R}^n)$ defines a regular distribution.
- $f(x) = 1/|x|$ (for $n \geq 2$) defines a regular distribution.
- Polynomials define tempered distributions.

13.3 The Dirac Delta Distribution

13.3.1 Definition and Properties

Definition 13.3.1 (Dirac delta). The **Dirac delta distribution** δ is defined by:

$$\langle \delta, \phi \rangle := \phi(0).$$

More generally, δ_a centered at a:

$$\langle \delta_a, \phi \rangle := \phi(a).$$

Proposition 13.3.1. *δ is a distribution (continuous linear functional on $\mathcal{D}$) but not a regular distribution (no function f satisfies $\int f\phi = \phi(0)$ for all ϕ).*

Theorem 13.3.2 (Sifting property). *For any continuous f:*

$$\langle \delta_a, f \rangle = f(a), \quad \text{“}\int \delta(x-a)f(x)\,dx = f(a)\text{”}.$$

13.3.2 Representations and Approximations

Theorem 13.3.3 (Delta as limit). *Let $\{\rho_\epsilon\}$ be an **approximate identity**: $\rho_\epsilon \geq 0$, $\int \rho_\epsilon = 1$, $\text{supp}(\rho_\epsilon) \to \{0\}$ as $\epsilon \to 0$. Then:*

$$\rho_\epsilon \to \delta \quad \textit{in } \mathcal{D}'.$$

Example 13.3.1 (Approximations to delta).

$$\delta_\epsilon(x) = \frac{1}{\epsilon\sqrt{\pi}} e^{-x^2/\epsilon^2} \qquad \text{(Gaussian)}$$

$$\delta_\epsilon(x) = \frac{1}{\pi} \frac{\epsilon}{x^2 + \epsilon^2} \qquad \text{(Lorentzian/Cauchy)}$$

$$\delta_\epsilon(x) = \frac{\sin(x/\epsilon)}{\pi x} \qquad \text{(sinc function)}$$

$$\delta_\epsilon(x) = \frac{1}{2\epsilon} \chi_{[-\epsilon,\epsilon]}(x) \qquad \text{(rectangular)}$$

All converge to δ as $\epsilon \to 0$.

Proposition 13.3.4 (Fourier representation).

$$\delta(x) = \frac{1}{2\pi} \int_{-\infty}^{\infty} e^{ikx}\, dk \quad (\text{in } \mathcal{S}').$$

13.3.3 Properties of the Delta Distribution

Proposition 13.3.5 (Basic properties).

(i) $x\delta(x) = 0$.

(ii) $\delta(-x) = \delta(x)$ *(even).*

(iii) $\delta(ax) = \frac{1}{|a|}\delta(x)$ *for* $a \neq 0$.

(iv) $f(x)\delta(x-a) = f(a)\delta(x-a)$ *for continuous* f.

Proposition 13.3.6 (Composition with functions). *If* g *has simple zeros at* $x_1, \ldots, x_n$*:*

$$\delta(g(x)) = \sum_{i=1}^{n} \frac{\delta(x - x_i)}{|g'(x_i)|}.$$

Example 13.3.2. $\delta(x^2 - a^2) = \frac{1}{2|a|}[\delta(x-a) + \delta(x+a)]$ for $a \neq 0$.

13.3.4 Multidimensional Delta

Definition 13.3.2. In $\mathbb{R}^n$:

$$\langle \delta^{(n)}, \phi \rangle := \phi(0), \qquad \delta^{(n)}(\mathbf{x}) = \delta(x_1)\delta(x_2)\cdots\delta(x_n).$$

Proposition 13.3.7 (Coordinate transformations). *Under diffeomorphism* $\mathbf{y} = \mathbf{g}(\mathbf{x})$*:*

$$\delta^{(n)}(\mathbf{g}(\mathbf{x})) = \frac{\delta^{(n)}(\mathbf{x} - \mathbf{x}_0)}{|\det J_g(\mathbf{x}_0)|}$$

where $\mathbf{g}(\mathbf{x}_0) = \mathbf{0}$.

Example 13.3.3 (Spherical coordinates). $\delta^{(3)}(\mathbf{r}) = \frac{\delta(r)}{4\pi r^2}$ (in the sense that $\int \delta^{(3)}(\mathbf{r}) f(\mathbf{r})\, d^3r = f(\mathbf{0})$).

13.4 Operations on Distributions

13.4.1 Differentiation

Definition 13.4.1 (Distributional derivative). The derivative of a distribution T is defined by:

$$\langle T', \phi \rangle := -\langle T, \phi' \rangle.$$

More generally:

$$\langle D^\alpha T, \phi \rangle := (-1)^{|\alpha|} \langle T, D^\alpha \phi \rangle.$$

Theorem 13.4.1. *Every distribution is infinitely differentiable (in the distributional sense).*

Example 13.4.1 (Derivative of Heaviside function). For $\theta(x) = \begin{cases} 1 & x > 0 \\ 0 & x < 0 \end{cases}$:

$$\langle \theta', \phi \rangle = -\langle \theta, \phi' \rangle = -\int_0^\infty \phi'(x)\, dx = \phi(0) = \langle \delta, \phi \rangle.$$

Thus $\theta' = \delta$.

Example 13.4.2 (Derivatives of delta).

$$\langle \delta', \phi \rangle = -\phi'(0), \quad \langle \delta^{(n)}, \phi \rangle = (-1)^n \phi^{(n)}(0).$$

Example 13.4.3 (Sign function). $\operatorname{sgn}(x) = 2\theta(x) - 1$, so $\operatorname{sgn}'(x) = 2\delta(x)$.

13.4.2 Multiplication

Definition 13.4.2 (Multiplication by smooth function). For $\psi \in C^\infty$ and distribution T:

$$\langle \psi T, \phi \rangle := \langle T, \psi \phi \rangle.$$

Remark 13.4.1. Products of arbitrary distributions are not always defined! $\delta(x)^2$ is meaningless.

Proposition 13.4.2 (Leibniz rule). *$(fT)' = f'T + fT'$ for smooth f.*

13.4.3 Convolution

Definition 13.4.3 (Convolution). For suitable distributions S, T:

$$\langle S * T, \phi \rangle := \langle S_x \otimes T_y, \phi(x + y) \rangle.$$

For regular distributions:

$$(f * g)(x) = \int f(x - y)g(y)\, dy.$$

Theorem 13.4.3 (Properties of convolution).

(i) $f * g = g * f$ *(commutative).*

(ii) $(f * g) * h = f * (g * h)$ *(associative, when defined).*

(iii) $D^\alpha(f * g) = (D^\alpha f) * g = f * (D^\alpha g)$.

(iv) $\delta * f = f$ *(delta is identity).*

Example 13.4.4. $\delta' * f = f'$ (differentiation by convolution with δ').

13.4.4 Fourier Transform of Distributions

Definition 13.4.4 (Fourier transform on $\mathcal{S}$). For $\phi \in \mathcal{S}(\mathbb{R}^n)$:

$$\hat{\phi}(k) = \mathcal{F}[\phi](k) := \int_{\mathbb{R}^n} \phi(x) e^{-ik \cdot x}\, dx.$$

Theorem 13.4.4. $\mathcal{F} : \mathcal{S} \to \mathcal{S}$ *is a bijection.*

Definition 13.4.5 (Fourier transform of tempered distribution). For $T \in \mathcal{S}'$:

$$\langle \hat{T}, \phi \rangle := \langle T, \hat{\phi} \rangle.$$

Theorem 13.4.5 (Fourier transform properties).

(i) $\mathcal{F}[\delta] = 1$ *(constant function).*

(ii) $\mathcal{F}[1] = (2\pi)^n \delta$.

(iii) $\mathcal{F}[D^\alpha T] = (ik)^\alpha \hat{T}$.

(iv) $\mathcal{F}[x^\alpha T] = i^{|\alpha|} D^\alpha \hat{T}$.

(v) $\mathcal{F}[T * S] = \hat{T} \cdot \hat{S}$.

Example 13.4.5 (Important transforms).

$$\mathcal{F}[e^{-a|x|}] = \frac{2a}{k^2 + a^2} \qquad (a > 0)$$

$$\mathcal{F}[e^{-ax^2}] = \sqrt{\frac{\pi}{a}} e^{-k^2/4a} \qquad (a > 0)$$

$$\mathcal{F}[\theta(x)e^{-ax}] = \frac{1}{a + ik} \qquad (a > 0)$$

$$\mathcal{F}[\text{sgn}(x)] = \frac{2}{ik} \qquad \text{(p.v.)}$$

13.5 Green's Functions

13.5.1 Definition and Motivation

Definition 13.5.1 (Green's function). The **Green's function** for a linear differential operator L is a distribution G satisfying:

$$LG = \delta.$$

Theorem 13.5.1 (Solution by Green's function). *If $LG = \delta$, then the solution to $Lu = f$ is:*

$$u = G * f$$

provided the convolution exists.

Proof. $Lu = L(G * f) = (LG) * f = \delta * f = f$. □

13.5.2 Green's Function for the Laplacian

Theorem 13.5.2 (Fundamental solution of Laplacian). *The Green's function for $-\Delta$ in $\mathbb{R}^n$ is:*

$$G(\mathbf{x}) = \begin{cases} -\frac{1}{2}|x| & n = 1 \\ \frac{1}{2\pi} \ln |\mathbf{x}| & n = 2 \\ \frac{1}{4\pi|\mathbf{x}|} & n = 3 \\ \frac{1}{(n-2)\omega_n |\mathbf{x}|^{n-2}} & n \geq 3 \end{cases}$$

where $\omega_n = 2\pi^{n/2}/\Gamma(n/2)$ is the surface area of S^{n-1}.

Corollary 13.5.3 (Solution of Poisson equation). *The solution to $-\Delta u = f$ in $\mathbb{R}^3$ is:*

$$u(\mathbf{x}) = \frac{1}{4\pi} \int_{\mathbb{R}^3} \frac{f(\mathbf{y})}{|\mathbf{x} - \mathbf{y}|} \, d^3y.$$

Example 13.5.1 (Electrostatics). For charge density ρ, the potential ϕ satisfies $-\epsilon_0 \Delta\phi = \rho$:

$$\phi(\mathbf{x}) = \frac{1}{4\pi\epsilon_0} \int \frac{\rho(\mathbf{y})}{|\mathbf{x}-\mathbf{y}|} \, d^3y.$$

For a point charge $\rho = q\delta^{(3)}(\mathbf{x})$: $\phi = \frac{q}{4\pi\epsilon_0 r}$ (Coulomb).

13.5.3 Green's Functions with Boundary Conditions

Definition 13.5.2 (Green's function for boundary value problem). For operator L on domain Ω with boundary conditions, the Green's function $G(\mathbf{x}, \mathbf{y})$ satisfies:

(i) $L_x G(\mathbf{x}, \mathbf{y}) = \delta(\mathbf{x} - \mathbf{y})$ in Ω.

(ii) G satisfies the boundary conditions in $\mathbf{x}$.

Theorem 13.5.4 (Symmetry). *For self-adjoint L with appropriate boundary conditions:*

$$G(\mathbf{x}, \mathbf{y}) = G(\mathbf{y}, \mathbf{x}).$$

Example 13.5.2 (Dirichlet problem on half-space). For $-\Delta u = f$ in $z > 0$ with $u = 0$ on $z = 0$:

$$G(\mathbf{x}, \mathbf{y}) = \frac{1}{4\pi|\mathbf{x}-\mathbf{y}|} - \frac{1}{4\pi|\mathbf{x}-\mathbf{y}'|}$$

where $\mathbf{y}' = (y_1, y_2, -y_3)$ is the image charge.

13.5.4 Helmholtz Equation

Theorem 13.5.5 (Green's function for Helmholtz). *For $(-\Delta - k^2)G = \delta$ in $\mathbb{R}^3$:*

$$G_\pm(\mathbf{x}) = \frac{e^{\pm ik|\mathbf{x}|}}{4\pi|\mathbf{x}|}.$$

G_+ (outgoing) and G_- (incoming) correspond to different boundary conditions at infinity.

Example 13.5.3 (Acoustic radiation). The outgoing wave from a point source at origin satisfies $(\Delta + k^2)u = -\delta$:

$$u(\mathbf{r}) = \frac{e^{ikr}}{4\pi r}.$$

13.6 Green's Functions in Time-Dependent Problems

13.6.1 Heat Equation

Theorem 13.6.1 (Heat kernel). *The Green's function for $\partial_t u - D\Delta u = f$ in $\mathbb{R}^n$ is:*

$$G(\mathbf{x}, t) = \frac{\theta(t)}{(4\pi Dt)^{n/2}} e^{-|\mathbf{x}|^2/4Dt}.$$

Corollary 13.6.2. *Solution with initial condition $u(\mathbf{x}, 0) = u_0(\mathbf{x})$:*

$$u(\mathbf{x}, t) = \int_{\mathbb{R}^n} G(\mathbf{x} - \mathbf{y}, t) u_0(\mathbf{y})\, d^n y.$$

13.6.2 Wave Equation

Theorem 13.6.3 (Retarded Green's function). *For $(\partial_t^2 - c^2\Delta)G = \delta(\mathbf{x})\delta(t)$ in $\mathbb{R}^3$:*

$$G_{\text{ret}}(\mathbf{x}, t) = \frac{\delta(t - |\mathbf{x}|/c)}{4\pi|\mathbf{x}|}.$$

Corollary 13.6.4 (Retarded potential). *Solution to $(\partial_t^2 - c^2\Delta)\phi = f$:*

$$\phi(\mathbf{x}, t) = \frac{1}{4\pi} \int \frac{f(\mathbf{y}, t - |\mathbf{x} - \mathbf{y}|/c)}{|\mathbf{x} - \mathbf{y}|}\, d^3 y.$$

This is the ***retarded solution****: effects propagate at speed c.*

Example 13.6.1 (Liénard-Wiechert potentials). For a moving point charge q at $\mathbf{r}_0(t)$:

$$\phi(\mathbf{x}, t) = \frac{q}{4\pi\epsilon_0} \frac{1}{|\mathbf{x} - \mathbf{r}_0(t_{\text{ret}})| - \mathbf{v}(t_{\text{ret}}) \cdot (\mathbf{x} - \mathbf{r}_0(t_{\text{ret}}))/c}$$

where t_{ret} is the retarded time.

13.6.3 Schrödinger Equation

Theorem 13.6.5 (Free particle propagator). *For $i\hbar\partial_t\psi = -\frac{\hbar^2}{2m}\Delta\psi$, the propagator is:*

$$K(\mathbf{x}, t; \mathbf{x}', 0) = \left(\frac{m}{2\pi i\hbar t}\right)^{n/2} e^{im|\mathbf{x}-\mathbf{x}'|^2/2\hbar t} \quad (t > 0).$$

Corollary 13.6.6. $\psi(\mathbf{x}, t) = \int K(\mathbf{x}, t; \mathbf{x}', 0)\psi(\mathbf{x}', 0)\, d^n x'$.

13.7 Propagators in Quantum Field Theory

13.7.1 The Klein-Gordon Propagator

Definition 13.7.1 (Klein-Gordon equation). $(\Box + m^2)\phi = 0$ where $\Box = \partial_t^2 - \nabla^2$ (with $c = \hbar = 1$).

Theorem 13.7.1 (Feynman propagator). *The **Feynman propagator** for $(\Box + m^2)G = -\delta^{(4)}$ is:*

$$G_F(x) = \int \frac{d^4k}{(2\pi)^4} \frac{e^{-ik\cdot x}}{k^2 - m^2 + i\epsilon}$$

where the $i\epsilon$ prescription defines how to handle poles.

Proposition 13.7.2 (Position space form).

$$G_F(x) = \begin{cases} -\frac{1}{4\pi}\delta(s) + \frac{m}{8\pi\sqrt{s}} H_1^{(1)}(m\sqrt{s}) & s > 0 \\ -\frac{m}{4\pi^2\sqrt{-s}} K_1(m\sqrt{-s}) & s < 0 \end{cases}$$

where $s = t^2 - |\mathbf{x}|^2$ and $H_1^{(1)}, K_1$ are Hankel/Bessel functions.

13.7.2 Physical Interpretation

The Feynman propagator $G_F(x - y)$ represents:

- Amplitude for a particle to propagate from y to x (if $t_x > t_y$).
- Amplitude for an antiparticle to propagate from x to y (if $t_y > t_x$).

Theorem 13.7.3 (Causal structure).

- $G_{\text{ret}}(x) = 0$ *for* $t < 0$ *(retarded: cause before effect).*
- $G_{\text{adv}}(x) = 0$ *for* $t > 0$ *(advanced: effect before cause).*
- $G_F = \theta(t)G_{\text{ret}} + \theta(-t)G_{\text{adv}}$ *(Feynman: positive frequency forward, negative backward).*

13.7.3 Dirac Propagator

Definition 13.7.2 (Dirac equation). $(i\gamma^\mu\partial_\mu - m)\psi = 0$.

Theorem 13.7.4 (Dirac propagator). *The Feynman propagator for $(i\gamma^\mu\partial_\mu - m)S_F = \delta^{(4)}$:*

$$S_F(x) = (i\gamma^\mu\partial_\mu + m)G_F(x) = \int \frac{d^4k}{(2\pi)^4} \frac{(\gamma^\mu k_\mu + m)e^{-ik\cdot x}}{k^2 - m^2 + i\epsilon}.$$

13.8 Principal Value and Regularization

13.8.1 Principal Value Distribution

Definition 13.8.1 (Principal value). The **principal value** of $1/x$ is the distribution:

$$\left\langle \text{p.v.}\frac{1}{x}, \phi \right\rangle := \lim_{\epsilon \to 0^+} \left(\int_{-\infty}^{-\epsilon} + \int_{\epsilon}^{\infty} \right) \frac{\phi(x)}{x}\, dx.$$

Theorem 13.8.1 (Sokhotski-Plemelj formula).

$$\lim_{\epsilon \to 0^+} \frac{1}{x \pm i\epsilon} = \text{p.v.}\frac{1}{x} \mp i\pi\delta(x).$$

Corollary 13.8.2. $\frac{1}{x+i\epsilon} - \frac{1}{x-i\epsilon} = -2\pi i\delta(x)$.

13.8.2 Regularization

Definition 13.8.2 (Hadamard regularization). For divergent integrals, subtract the divergent part:

$$\text{Pf} \int_0^\infty \frac{f(x)}{x^{n+1}}\, dx := \lim_{\epsilon \to 0^+} \left[\int_\epsilon^\infty \frac{f(x)}{x^{n+1}}\, dx - \sum_{k=0}^{n-1} \frac{f^{(k)}(0)}{k!} \frac{\epsilon^{k-n}}{k-n} \right].$$

Example 13.8.1 (Dimensional regularization). In QFT, divergent loop integrals are regularized by computing in $d = 4 - \epsilon$ dimensions:

$$\int \frac{d^d k}{(k^2 + m^2)^2} = \frac{\pi^{d/2}\Gamma(2 - d/2)}{(m^2)^{2-d/2}\Gamma(2)}.$$

The pole at $d = 4$ is isolated and removed by renormalization.

13.9 Applications to Physics

13.9.1 Electrodynamics

Proposition 13.9.1 (Maxwell's equations with sources).

$$\nabla \cdot \mathbf{E} = \rho/\epsilon_0$$

$$\nabla \times \mathbf{B} - \frac{1}{c^2}\partial_t \mathbf{E} = \mu_0 \mathbf{J}$$

With potentials $\mathbf{E} = -\nabla\phi - \partial_t \mathbf{A}$, $\mathbf{B} = \nabla \times \mathbf{A}$ *in Lorenz gauge:*

$$\Box\phi = -\rho/\epsilon_0, \quad \Box\mathbf{A} = -\mu_0 \mathbf{J}.$$

Corollary 13.9.2 (Retarded potentials).

$$\phi(\mathbf{x}, t) = \frac{1}{4\pi\epsilon_0} \int \frac{\rho(\mathbf{y}, t_{\text{ret}})}{|\mathbf{x} - \mathbf{y}|} \, d^3y$$

where $t_{\text{ret}} = t - |\mathbf{x} - \mathbf{y}|/c$.

13.9.2 Quantum Scattering

Theorem 13.9.3 (Lippmann-Schwinger equation). *For scattering from potential* V*, the scattering state satisfies:*

$$|\psi^{(+)}\rangle = |\phi\rangle + G_0^{(+)}(E)V|\psi^{(+)}\rangle$$

where $G_0^{(+)}(E) = (E - H_0 + i\epsilon)^{-1}$ *is the free Green's operator.*

Corollary 13.9.4 (Born series). *Iterating:* $|\psi^{(+)}\rangle = |\phi\rangle + G_0^{(+)}V|\phi\rangle + G_0^{(+)}VG_0^{(+)}V|\phi\rangle + \ldots$

Theorem 13.9.5 (Optical theorem).

$$\text{Im}\, f(0) = \frac{k}{4\pi}\sigma_{\text{tot}}$$

relating forward scattering amplitude to total cross section.

13.9.3 Statistical Mechanics

Definition 13.9.1 (Matsubara Green's function). At finite temperature $T = 1/\beta$:

$$\mathcal{G}(\mathbf{x}, \tau) = -\langle T_\tau \psi(\mathbf{x}, \tau)\psi^\dagger(\mathbf{0}, 0)\rangle$$

where $\tau \in [0, \beta]$ is imaginary time.

Proposition 13.9.6. *Matsubara Green's functions have discrete frequencies* $\omega_n = \pi n/\beta$ *(bosons) or* $\omega_n = \pi(2n + 1)/\beta$ *(fermions).*

13.10 Summary

Distributions extend classical analysis to handle singularities:

Distributions: Linear functionals on test functions; include delta functions and derivatives of discontinuous functions.

Dirac delta: $\langle \delta, \phi \rangle = \phi(0)$; fundamental for point sources and sifting.

Operations: Differentiation always defined; Fourier transform extends to $\mathcal{S}'$.

Green's functions: Solutions to $LG = \delta$; general solutions via convolution $u = G * f$.

Propagators: Green's functions for time-dependent problems; causal structure encoded in $i\epsilon$ prescription.

Applications: Electrodynamics, quantum mechanics, QFT, statistical mechanics.

Exercises

Exercise 13.10.1. Show that $\langle \delta, \phi \rangle = \phi(0)$ defines a continuous linear functional on $\mathcal{D}(\mathbb{R})$.

Exercise 13.10.2. Prove that $x\delta(x) = 0$ in $\mathcal{D}'$.

Exercise 13.10.3. Compute the distributional derivative of $|x|$.

Exercise 13.10.4. Verify that $\delta(x^2 - a^2) = \frac{1}{2|a|}[\delta(x-a) + \delta(x+a)]$ for $a \neq 0$.

Exercise 13.10.5. Show that $\mathcal{F}[\delta] = 1$ and $\mathcal{F}[1] = 2\pi\delta$.

Exercise 13.10.6. Derive the Green's function for $-\frac{d^2}{dx^2} + k^2$ on $\mathbb{R}$ and verify it satisfies the equation.

Exercise 13.10.7. Use the method of images to find the Green's function for the Laplacian in the first quadrant with Dirichlet boundary conditions.

Exercise 13.10.8. Verify that the heat kernel $G(\mathbf{x}, t) = (4\pi Dt)^{-n/2} e^{-|\mathbf{x}|^2/4Dt}$ satisfies $\partial_t G = D\Delta G$ and $G \to \delta$ as $t \to 0^+$.

Exercise 13.10.9. Prove the Sokhotski-Plemelj formula.

Exercise 13.10.10. Show that the retarded Green's function for the wave equation satisfies $G_{\text{ret}}(\mathbf{x}, t) = 0$ for $t < |\mathbf{x}|/c$.

Exercise 13.10.11. Derive the Born approximation for the scattering amplitude from the Lippmann-Schwinger equation.

Exercise 13.10.12. Show that the Feynman propagator $G_F(x - y)$ equals the time-ordered product $\langle 0|T\phi(x)\phi(y)|0\rangle$ for the free scalar field.

Part V

Differential Equations

Chapter 14

Ordinary Differential Equations

Differential equations are the language of dynamics. From Newton's laws to quantum mechanics, physical systems are described by equations relating quantities to their rates of change. This chapter develops the theory of ordinary differential equations, with emphasis on the analytical and qualitative methods most useful in physics.

14.1 First-Order Equations

14.1.1 Basic Concepts

Definition 14.1.1 (Ordinary differential equation). An **ordinary differential equation** (ODE) of order n is an equation:

$$F(x, y, y', y'', \ldots, y^{(n)}) = 0$$

where $y = y(x)$ is the unknown function.

Definition 14.1.2 (Solution). A **solution** on interval I is a function $y : I \to \mathbb{R}$ that satisfies the equation.

The **general solution** contains arbitrary constants; a **particular solution** has specific values.

Definition 14.1.3 (Initial value problem). An **initial value problem** (IVP) consists of an ODE plus initial conditions:

$$y^{(n)} = f(x, y, y', \ldots, y^{(n-1)}), \quad y(x_0) = y_0, y'(x_0) = y_1, \ldots$$

14.1.2 Separable Equations

Definition 14.1.4 (Separable equation). $\frac{dy}{dx} = g(x)h(y)$ is **separable**: variables can be separated.

Theorem 14.1.1 (Solution method). *If $h(y) \neq 0$:*

$$\int \frac{dy}{h(y)} = \int g(x)\, dx + C.$$

Example 14.1.1. $\frac{dy}{dx} = xy$: $\int \frac{dy}{y} = \int x\, dx$ gives $\ln|y| = \frac{x^2}{2} + C$, so $y = Ae^{x^2/2}$.

14.1.3 Linear First-Order Equations

Definition 14.1.5 (Linear first-order ODE). $y' + p(x)y = q(x)$ is **linear** (in y).

Theorem 14.1.2 (Integrating factor method). *Multiply by integrating factor $\mu(x) = e^{\int p(x)\, dx}$:*

$$(\mu y)' = \mu q \implies y = \frac{1}{\mu}\left(\int \mu q\, dx + C\right).$$

Example 14.1.2. $y' + 2xy = x$: $\mu = e^{x^2}$, so $y = e^{-x^2} \int xe^{x^2} dx = e^{-x^2}(\frac{1}{2}e^{x^2} + C) = \frac{1}{2} + Ce^{-x^2}$.

14.1.4 Exact Equations

Definition 14.1.6 (Exact equation). $M(x,y)dx + N(x,y)dy = 0$ is **exact** if $\frac{\partial M}{\partial y} = \frac{\partial N}{\partial x}$.

Theorem 14.1.3. *If exact, there exists $F(x,y)$ with $\frac{\partial F}{\partial x} = M$, $\frac{\partial F}{\partial y} = N$, and solutions satisfy $F(x,y) = C$.*

14.1.5 Existence and Uniqueness

Theorem 14.1.4 (Picard-Lindelöf theorem). *Consider $y' = f(x,y)$, $y(x_0) = y_0$. If f is continuous and Lipschitz in y on a rectangle containing (x_0, y_0), then there exists a unique local solution.*

Definition 14.1.7 (Lipschitz condition). f is **Lipschitz in** y if $|f(x,y_1) - f(x,y_2)| \leq L|y_1 - y_2|$ for some $L > 0$.

Example 14.1.3 (Non-uniqueness). $y' = y^{1/3}$, $y(0) = 0$ has solutions $y = 0$ and $y = (\frac{2x}{3})^{3/2}$. Here $f(y) = y^{1/3}$ is not Lipschitz at $y = 0$.

14.2 Second-Order Linear Equations

14.2.1 Homogeneous Equations

Definition 14.2.1 (Second-order linear ODE).

$$y'' + p(x)y' + q(x)y = r(x).$$

Homogeneous if $r(x) = 0$; **inhomogeneous** otherwise.

Theorem 14.2.1 (Superposition principle). *If y_1, y_2 are solutions to the homogeneous equation, so is $c_1y_1 + c_2y_2$.*

Definition 14.2.2 (Wronskian). The **Wronskian** of y_1, y_2 is:

$$W(y_1, y_2) := \begin{vmatrix} y_1 & y_2 \\ y_1' & y_2' \end{vmatrix} = y_1y_2' - y_2y_1'.$$

Theorem 14.2.2 (Linear independence). *y_1, y_2 are linearly independent solutions iff $W(y_1, y_2) \neq 0$.*

Theorem 14.2.3 (Abel's identity). $W(x) = W(x_0)\exp\left(-\int_{x_0}^{x} p(t)\,dt\right)$.

Theorem 14.2.4 (General solution). *If y_1, y_2 are linearly independent solutions to the homogeneous equation:*

$$y_{\text{gen}} = c_1y_1 + c_2y_2.$$

14.2.2 Constant Coefficients

Theorem 14.2.5 (Constant coefficient homogeneous). *For $ay''+by'+cy = 0$, the **characteristic equation** is $ar^2 + br + c = 0$.*

- *Two real roots $r_1 \neq r_2$: $y = c_1e^{r_1x} + c_2e^{r_2x}$.*
- *Repeated root r: $y = (c_1 + c_2x)e^{rx}$.*
- *Complex roots $\alpha \pm i\beta$: $y = e^{\alpha x}(c_1\cos\beta x + c_2\sin\beta x)$.*

Example 14.2.1 (Harmonic oscillator). $y'' + \omega^2y = 0$: roots $\pm i\omega$, so $y = A\cos\omega x + B\sin\omega x = C\cos(\omega x - \phi)$.

Example 14.2.2 (Damped oscillator). $y'' + 2\gamma y' + \omega_0^2y = 0$: roots $-\gamma \pm \sqrt{\gamma^2 - \omega_0^2}$.

- Underdamped ($\gamma < \omega_0$): $y = e^{-\gamma x}(A\cos\omega_dx + B\sin\omega_dx)$, $\omega_d = \sqrt{\omega_0^2 - \gamma^2}$.
- Critically damped ($\gamma = \omega_0$): $y = (A + Bx)e^{-\gamma x}$.
- Overdamped ($\gamma > \omega_0$): $y = Ae^{r_1x} + Be^{r_2x}$, both $r_i < 0$.

14.2.3 Inhomogeneous Equations

Theorem 14.2.6 (Structure of general solution). *$y_{gen} = y_h + y_p$ where y_h is general homogeneous solution and y_p is any particular solution.*

Theorem 14.2.7 (Method of undetermined coefficients). *For $ay'' + by' + cy = r(x)$ with $r(x)$ a polynomial, exponential, sine/cosine, or product thereof, guess y_p of similar form and solve for coefficients.*

Example 14.2.3. $y''-3y'+2y = e^{3x}$: guess $y_p = Ae^{3x}$, get $9A-9A+2A = 1$, so $A = 1/2$.

Theorem 14.2.8 (Variation of parameters). *For $y'' + p(x)y' + q(x)y = r(x)$ with homogeneous solutions y_1, y_2:*

$$y_p = -y_1 \int \frac{y_2 r}{W}\, dx + y_2 \int \frac{y_1 r}{W}\, dx.$$

14.2.4 Reduction of Order

Theorem 14.2.9 (Reduction of order). *If y_1 is a solution to $y'' + p(x)y' + q(x)y = 0$, a second independent solution is:*

$$y_2 = y_1 \int \frac{e^{-\int p\,dx}}{y_1^2}\, dx.$$

14.3 Series Solutions

14.3.1 Power Series Method

Definition 14.3.1 (Ordinary and singular points). For $y''+P(x)y'+Q(x)y = 0$, point x_0 is:

- **Ordinary** if P, Q are analytic at x_0.
- **Singular** otherwise.
- **Regular singular** if $(x - x_0)P$ and $(x - x_0)^2 Q$ are analytic.
- **Irregular singular** otherwise.

Theorem 14.3.1 (Power series solution). *Near an ordinary point x_0, solutions have the form:*

$$y = \sum_{n=0}^{\infty} a_n (x - x_0)^n$$

with radius of convergence at least to the nearest singular point.

Example 14.3.1 (Airy equation). $y'' - xy = 0$: substitute $y = \sum a_n x^n$, get recurrence $a_{n+2} = \frac{a_{n-1}}{(n+1)(n+2)}$.

Solutions: Airy functions $\mathrm{Ai}(x)$, $\mathrm{Bi}(x)$.

14.3.2 Frobenius Method

Theorem 14.3.2 (Frobenius method). *Near a regular singular point $x_0 = 0$, seek solutions:*

$$y = x^r \sum_{n=0}^{\infty} a_n x^n = \sum_{n=0}^{\infty} a_n x^{n+r}.$$

The ***indicial equation*** *determines r.*

Theorem 14.3.3 (Fuchs' theorem). *Near a regular singular point:*

- *If indicial roots r_1, r_2 differ by non-integer: two Frobenius solutions.*
- *If $r_1 = r_2$: one Frobenius, one with $\ln x$ term.*
- *If $r_1 - r_2 \in \mathbb{Z}^+$: may need $\ln x$ term for second solution.*

Example 14.3.2 (Bessel's equation). $x^2 y'' + xy' + (x^2 - \nu^2)y = 0$: regular singular at $x = 0$.

Indicial equation: $r^2 - \nu^2 = 0$, so $r = \pm\nu$.

Solutions: Bessel functions $J_\nu(x)$, $Y_\nu(x)$ (or $J_{-\nu}$ if $\nu \notin \mathbb{Z}$).

14.4 Special Functions

14.4.1 Bessel Functions

Definition 14.4.1 (Bessel functions of first kind).

$$J_\nu(x) = \sum_{k=0}^{\infty} \frac{(-1)^k}{k!\,\Gamma(k+\nu+1)} \left(\frac{x}{2}\right)^{2k+\nu}.$$

Proposition 14.4.1 (Properties). *(i) $J_{-n}(x) = (-1)^n J_n(x)$ for integer n.*

(ii) $\frac{d}{dx}[x^\nu J_\nu(x)] = x^\nu J_{\nu-1}(x)$.

(iii) $\frac{d}{dx}[x^{-\nu} J_\nu(x)] = -x^{-\nu} J_{\nu+1}(x)$.

(iv) $J_{\nu-1}(x) + J_{\nu+1}(x) = \frac{2\nu}{x} J_\nu(x)$.

Definition 14.4.2 (Bessel functions of second kind). $Y_\nu(x) = \frac{J_\nu(x)\cos\nu\pi - J_{-\nu}(x)}{\sin\nu\pi}$ (limit for integer ν).

Example 14.4.1 (Physical applications).

- Vibrating circular membrane: $J_n(k_{nm}r)\cos(n\theta)$.
- Cylindrical waveguides: J_n, Y_n for TE/TM modes.
- Diffraction: J_1 in Airy disk pattern.

14.4.2 Legendre Functions

Definition 14.4.3 (Legendre's equation). $(1-x^2)y'' - 2xy' + \ell(\ell+1)y = 0$.

Definition 14.4.4 (Legendre polynomials). For non-negative integer ℓ:

$$P_\ell(x) = \frac{1}{2^\ell \ell!}\frac{d^\ell}{dx^\ell}(x^2-1)^\ell \quad \text{(Rodrigues' formula)}.$$

Proposition 14.4.2 (Properties).

(i) $P_\ell(1) = 1$, $P_\ell(-1) = (-1)^\ell$.

(ii) Orthogonality: $\int_{-1}^{1} P_\ell(x)P_m(x)\,dx = \frac{2}{2\ell+1}\delta_{\ell m}$.

(iii) Recurrence: $(\ell+1)P_{\ell+1} = (2\ell+1)xP_\ell - \ell P_{\ell-1}$.

(iv) Generating function: $\frac{1}{\sqrt{1-2xt+t^2}} = \sum_{\ell=0}^{\infty} P_\ell(x)t^\ell$.

Definition 14.4.5 (Associated Legendre functions). $P_\ell^m(x) = (-1)^m(1-x^2)^{m/2}\frac{d^m}{dx^m}P_\ell(x)$ for $0 \le m \le \ell$.

14.4.3 Hermite and Laguerre Functions

Definition 14.4.6 (Hermite's equation). $y'' - 2xy' + 2ny = 0$.

Definition 14.4.7 (Hermite polynomials).

$$H_n(x) = (-1)^n e^{x^2}\frac{d^n}{dx^n}e^{-x^2}.$$

Proposition 14.4.3. *Orthogonality:* $\int_{-\infty}^{\infty} H_m(x)H_n(x)e^{-x^2}\,dx = \sqrt{\pi}2^n n!\delta_{mn}$.

Example 14.4.2. Quantum harmonic oscillator eigenfunctions: $\psi_n(x) \propto H_n(x)e^{-x^2/2}$.

Definition 14.4.8 (Laguerre polynomials).

$$L_n(x) = e^x\frac{d^n}{dx^n}(x^n e^{-x}) = \sum_{k=0}^{n}\binom{n}{k}\frac{(-x)^k}{k!}.$$

Definition 14.4.9 (Associated Laguerre polynomials). $L_n^k(x) = (-1)^k\frac{d^k}{dx^k}L_{n+k}(x)$.

Example 14.4.3. Hydrogen atom radial functions: $R_{n\ell}(r) \propto e^{-r/na_0}(r/a_0)^\ell L_{n-\ell-1}^{2\ell+1}(2r/n$

14.4.4 Hypergeometric Functions

Definition 14.4.10 (Hypergeometric equation). $x(1-x)y'' + [c-(a+b+1)x]y' - aby = 0$.

Definition 14.4.11 (Hypergeometric function).

$$ {}_2F_1(a,b;c;x) = \sum_{n=0}^{\infty} \frac{(a)_n(b)_n}{(c)_n n!} x^n $$

where $(a)_n = a(a+1)\cdots(a+n-1)$ is the Pochhammer symbol.

Proposition 14.4.4 (Special cases). *Many special functions are hypergeometric:*

$$ \begin{aligned} P_\ell(x) &= {}_2F_1(-\ell, \ell+1; 1; \frac{1-x}{2}) \\ \ln(1+x) &= x \cdot {}_2F_1(1,1;2;-x) \\ \arcsin x &= x \cdot {}_2F_1(\frac{1}{2}, \frac{1}{2}; \frac{3}{2}; x^2) \end{aligned} $$

14.5 Systems of ODEs

14.5.1 First-Order Linear Systems

Definition 14.5.1 (Linear system).

$$ \mathbf{x}' = A(t)\mathbf{x} + \mathbf{b}(t) $$

where $\mathbf{x} = (x_1, \ldots, x_n)^T$ and A is an $n \times n$ matrix.

Theorem 14.5.1 (Reduction to first-order system). *Any n-th order ODE $y^{(n)} = f(t, y, y', \ldots, y^{(n-1)})$ is equivalent to a first-order system via $x_1 = y, x_2 = y', \ldots, x_n = y^{(n-1)}$.*

14.5.2 Constant Coefficient Systems

Theorem 14.5.2 (Solution via eigenvalues). *For $\mathbf{x}' = A\mathbf{x}$ with constant A:*

- *If A has n independent eigenvectors $\mathbf{v}_i$ with eigenvalues λ_i:*

$$ \mathbf{x}(t) = \sum_{i=1}^{n} c_i e^{\lambda_i t} \mathbf{v}_i. $$

- *Complex eigenvalues $\alpha \pm i\beta$ give real solutions involving $e^{\alpha t}\cos\beta t$, $e^{\alpha t}\sin\beta t$.*

- *Repeated eigenvalues may require generalized eigenvectors.*

Definition 14.5.2 (Matrix exponential). $e^{At} := \sum_{n=0}^{\infty} \frac{(At)^n}{n!}$.

Theorem 14.5.3. *The solution to* $\mathbf{x}' = A\mathbf{x}$, $\mathbf{x}(0) = \mathbf{x}_0$ *is* $\mathbf{x}(t) = e^{At}\mathbf{x}_0$.

Proposition 14.5.4 (Computing e^{At}).

- *If* $A = PDP^{-1}$ *diagonalizable:* $e^{At} = Pe^{Dt}P^{-1}$.
- *If* A *is* 2×2*: use Cayley-Hamilton or direct series.*
- *Laplace transform:* $e^{At} = \mathcal{L}^{-1}[(sI - A)^{-1}]$.

14.5.3 Phase Plane Analysis

Definition 14.5.3 (Phase portrait). The **phase portrait** shows solution trajectories in the (x_1, x_2) plane.

Theorem 14.5.5 (Classification of 2D linear systems). *For* $\mathbf{x}' = A\mathbf{x}$ *with eigenvalues* λ_1, λ_2*:*

Eigenvalues	*Fixed Point Type*
$\lambda_1, \lambda_2 > 0$	*Unstable node*
$\lambda_1, \lambda_2 < 0$	*Stable node*
$\lambda_1 > 0 > \lambda_2$	*Saddle*
$\alpha \pm i\beta$, $\alpha > 0$	*Unstable spiral*
$\alpha \pm i\beta$, $\alpha < 0$	*Stable spiral*
$\pm i\beta$ *(purely imaginary)*	*Center*

14.6 Qualitative Theory

14.6.1 Existence and Uniqueness

Theorem 14.6.1 (Picard-Lindelöf for systems). *For* $\mathbf{x}' = \mathbf{f}(t, \mathbf{x})$ *with* $\mathbf{f}$ *continuous and locally Lipschitz in* $\mathbf{x}$*, local existence and uniqueness hold.*

Theorem 14.6.2 (Maximal solutions). *Solutions extend until they leave every compact set (reach infinity or boundary in finite time).*

14.6.2 Stability

Definition 14.6.1 (Stability). An equilibrium $\mathbf{x}^*$ (where $\mathbf{f}(\mathbf{x}^*) = 0$) is:

- **Stable** if nearby solutions stay nearby.
- **Asymptotically stable** if stable and nearby solutions converge to $\mathbf{x}^*$.
- **Unstable** if not stable.

Theorem 14.6.3 (Linearization). *For $\mathbf{x}' = \mathbf{f}(\mathbf{x})$ near equilibrium $\mathbf{x}^*$, let $A = D\mathbf{f}(\mathbf{x}^*)$ be the Jacobian.*

- *If all eigenvalues of A have $\text{Re}(\lambda) < 0$: asymptotically stable.*
- *If any eigenvalue has $\text{Re}(\lambda) > 0$: unstable.*
- *If all $\text{Re}(\lambda) \leq 0$ with some $= 0$: linearization inconclusive.*

Definition 14.6.2 (Lyapunov function). A **Lyapunov function** for $\mathbf{x}' = \mathbf{f}(\mathbf{x})$ near $\mathbf{x}^*$ is $V : U \to \mathbb{R}$ with:

(i) $V(\mathbf{x}^*) = 0$, $V(\mathbf{x}) > 0$ for $\mathbf{x} \neq \mathbf{x}^*$.

(ii) $\dot{V} = \nabla V \cdot \mathbf{f} \leq 0$.

Theorem 14.6.4 (Lyapunov stability theorem). *If a Lyapunov function exists with $\dot{V} \leq 0$: stable.*

If $\dot{V} < 0$ for $\mathbf{x} \neq \mathbf{x}^*$: asymptotically stable.

Example 14.6.1 (Energy as Lyapunov function). For $\ddot{x} + \gamma\dot{x} + \omega^2 x = 0$ (damped oscillator), $V = \frac{1}{2}\dot{x}^2 + \frac{1}{2}\omega^2 x^2$ gives $\dot{V} = -\gamma\dot{x}^2 \leq 0$.

14.6.3 Limit Cycles and Bifurcations

Definition 14.6.3 (Limit cycle). A **limit cycle** is an isolated periodic orbit.

Theorem 14.6.5 (Poincaré-Bendixson). *For a planar system, if a trajectory stays in a bounded region with no equilibria, it approaches a limit cycle.*

Definition 14.6.4 (Bifurcation). A **bifurcation** occurs when the qualitative behavior changes as a parameter varies.

Example 14.6.2 (Common bifurcations).

- **Saddle-node**: $\dot{x} = r - x^2$ (equilibria appear/disappear at $r = 0$).
- **Pitchfork**: $\dot{x} = rx - x^3$ (symmetry breaking at $r = 0$).
- **Hopf**: equilibrium loses stability, limit cycle appears.

14.7 Sturm-Liouville Theory

14.7.1 Sturm-Liouville Problems

Definition 14.7.1 (Sturm-Liouville equation).

$$\frac{d}{dx}\left[p(x)\frac{dy}{dx}\right] + [q(x) + \lambda w(x)]y = 0$$

on $[a, b]$ with $p > 0$, $w > 0$ (weight function), and boundary conditions.

Theorem 14.7.1 (Sturm-Liouville eigenvalue properties). *(i) Eigenvalues λ_n are real and can be ordered $\lambda_1 < \lambda_2 < \cdots \to \infty$.*

(ii) Eigenfunctions y_n are orthogonal with weight w: $\int_a^b y_m y_n w\, dx = 0$ for $m \neq n$.

(iii) Eigenfunctions form a complete basis for $L^2([a, b], w)$.

(iv) The n-th eigenfunction has exactly $n - 1$ zeros in (a, b).

Example 14.7.1 (Fourier series). $y'' + \lambda y = 0$ on $[0, L]$ with $y(0) = y(L) = 0$: $\lambda_n = (n\pi/L)^2$, $y_n = \sin(n\pi x/L)$.

Example 14.7.2 (Bessel equation as S-L). $\frac{d}{dx}[x\frac{dy}{dx}] + (\lambda x - \frac{\nu^2}{x})y = 0$ with appropriate BCs gives Bessel function eigenvalue problems.

14.7.2 Eigenfunction Expansions

Theorem 14.7.2 (Eigenfunction expansion). *Any $f \in L^2([a, b], w)$ has expansion:*

$$f(x) = \sum_{n=1}^{\infty} c_n y_n(x), \quad c_n = \frac{\int_a^b f(x) y_n(x) w(x)\, dx}{\int_a^b y_n(x)^2 w(x)\, dx}.$$

14.8 Applications to Physics

14.8.1 Classical Mechanics

Example 14.8.1 (Simple pendulum). $\ddot{\theta} + \frac{g}{L}\sin\theta = 0$.
Small angle: $\ddot{\theta} + \omega^2\theta = 0$, $\omega = \sqrt{g/L}$.
Phase portrait has center at origin, separatrix through $(\pm\pi, 0)$.

Example 14.8.2 (Kepler problem). $\ddot{\mathbf{r}} = -\frac{GM}{r^3}\mathbf{r}$.
Conservation of $\mathbf{L} = \mathbf{r} \times \dot{\mathbf{r}}$ and E gives conic section orbits.

14.8.2 Electrical Circuits

Example 14.8.3 (RLC circuit). $L\ddot{Q} + R\dot{Q} + \frac{Q}{C} = V(t)$.

Same as damped driven oscillator with $\gamma = R/2L$, $\omega_0 = 1/\sqrt{LC}$.

14.8.3 Quantum Mechanics

Example 14.8.4 (Time-independent Schrödinger equation). $-\frac{\hbar^2}{2m}\psi'' + V(x)\psi = E\psi$.

Sturm-Liouville problem; bound states form discrete spectrum.

14.9 Summary

Ordinary differential equations describe dynamical systems:

First-order: Separable, linear, exact equations; Picard-Lindelöf existence/uniqueness.

Second-order linear: Superposition, Wronskian, constant coefficients via characteristic equation, variation of parameters.

Series solutions: Power series at ordinary points, Frobenius at regular singular points; special functions (Bessel, Legendre, Hermite, Laguerre).

Systems: Matrix exponential, eigenvalue methods, phase portraits.

Qualitative theory: Stability, Lyapunov functions, linearization, bifurcations, limit cycles.

Sturm-Liouville: Eigenvalue problems, orthogonal eigenfunctions, completeness.

Exercises

Exercise 14.9.1. Solve $y' = y^2$ with $y(0) = 1$ and find the maximal interval of existence.

Exercise 14.9.2. Find the general solution of $y'' - 4y' + 4y = e^{2x}$.

Exercise 14.9.3. Use the Frobenius method to find solutions to Bessel's equation $x^2y'' + xy' + (x^2 - \nu^2)y = 0$ near $x = 0$.

Exercise 14.9.4. Verify that $P_2(x) = \frac{1}{2}(3x^2 - 1)$ satisfies Legendre's equation with $\ell = 2$.

Exercise 14.9.5. Find the general solution of $\mathbf{x}' = \begin{pmatrix} 1 & 1 \\ 4 & 1 \end{pmatrix} \mathbf{x}$.

Exercise 14.9.6. Classify the equilibrium at the origin for $\dot{x} = y$, $\dot{y} = -x - 0.1y$.

Exercise 14.9.7. Show that $V = x^2 + y^2$ is a Lyapunov function for $\dot{x} = -x + y^2$, $\dot{y} = -y$ near the origin.

Exercise 14.9.8. Find the eigenvalues and eigenfunctions for $y'' + \lambda y = 0$ with $y(0) = 0$, $y'(\pi) = 0$.

Exercise 14.9.9. Analyze the bifurcations in $\dot{x} = rx + x^3 - x^5$ as r varies.

Exercise 14.9.10. Solve the damped driven oscillator $\ddot{x} + 2\gamma\dot{x} + \omega_0^2 x = F_0 \cos \omega t$ and find the resonance frequency.

Exercise 14.9.11. Show that the Legendre polynomials satisfy $\int_{-1}^{1} P_m(x)P_n(x)\,dx = \frac{2}{2n+1}\delta_{mn}$.

Exercise 14.9.12. For the nonlinear pendulum $\ddot{\theta} + \sin\theta = 0$, sketch the phase portrait and identify all equilibria and their stability.

Chapter 15

Partial Differential Equations

Partial differential equations (PDEs) describe systems with multiple independent variables—typically space and time. The fundamental equations of physics are PDEs: Maxwell's equations, the Schrödinger equation, Einstein's field equations, the Navier-Stokes equations. This chapter develops the theory and solution methods for the most important classes of PDEs in theoretical physics.

15.1 Introduction and Classification

15.1.1 Basic Definitions

Definition 15.1.1 (Partial differential equation). A **partial differential equation** (PDE) is an equation involving a function $u(x_1, \ldots, x_n)$ and its partial derivatives:

$$F\left(x_1, \ldots, x_n, u, \frac{\partial u}{\partial x_1}, \ldots, \frac{\partial^2 u}{\partial x_1^2}, \ldots\right) = 0.$$

The **order** is the highest derivative appearing.

Definition 15.1.2 (Linear PDE). A PDE is **linear** if it is linear in u and all its derivatives:

$$\sum_{|\alpha| \leq k} a_\alpha(x) D^\alpha u = f(x).$$

It is **homogeneous** if $f = 0$.

Definition 15.1.3 (Quasilinear and semilinear). A PDE is **quasilinear** if linear in highest-order derivatives (coefficients may depend on lower derivatives). It is **semilinear** if coefficients of highest derivatives depend only on independent variables.

15.1.2 Classification of Second-Order Linear PDEs

Definition 15.1.4 (Second-order linear PDE). The general second-order linear PDE in two variables:

$$a(x,y)u_{xx}+2b(x,y)u_{xy}+c(x,y)u_{yy}+d(x,y)u_x+e(x,y)u_y+f(x,y)u=g(x,y).$$

Definition 15.1.5 (Classification by discriminant). The discriminant $\Delta = b^2 - ac$ determines the type:

- $\Delta < 0$: **Elliptic** (e.g., Laplace equation).
- $\Delta = 0$: **Parabolic** (e.g., heat equation).
- $\Delta > 0$: **Hyperbolic** (e.g., wave equation).

Theorem 15.1.1 (Canonical forms). *By change of variables, second-order PDEs reduce to:*

- *Elliptic:* $u_{\xi\xi} + u_{\eta\eta} + \cdots = 0$.
- *Parabolic:* $u_{\xi\xi} + \cdots = u_\tau$.
- *Hyperbolic:* $u_{\xi\xi} - u_{\eta\eta} + \cdots = 0$ *or* $u_{\xi\eta} + \cdots = 0$.

Example 15.1.1 (The fundamental PDEs of physics).

Equation	Form	Type	Physics
Laplace	$\nabla^2 u = 0$	Elliptic	Electrostatics, steady state
Poisson	$\nabla^2 u = f$	Elliptic	Gravity, electrostatics
Heat	$u_t = \alpha\nabla^2 u$	Parabolic	Diffusion, conduction
Wave	$u_{tt} = c^2\nabla^2 u$	Hyperbolic	Waves, vibrations
Schrödinger	$i\hbar u_t = -\frac{\hbar^2}{2m}\nabla^2 u + Vu$	Parabolic	Quantum mechanics

15.1.3 Well-Posed Problems

Definition 15.1.6 (Well-posed problem). A problem is **well-posed** (in the sense of Hadamard) if:

(i) A solution exists.

(ii) The solution is unique.

(iii) The solution depends continuously on the data.

Example 15.1.2 (Appropriate boundary conditions).

- **Elliptic**: boundary conditions on entire boundary (Dirichlet, Neumann, Robin).
- **Parabolic**: initial condition + boundary conditions.
- **Hyperbolic**: initial conditions (position and velocity) + boundary conditions.

15.2 First-Order PDEs

15.2.1 Method of Characteristics

Definition 15.2.1 (First-order quasilinear PDE). $a(x, y, u)u_x + b(x, y, u)u_y = c(x, y, u)$.

Theorem 15.2.1 (Method of characteristics). *Solutions are found by solving the **characteristic equations**:*

$$\frac{dx}{a} = \frac{dy}{b} = \frac{du}{c}.$$

The solution surface $u(x, y)$ is generated by characteristic curves.

Example 15.2.1 (Transport equation). $u_t + cu_x = 0$ with $u(x, 0) = f(x)$.
Characteristics: $\frac{dx}{c} = \frac{dt}{1}$, so $x - ct = \text{const}$.
Solution: $u(x, t) = f(x - ct)$ (wave traveling with speed c).

Example 15.2.2 (Inviscid Burgers equation). $u_t + uu_x = 0$ (nonlinear).
Characteristics: $\frac{dx}{u} = \frac{dt}{1}$, so $x = ut + x_0$.
Characteristics can cross, forming **shock waves**.

15.2.2 General Solution

Theorem 15.2.2 (General solution of linear first-order). *For $a(x, y)u_x + b(x, y)u_y = 0$, if $\phi(x, y) = c$ defines characteristics, then $u = F(\phi)$ for arbitrary F.*

Theorem 15.2.3 (Cauchy problem). *Given initial data $u = h(s)$ on curve $\Gamma : (x_0(s), y_0(s))$, the solution is determined if Γ is not tangent to characteristics (non-characteristic).*

15.3 The Wave Equation

15.3.1 One-Dimensional Wave Equation

Definition 15.3.1 (Wave equation).

$$u_{tt} = c^2 u_{xx}$$

describes waves propagating with speed c.

Theorem 15.3.1 (d'Alembert's solution). *The general solution of $u_{tt} = c^2 u_{xx}$ is:*

$$u(x,t) = F(x-ct) + G(x+ct)$$

representing right- and left-traveling waves.

Theorem 15.3.2 (Initial value problem). *For $u(x,0) = f(x)$, $u_t(x,0) = g(x)$ on $\mathbb{R}$:*

$$u(x,t) = \frac{1}{2}[f(x-ct) + f(x+ct)] + \frac{1}{2c}\int_{x-ct}^{x+ct} g(s)\,ds.$$

Proposition 15.3.3 (Domain of dependence). *The value $u(x_0, t_0)$ depends only on initial data in $[x_0 - ct_0, x_0 + ct_0]$.*

*The **domain of influence** of a point is the forward cone bounded by characteristics.*

15.3.2 Higher Dimensions

Theorem 15.3.4 (Spherical means). *In 3D, the solution to $u_{tt} = c^2\nabla^2 u$ with $u(x,0) = f(x)$, $u_t(x,0) = g(x)$:*

$$u(\mathbf{x},t) = \frac{\partial}{\partial t}\left[\frac{t}{4\pi}\int_{|\mathbf{y}|=1} f(\mathbf{x} + ct\mathbf{y})\,dS\right] + \frac{t}{4\pi}\int_{|\mathbf{y}|=1} g(\mathbf{x} + ct\mathbf{y})\,dS.$$

Theorem 15.3.5 (Huygens' principle). *In odd dimensions $n \geq 3$, signals propagate on the light cone only (sharp propagation). In even dimensions, signals persist inside the cone (diffusion of waves).*

15.3.3 Boundary Value Problems

Example 15.3.1 (Vibrating string). $u_{tt} = c^2 u_{xx}$ on $[0, L]$ with $u(0,t) = u(L,t) = 0$.

By separation of variables: $u(x,t) = \sum_{n=1}^{\infty} \sin\frac{n\pi x}{L}\left(A_n \cos\frac{n\pi ct}{L} + B_n \sin\frac{n\pi ct}{L}\right)$.

Normal modes with frequencies $\omega_n = n\pi c/L$.

15.4 The Heat Equation

15.4.1 Physical Derivation

Definition 15.4.1 (Heat equation).

$$u_t = \alpha \nabla^2 u$$

where $\alpha = k/(\rho c_p)$ is thermal diffusivity.

Proposition 15.4.1 (Conservation form). *Derived from conservation of energy:* $\rho c_p \frac{\partial T}{\partial t} = -\nabla \cdot \mathbf{q}$ *with Fourier's law* $\mathbf{q} = -k\nabla T$.

15.4.2 Solutions on Infinite Domain

Theorem 15.4.2 (Fundamental solution). *The fundamental solution (heat kernel) in* $\mathbb{R}^n$*:*

$$\Phi(\mathbf{x}, t) = \frac{1}{(4\pi\alpha t)^{n/2}} e^{-|\mathbf{x}|^2/4\alpha t}, \quad t > 0.$$

Satisfies $\Phi_t = \alpha\nabla^2\Phi$ *and* $\Phi \to \delta(\mathbf{x})$ *as* $t \to 0^+$.

Theorem 15.4.3 (Solution by convolution). *For initial condition* $u(\mathbf{x}, 0) = f(\mathbf{x})$*:*

$$u(\mathbf{x}, t) = \int_{\mathbb{R}^n} \Phi(\mathbf{x} - \mathbf{y}, t) f(\mathbf{y})\, d^n y.$$

Proposition 15.4.4 (Properties). *(i)* ***Smoothing****:* $u(\mathbf{x}, t)$ *is* C^∞ *for* $t > 0$ *even if* f *is rough.*

(ii) ***Infinite speed****: Effects propagate instantaneously (non-relativistic).*

(iii) ***Maximum principle****:* $\max u$ *occurs on boundary or initial surface.*

(iv) ***Irreversibility****: Information is lost; backward heat equation is ill-posed.*

15.4.3 Boundary Value Problems

Example 15.4.1 (Heat conduction in a rod). $u_t = \alpha u_{xx}$ on $[0, L]$ with $u(0, t) = u(L, t) = 0$, $u(x, 0) = f(x)$.

Solution: $u(x, t) = \sum_{n=1}^{\infty} B_n \sin \frac{n\pi x}{L} e^{-\alpha(n\pi/L)^2 t}$
where $B_n = \frac{2}{L}\int_0^L f(x) \sin \frac{n\pi x}{L}\, dx$.

Proposition 15.4.5 (Decay to equilibrium). *Higher modes decay faster (*$\propto e^{-n^2 t}$*). Solution approaches steady state (or zero for homogeneous BC).*

15.5 Laplace and Poisson Equations

15.5.1 Laplace's Equation

Definition 15.5.1 (Laplace equation).

$$\nabla^2 u = 0.$$

Solutions are called **harmonic functions**.

Theorem 15.5.1 (Mean value property). *u is harmonic iff for every ball $B_r(\mathbf{x}_0)$ in the domain:*

$$u(\mathbf{x}_0) = \frac{1}{|S_r|}\int_{S_r(\mathbf{x}_0)} u\,dS = \frac{1}{|B_r|}\int_{B_r(\mathbf{x}_0)} u\,dV.$$

Theorem 15.5.2 (Maximum principle). *A harmonic function on a bounded domain attains its maximum and minimum on the boundary.*

Corollary 15.5.3 (Uniqueness). *The Dirichlet problem $\nabla^2 u = 0$ with $u|_{\partial\Omega} = f$ has at most one solution.*

15.5.2 Poisson's Equation

Definition 15.5.2 (Poisson equation).

$$\nabla^2 u = f.$$

Theorem 15.5.4 (Solution via Green's function).

$$u(\mathbf{x}) = \int_\Omega G(\mathbf{x}, \mathbf{y}) f(\mathbf{y})\, d^n y + \textit{boundary terms}$$

where G is the Green's function for the domain.

Example 15.5.1 (Fundamental solution). In $\mathbb{R}^3$: $\nabla^2 u = \delta(\mathbf{x})$ has solution $u = -\frac{1}{4\pi|\mathbf{x}|}$.

Thus $\nabla^2 u = f$ gives $u(\mathbf{x}) = -\frac{1}{4\pi}\int \frac{f(\mathbf{y})}{|\mathbf{x}-\mathbf{y}|}\, d^3y$.

15.5.3 Boundary Conditions

Definition 15.5.3 (Types of boundary conditions).

- **Dirichlet**: $u = f$ on $\partial\Omega$.
- **Neumann**: $\frac{\partial u}{\partial n} = g$ on $\partial\Omega$.

- **Robin** (mixed): $au + b\frac{\partial u}{\partial n} = h$ on $\partial\Omega$.

Theorem 15.5.5 (Existence and uniqueness).

- *Dirichlet: unique solution exists.*

- *Neumann: solution exists iff* $\int_{\partial\Omega} g\,dS = \int_\Omega f\,dV$ *(compatibility); unique up to constant.*

15.6 Separation of Variables

15.6.1 The Method

Theorem 15.6.1 (Separation of variables). *For linear homogeneous PDEs with homogeneous BC, seek solutions of the form:*

$$u(x_1, \ldots, x_n) = X_1(x_1) \cdots X_n(x_n).$$

The PDE separates into ODEs connected by separation constants.

Example 15.6.1 (Heat equation on rectangle). $u_t = \alpha(u_{xx} + u_{yy})$ on $[0, a] \times [0, b]$ with $u = 0$ on boundary.

Try $u = X(x)Y(y)T(t)$:

$$\frac{T'}{\alpha T} = \frac{X''}{X} + \frac{Y''}{Y} = -\lambda.$$

Further separation: $\frac{X''}{X} = -\mu$, $\frac{Y''}{Y} = -(\lambda - \mu)$.
Eigenfunctions: $X_m = \sin\frac{m\pi x}{a}$, $Y_n = \sin\frac{n\pi y}{b}$.
Solution: $u = \sum_{m,n} B_{mn} \sin\frac{m\pi x}{a} \sin\frac{n\pi y}{b} e^{-\alpha\lambda_{mn} t}$
where $\lambda_{mn} = \pi^2(m^2/a^2 + n^2/b^2)$.

15.6.2 Separation in Various Coordinate Systems

Theorem 15.6.2 (Separable coordinates). *The Laplacian separates in 11 coordinate systems in 3D, including:*

- *Cartesian: products of exponentials/trigonometric functions.*

- *Cylindrical: Bessel functions* $\times$ *trigonometric* $\times$ *exponential.*

- *Spherical: spherical harmonics* $\times$ *powers/Bessel.*

Example 15.6.2 (Laplace equation in spherical coordinates). $\nabla^2 u = 0$ with $u = R(r)\Theta(\theta)\Phi(\phi)$:

$$\Phi'' + m^2\Phi = 0$$
$$\frac{1}{\sin\theta}\frac{d}{d\theta}\left(\sin\theta\frac{d\Theta}{d\theta}\right) + \left[l(l+1) - \frac{m^2}{\sin^2\theta}\right]\Theta = 0$$
$$r^2R'' + 2rR' - l(l+1)R = 0$$

Solutions: $\Phi = e^{im\phi}$, $\Theta = P_l^m(\cos\theta)$, $R = r^l$ or $r^{-(l+1)}$.

General solution: $u = \sum_{l=0}^{\infty}\sum_{m=-l}^{l}(A_{lm}r^l + B_{lm}r^{-(l+1)})Y_l^m(\theta,\phi)$.

15.7 Fourier Transform Methods

15.7.1 Fourier Transform for PDEs

Theorem 15.7.1 (Fourier transform approach). *For PDEs on $\mathbb{R}^n$, apply Fourier transform to convert to algebraic or ODE:*

$$\mathcal{F}[\nabla^2 u] = -|\mathbf{k}|^2\hat{u}, \quad \mathcal{F}[\partial_t u] = \partial_t\hat{u}.$$

Example 15.7.1 (Heat equation on $\mathbb{R}$). $u_t = \alpha u_{xx}$, $u(x,0) = f(x)$.

Fourier transform: $\hat{u}_t = -\alpha k^2\hat{u}$, so $\hat{u}(k,t) = \hat{f}(k)e^{-\alpha k^2 t}$.

Inverse: $u(x,t) = \mathcal{F}^{-1}[\hat{f}(k)e^{-\alpha k^2 t}] = f * \Phi$ where Φ is the heat kernel.

Example 15.7.2 (Wave equation on $\mathbb{R}$). $u_{tt} = c^2u_{xx}$, $u(x,0) = f(x)$, $u_t(x,0) = g(x)$.

Fourier: $\hat{u}_{tt} = -c^2k^2\hat{u}$, general solution $\hat{u} = A(k)\cos(ckt) + B(k)\sin(ckt)$.

Initial conditions: $A = \hat{f}$, $B = \hat{g}/(ck)$.

Recovers d'Alembert's solution.

15.7.2 Dispersion Relations

Definition 15.7.1 (Dispersion relation). For wave-like solutions $u \sim e^{i(kx-\omega t)}$, the **dispersion relation** $\omega(k)$ relates frequency to wavenumber.

Example 15.7.3 (Common dispersion relations).

- Wave equation: $\omega = \pm ck$ (non-dispersive, $v_p = v_g = c$).
- Schrödinger: $\omega = \hbar k^2/2m$ (dispersive, $v_g = 2v_p$).
- Klein-Gordon: $\omega = \sqrt{c^2k^2 + m^2c^4/\hbar^2}$ (relativistic).
- Deep water waves: $\omega = \sqrt{gk}$.

Definition 15.7.2 (Phase and group velocity).

$$v_p = \frac{\omega}{k}, \quad v_g = \frac{d\omega}{dk}.$$

Wave packets travel at group velocity.

15.8 Green's Functions for PDEs

15.8.1 General Theory

Definition 15.8.1 (Green's function for PDE). For linear operator L, the Green's function satisfies:

$$L_x G(\mathbf{x}, \mathbf{x}') = \delta(\mathbf{x} - \mathbf{x}')$$

with appropriate boundary/initial conditions.

Theorem 15.8.1 (Solution via Green's function). *Solution to $Lu = f$ with homogeneous BC:*

$$u(\mathbf{x}) = \int G(\mathbf{x}, \mathbf{x}') f(\mathbf{x}')\, d^n x'.$$

15.8.2 Examples

Example 15.8.1 (Green's function for Laplacian). Free space ($\mathbb{R}^3$): $G(\mathbf{x}, \mathbf{x}') = -\frac{1}{4\pi|\mathbf{x}-\mathbf{x}'|}$.

Half-space $z > 0$ with Dirichlet BC: method of images gives

$$G = -\frac{1}{4\pi|\mathbf{x} - \mathbf{x}'|} + \frac{1}{4\pi|\mathbf{x} - \mathbf{x}''|}$$

where $\mathbf{x}'' = (x', y', -z')$.

Example 15.8.2 (Green's function for heat equation).

$$G(\mathbf{x}, t; \mathbf{x}', t') = \frac{\theta(t - t')}{[4\pi\alpha(t - t')]^{n/2}} \exp\left(-\frac{|\mathbf{x} - \mathbf{x}'|^2}{4\alpha(t - t')}\right).$$

Example 15.8.3 (Retarded Green's function for wave equation). In 3D:

$$G(\mathbf{x}, t; \mathbf{x}', t') = \frac{\delta(t - t' - |\mathbf{x} - \mathbf{x}'|/c)}{4\pi|\mathbf{x} - \mathbf{x}'|}.$$

15.9 Variational Methods

15.9.1 Calculus of Variations

Theorem 15.9.1 (Euler-Lagrange equation). *The functional $J[u] = \int_\Omega L(x, u, \nabla u)\, dx$ is stationary iff:*

$$\frac{\partial L}{\partial u} - \nabla \cdot \frac{\partial L}{\partial(\nabla u)} = 0.$$

Example 15.9.1 (Laplace equation as variational problem). $\nabla^2 u = 0$ is Euler-Lagrange for $J[u] = \frac{1}{2}\int |\nabla u|^2\, dx$ (Dirichlet energy).

Example 15.9.2 (Poisson equation). $\nabla^2 u = f$ is Euler-Lagrange for $J[u] = \frac{1}{2}\int |\nabla u|^2\, dx - \int f u\, dx$.

15.9.2 Rayleigh-Ritz Method

Theorem 15.9.2 (Rayleigh-Ritz method). *Approximate solution by $u_N = \sum_{i=1}^N c_i \phi_i(x)$ where ϕ_i satisfy boundary conditions. Minimize $J[u_N]$ over coefficients c_i.*

Theorem 15.9.3 (Eigenvalue estimation). *For $-\nabla^2 u = \lambda u$, the Rayleigh quotient:*

$$R[u] = \frac{\int |\nabla u|^2\, dx}{\int u^2\, dx}$$

satisfies $R[u] \geq \lambda_1$ with equality iff u is the first eigenfunction.

15.10 Numerical Methods

15.10.1 Finite Difference Methods

Definition 15.10.1 (Finite differences). Approximate derivatives on a grid with spacing h:

$$u'(x) \approx \frac{u(x+h) - u(x-h)}{2h} + O(h^2)$$
$$u''(x) \approx \frac{u(x+h) - 2u(x) + u(x-h)}{h^2} + O(h^2)$$

Example 15.10.1 (Heat equation). $u_t = \alpha u_{xx}$ with explicit scheme:

$$\frac{u_j^{n+1} - u_j^n}{\Delta t} = \alpha \frac{u_{j+1}^n - 2u_j^n + u_{j-1}^n}{(\Delta x)^2}.$$

Stable if $r = \alpha \Delta t / (\Delta x)^2 \leq 1/2$ (CFL condition).

Example 15.10.2 (Wave equation). $u_{tt} = c^2 u_{xx}$ with explicit scheme:

$$u_j^{n+1} = 2u_j^n - u_j^{n-1} + r^2(u_{j+1}^n - 2u_j^n + u_{j-1}^n)$$

where $r = c\Delta t/\Delta x$. Stable if $r \leq 1$.

15.10.2 Finite Element Methods

Theorem 15.10.1 (Weak formulation). *Poisson $-\nabla^2 u = f$ with Dirichlet BC is equivalent to:*

Find $u \in H_0^1(\Omega)$ such that $\int \nabla u \cdot \nabla v \, dx = \int fv \, dx$ for all $v \in H_0^1(\Omega)$.

Theorem 15.10.2 (Galerkin method). *Approximate $u \approx u_h = \sum_i c_i \phi_i$ where ϕ_i are basis functions (e.g., piecewise linear on triangulation). Solve:*

$$\sum_j K_{ij} c_j = F_i, \quad K_{ij} = \int \nabla\phi_i \cdot \nabla\phi_j \, dx, \quad F_i = \int f\phi_i \, dx.$$

15.11 Nonlinear PDEs

15.11.1 Important Nonlinear Equations

Example 15.11.1 (Nonlinear wave equations).

- **Korteweg-de Vries** (KdV): $u_t + 6uu_x + u_{xxx} = 0$ (shallow water, solitons).
- **Nonlinear Schrödinger**: $iu_t + u_{xx} + |u|^2 u = 0$ (optics, BEC).
- **Sine-Gordon**: $u_{tt} - u_{xx} + \sin u = 0$ (kinks, Josephson junctions).

Example 15.11.2 (Fluid dynamics).

- **Euler equations**: $\rho(\partial_t \mathbf{v} + \mathbf{v} \cdot \nabla\mathbf{v}) = -\nabla p$ (inviscid).
- **Navier-Stokes**: adds $\mu\nabla^2\mathbf{v}$ (viscous).
- **Burgers**: $u_t + uu_x = \nu u_{xx}$ (model for shocks).

15.11.2 Solitons

Definition 15.11.1 (Soliton). A **soliton** is a localized wave that maintains its shape and speed, even after collisions with other solitons.

Example 15.11.3 (KdV soliton)**.** The KdV equation has soliton solution:

$$u(x,t) = \frac{c}{2}\operatorname{sech}^2\left(\frac{\sqrt{c}}{2}(x - ct - x_0)\right).$$

Amplitude and speed are related: faster solitons are taller.

Theorem 15.11.1 (Inverse scattering transform)**.** *KdV and related equations are integrable: solvable by inverse scattering, with infinitely many conserved quantities.*

15.12 Applications to Physics

15.12.1 Electromagnetism

Proposition 15.12.1 (Maxwell's equations as PDEs)**.** *In vacuum with sources* $(\rho, \mathbf{J})$*:*

$$\begin{aligned} \nabla^2\phi - \frac{1}{c^2}\frac{\partial^2\phi}{\partial t^2} &= -\frac{\rho}{\epsilon_0} \\ \nabla^2\mathbf{A} - \frac{1}{c^2}\frac{\partial^2\mathbf{A}}{\partial t^2} &= -\mu_0\mathbf{J} \end{aligned}$$

(in Lorenz gauge). Wave equations with sources.

Example 15.12.1 (Electromagnetic waves)**.** In vacuum without sources: $\nabla^2\mathbf{E} = \frac{1}{c^2}\frac{\partial^2\mathbf{E}}{\partial t^2}$.

Plane wave solutions: $\mathbf{E} = \mathbf{E}_0 e^{i(\mathbf{k}\cdot\mathbf{x} - \omega t)}$ with $\omega = c|\mathbf{k}|$.

15.12.2 Quantum Mechanics

Example 15.12.2 (Time-dependent Schrödinger equation)**.** $i\hbar\frac{\partial\psi}{\partial t} = -\frac{\hbar^2}{2m}\nabla^2\psi + V\psi$.

Parabolic PDE; separates into time-independent equation for stationary states.

Example 15.12.3 (Hydrogen atom)**.** Separation in spherical coordinates gives radial equation:

$$-\frac{\hbar^2}{2m}\left[\frac{d^2R}{dr^2} + \frac{2}{r}\frac{dR}{dr}\right] + \left[V(r) + \frac{\hbar^2 l(l+1)}{2mr^2}\right]R = ER.$$

15.12.3 General Relativity

Example 15.12.4 (Einstein field equations). $R_{\mu\nu} - \frac{1}{2}Rg_{\mu\nu} + \Lambda g_{\mu\nu} = \frac{8\pi G}{c^4}T_{\mu\nu}$. Second-order nonlinear PDEs for the metric $g_{\mu\nu}$.

Example 15.12.5 (Gravitational waves). Linearized gravity in vacuum: $\Box\bar{h}_{\mu\nu} = 0$ (wave equation for metric perturbations).

15.12.4 Fluid Dynamics

Example 15.12.6 (Incompressible Navier-Stokes).

$$\frac{\partial \mathbf{v}}{\partial t} + (\mathbf{v} \cdot \nabla)\mathbf{v} = -\frac{1}{\rho}\nabla p + \nu\nabla^2\mathbf{v}$$
$$\nabla \cdot \mathbf{v} = 0$$

Nonlinear PDEs; existence and smoothness is a Millennium Prize problem.

15.13 Summary

PDEs are the mathematical language of continuum physics:

Classification: Elliptic (steady-state), parabolic (diffusion), hyperbolic (waves) have different mathematical properties and require different solution methods.

Well-posedness: Appropriate boundary/initial conditions depend on equation type.

Solution methods: Separation of variables for bounded domains with symmetry; Fourier transforms for infinite domains; Green's functions for inhomogeneous problems.

Physical equations: Wave equation (finite propagation speed), heat equation (smoothing, infinite speed), Laplace/Poisson (equilibrium).

Nonlinear PDEs: Shocks, solitons, turbulence—rich phenomena beyond linear theory.

Numerical methods: Finite differences and finite elements for computational solutions.

Exercises

Exercise 15.13.1. Classify $u_{xx} + 2u_{xy} - 3u_{yy} = 0$ and find its canonical form.

Exercise 15.13.2. Solve $u_t + cu_x = 0$ with $u(x, 0) = e^{-x^2}$ using characteristics.

Exercise 15.13.3. Derive d'Alembert's solution for the wave equation from the factorization $(\partial_t - c\partial_x)(\partial_t + c\partial_x)u = 0$.

Exercise 15.13.4. Solve the heat equation $u_t = u_{xx}$ on $[0, \pi]$ with $u(0,t) = u(\pi, t) = 0$ and $u(x, 0) = \sin x + 2\sin 3x$.

Exercise 15.13.5. Show that the mean value property implies the maximum principle for harmonic functions.

Exercise 15.13.6. Find the steady-state temperature in a disk of radius a with boundary temperature $f(\theta) = \cos^2\theta$.

Exercise 15.13.7. Use separation of variables to solve Laplace's equation inside a sphere with $u(a, \theta) = \cos\theta$.

Exercise 15.13.8. Derive the dispersion relation for the linearized KdV equation $u_t + u_{xxx} = 0$.

Exercise 15.13.9. Find the Green's function for Laplace's equation inside a sphere of radius a with Dirichlet boundary conditions.

Exercise 15.13.10. Show that the wave equation conserves energy: $E = \frac{1}{2}\int(u_t^2 + c^2u_x^2)\,dx$.

Exercise 15.13.11. Verify that the KdV soliton $u = \frac{c}{2}\text{sech}^2(\frac{\sqrt{c}}{2}(x - ct))$ satisfies $u_t + 6uu_x + u_{xxx} = 0$.

Exercise 15.13.12. Derive the CFL stability condition $c\Delta t/\Delta x \leq 1$ for the explicit wave equation scheme.

Chapter 16

Variational Methods

Variational principles lie at the heart of theoretical physics. From Fermat's principle in optics to the principle of least action in mechanics, from the Rayleigh-Ritz method in quantum mechanics to variational formulations of field theories, the idea that physical systems extremize certain quantities provides both deep insights and powerful computational tools. This chapter develops the calculus of variations and its applications throughout physics.

16.1 Calculus of Variations: Fundamentals

16.1.1 The Basic Problem

Definition 16.1.1 (Functional). A **functional** is a map from a space of functions to the real numbers:

$$J : \mathcal{F} \to \mathbb{R}, \quad y(\cdot) \mapsto J[y].$$

Example 16.1.1 (Typical functionals).

- Arc length: $J[y] = \int_a^b \sqrt{1 + y'^2}\, dx$.
- Action: $S[q] = \int_{t_1}^{t_2} L(q, \dot{q}, t)\, dt$.
- Energy: $E[u] = \int_\Omega |\nabla u|^2\, dx$.

Definition 16.1.2 (Extremal problem). Find function $y(x)$ that makes $J[y]$ stationary (extremal) subject to boundary conditions $y(a) = y_a$, $y(b) = y_b$.

16.1.2 First Variation and Euler-Lagrange Equation

Definition 16.1.3 (Variation). For $y \in \mathcal{F}$ and admissible perturbation η with $\eta(a) = \eta(b) = 0$, the **first variation** is:

$$\delta J[y;\eta] := \left.\frac{d}{d\epsilon}\right|_{\epsilon=0} J[y+\epsilon\eta] = \lim_{\epsilon\to 0}\frac{J[y+\epsilon\eta]-J[y]}{\epsilon}.$$

Definition 16.1.4 (Stationary point). y is a **stationary point** of J if $\delta J[y;\eta] = 0$ for all admissible η.

Theorem 16.1.1 (Euler-Lagrange equation). *For $J[y] = \int_a^b F(x,y,y')\,dx$, stationary points satisfy:*

$$\frac{\partial F}{\partial y} - \frac{d}{dx}\frac{\partial F}{\partial y'} = 0.$$

Proof. Compute $\delta J = \int_a^b \left(\frac{\partial F}{\partial y}\eta + \frac{\partial F}{\partial y'}\eta'\right) dx$. Integrate by parts:

$$\delta J = \int_a^b \left(\frac{\partial F}{\partial y} - \frac{d}{dx}\frac{\partial F}{\partial y'}\right)\eta\,dx + \left[\frac{\partial F}{\partial y'}\eta\right]_a^b.$$

Boundary terms vanish since $\eta(a) = \eta(b) = 0$. By the fundamental lemma of calculus of variations, the integrand must vanish. □

Lemma 16.1.2 (Fundamental lemma). *If $\int_a^b f(x)\eta(x)\,dx = 0$ for all smooth η with $\eta(a) = \eta(b) = 0$, and f is continuous, then $f \equiv 0$.*

16.1.3 Special Cases and First Integrals

Theorem 16.1.3 (Beltrami identity). *If F does not depend explicitly on x:*

$$F - y'\frac{\partial F}{\partial y'} = const.$$

Proof. $\frac{d}{dx}\left(F - y'\frac{\partial F}{\partial y'}\right) = \frac{\partial F}{\partial y}y' + \frac{\partial F}{\partial y'}y'' - y''\frac{\partial F}{\partial y'} - y'\frac{d}{dx}\frac{\partial F}{\partial y'} = y'\left(\frac{\partial F}{\partial y} - \frac{d}{dx}\frac{\partial F}{\partial y'}\right) = 0.$ □

Theorem 16.1.4 (No y dependence). *If $F = F(x,y')$ only, then $\frac{\partial F}{\partial y'} = const.$*

Example 16.1.2 (Geodesics). Shortest path: $J = \int \sqrt{1+y'^2}\,dx$. Since $F = \sqrt{1+y'^2}$ has no x or y:

$$\frac{\partial F}{\partial y'} = \frac{y'}{\sqrt{1+y'^2}} = \text{const} \Rightarrow y' = \text{const} \Rightarrow y = mx + c.$$

Straight lines are geodesics in Euclidean space.

Example 16.1.3 (Brachistochrone). Find curve of fastest descent from $(0, 0)$ to (x_1, y_1) under gravity.

Time: $T = \int \frac{ds}{v} = \int_0^{x_1} \frac{\sqrt{1+y'^2}}{\sqrt{2gy}}\, dx$.

Beltrami identity gives cycloid: $x = a(\theta - \sin\theta)$, $y = a(1 - \cos\theta)$.

16.2 Extensions and Generalizations

16.2.1 Higher Derivatives

Theorem 16.2.1 (Higher-order Euler-Lagrange). *For* $J[y] = \int F(x, y, y', y'', \ldots, y^{(n)})\, dx$

$$\frac{\partial F}{\partial y} - \frac{d}{dx}\frac{\partial F}{\partial y'} + \frac{d^2}{dx^2}\frac{\partial F}{\partial y''} - \cdots + (-1)^n \frac{d^n}{dx^n}\frac{\partial F}{\partial y^{(n)}} = 0.$$

Example 16.2.1 (Beam bending). Elastic energy of a bent beam: $J = \int \frac{1}{2} EI\, y''^2\, dx$.

Euler-Lagrange: $\frac{d^2}{dx^2}(EI\, y'') = 0$, giving $y'''' = 0$ for constant EI.

16.2.2 Multiple Functions

Theorem 16.2.2 (System of Euler-Lagrange equations). *For* $J[y_1, \ldots, y_n] = \int F(x, y_1, \ldots, y_n, y_1', \ldots, y_n')\, dx$*:*

$$\frac{\partial F}{\partial y_i} - \frac{d}{dx}\frac{\partial F}{\partial y_i'} = 0, \quad i = 1, \ldots, n.$$

16.2.3 Multiple Independent Variables

Theorem 16.2.3 (Multidimensional Euler-Lagrange). *For* $J[u] = \int_\Omega F(x_1, \ldots, x_n, u, u_{x_1}$

$$\frac{\partial F}{\partial u} - \sum_{i=1}^{n} \frac{\partial}{\partial x_i}\frac{\partial F}{\partial u_{x_i}} = 0.$$

In compact notation: $\frac{\partial F}{\partial u} - \nabla \cdot \frac{\partial F}{\partial(\nabla u)} = 0$.

Example 16.2.2 (Minimal surface). Area functional: $A[u] = \int \sqrt{1 + |\nabla u|^2}\, dx\, dy$.

Euler-Lagrange: $\nabla \cdot \frac{\nabla u}{\sqrt{1+|\nabla u|^2}} = 0$ (mean curvature $H = 0$).

16.2.4 Natural Boundary Conditions

Theorem 16.2.4 (Free endpoint conditions). *If $y(b)$ is not prescribed, the natural boundary condition is:*

$$\left.\frac{\partial F}{\partial y'}\right|_{x=b} = 0.$$

Example 16.2.3. For arc length with free endpoint: $\frac{y'}{\sqrt{1+y'^2}}\big|_{x=b} = 0$, so $y'(b) = 0$ (curve arrives horizontally).

16.3 Constraints

16.3.1 Lagrange Multipliers

Theorem 16.3.1 (Isoperimetric constraint). *Extremize $J[y] = \int F(x, y, y')\,dx$ subject to $K[y] = \int G(x, y, y')\,dx = \ell$:*

Solve Euler-Lagrange for $\tilde{F} = F - \lambda G$ where λ is a Lagrange multiplier.

Example 16.3.1 (Isoperimetric problem). Maximize area $A = \frac{1}{2}\oint(x\,dy - y\,dx)$ for fixed perimeter $L = \oint ds$.

Solution: circle (the curve enclosing maximum area for given perimeter).

Example 16.3.2 (Catenary). Hanging chain minimizes potential energy $\int \rho g y\,ds$ for fixed length $\int ds = L$.

Euler-Lagrange for $F = (y - \lambda)\sqrt{1 + y'^2}$ gives $y = a\cosh\frac{x-b}{a}$.

16.3.2 Holonomic Constraints

Definition 16.3.1 (Holonomic constraint). A constraint of the form $g(x, y_1, \ldots, y_n) = 0$.

Theorem 16.3.2 (Constrained extrema). *Extremize J subject to $g(x, y_1, \ldots, y_n) = 0$:*

$$\frac{\partial F}{\partial y_i} - \frac{d}{dx}\frac{\partial F}{\partial y_i'} = \lambda(x)\frac{\partial g}{\partial y_i}.$$

16.3.3 Non-holonomic Constraints

Definition 16.3.2 (Non-holonomic constraint). A constraint involving derivatives: $g(x, y, y') = 0$.

Non-holonomic constraints require special treatment and often cannot be incorporated via simple Lagrange multipliers.

16.4 Second Variation and Stability

16.4.1 Second Variation

Definition 16.4.1 (Second variation).

$$\delta^2 J[y;\eta] := \left.\frac{d^2}{d\epsilon^2}\right|_{\epsilon=0} J[y+\epsilon\eta].$$

Theorem 16.4.1 (Second variation formula). *For $J = \int F(x, y, y')\,dx$:*

$$\delta^2 J = \int_a^b (P\eta^2 + Q\eta'^2)\,dx$$

where $P = \frac{\partial^2 F}{\partial y^2} - \frac{d}{dx}\frac{\partial^2 F}{\partial y \partial y'}$ and $Q = \frac{\partial^2 F}{\partial y'^2}$.

Theorem 16.4.2 (Legendre condition). *For a minimum: $\frac{\partial^2 F}{\partial y'^2} \geq 0$ (weak condition) or > 0 (strong condition).*

16.4.2 Jacobi Equation and Conjugate Points

Definition 16.4.2 (Jacobi equation). The **Jacobi equation** (or accessory equation) is:

$$\frac{d}{dx}\left(Q\frac{d\eta}{dx}\right) - P\eta = 0.$$

Definition 16.4.3 (Conjugate point). Point x^* is **conjugate** to a if a non-trivial solution η of the Jacobi equation satisfies $\eta(a) = \eta(x^*) = 0$.

Theorem 16.4.3 (Jacobi condition). *An extremal is a local minimum only if there are no conjugate points in (a, b).*

Example 16.4.1 (Geodesics on sphere). Great circles are geodesics. Conjugate points are antipodal points. A great circle arc longer than a semicircle is not a minimum (can be shortened by going the other way).

16.5 Lagrangian Mechanics

16.5.1 Principle of Least Action

Definition 16.5.1 (Action). The **action** is the functional:

$$S[q] = \int_{t_1}^{t_2} L(q, \dot{q}, t)\,dt$$

where $L = T - V$ is the **Lagrangian**.

Theorem 16.5.1 (Hamilton's principle). *The physical trajectory between fixed endpoints extremizes the action. Euler-Lagrange equations:*

$$\frac{d}{dt}\frac{\partial L}{\partial \dot{q}_i} - \frac{\partial L}{\partial q_i} = 0.$$

Example 16.5.1 (Free particle). $L = \frac{1}{2}m\dot{x}^2$. Euler-Lagrange: $\frac{d}{dt}(m\dot{x}) = 0$, so $\ddot{x} = 0$ (uniform motion).

Example 16.5.2 (Particle in potential). $L = \frac{1}{2}m\dot{\mathbf{r}}^2 - V(\mathbf{r})$. Euler-Lagrange: $m\ddot{\mathbf{r}} = -\nabla V$ (Newton's second law).

16.5.2 Symmetries and Conservation Laws

Definition 16.5.2 (Cyclic coordinate). q_i is **cyclic** (or ignorable) if $\frac{\partial L}{\partial q_i} = 0$.

Theorem 16.5.2 (Conservation of conjugate momentum). *If q_i is cyclic, then $p_i = \frac{\partial L}{\partial \dot{q}_i}$ is conserved.*

Theorem 16.5.3 (Noether's theorem). *Every continuous symmetry of the Lagrangian corresponds to a conserved quantity.*

If L is invariant under $q_i \to q_i + \epsilon\xi_i(q,t)$, then:

$$Q = \sum_i \frac{\partial L}{\partial \dot{q}_i}\xi_i$$

is conserved.

Example 16.5.3 (Symmetries and conservation).

Symmetry	Transformation	Conserved Quantity
Time translation	$t \to t + \epsilon$	Energy H
Space translation	$\mathbf{r} \to \mathbf{r} + \epsilon$	Momentum $\mathbf{p}$
Rotation	$\mathbf{r} \to R_\epsilon \mathbf{r}$	Angular momentum $\mathbf{L}$

16.5.3 Energy Conservation

Definition 16.5.3 (Hamiltonian). $H = \sum_i \dot{q}_i \frac{\partial L}{\partial \dot{q}_i} - L$.

Theorem 16.5.4 (Energy theorem). *If L has no explicit time dependence, then H is conserved. If additionally the kinetic energy is a homogeneous quadratic function of velocities, $H = T + V = E$.*

16.6 Hamiltonian Mechanics

16.6.1 Legendre Transform

Definition 16.6.1 (Canonical momentum). $p_i = \frac{\partial L}{\partial \dot{q}_i}$.

Definition 16.6.2 (Hamiltonian via Legendre transform). $H(q, p, t) = \sum_i p_i \dot{q}_i - L(q, \dot{q}, t)$
where $\dot{q}$ is expressed in terms of p by inverting $p_i = \frac{\partial L}{\partial \dot{q}_i}$.

Theorem 16.6.1 (Hamilton's equations).

$$\dot{q}_i = \frac{\partial H}{\partial p_i}, \quad \dot{p}_i = -\frac{\partial H}{\partial q_i}.$$

Proposition 16.6.2 (Symplectic structure). *Hamilton's equations can be written as* $\dot{\mathbf{z}} = J\nabla H$ *where* $\mathbf{z} = (q, p)^T$ *and* $J = \begin{pmatrix} 0 & I \\ -I & 0 \end{pmatrix}$.

16.6.2 Poisson Brackets

Definition 16.6.3 (Poisson bracket).

$$\{f, g\} = \sum_i \left(\frac{\partial f}{\partial q_i} \frac{\partial g}{\partial p_i} - \frac{\partial f}{\partial p_i} \frac{\partial g}{\partial q_i} \right).$$

Proposition 16.6.3 (Properties). *(i) Antisymmetry:* $\{f, g\} = -\{g, f\}$.

(ii) Linearity: $\{af + bg, h\} = a\{f, h\} + b\{g, h\}$.

(iii) Leibniz: $\{fg, h\} = f\{g, h\} + g\{f, h\}$.

(iv) Jacobi: $\{f, \{g, h\}\} + \{g, \{h, f\}\} + \{h, \{f, g\}\} = 0$.

Theorem 16.6.4 (Equations of motion). $\dot{f} = \{f, H\} + \frac{\partial f}{\partial t}$.
Conservation: f *is conserved iff* $\{f, H\} = 0$ *(for explicit time-independent* f*).*

Proposition 16.6.5 (Canonical brackets). $\{q_i, q_j\} = 0$, $\{p_i, p_j\} = 0$, $\{q_i, p_j\} = \delta_{ij}$.

16.6.3 Canonical Transformations

Definition 16.6.4 (Canonical transformation). A transformation $(q, p) \to (Q, P)$ is **canonical** if it preserves the form of Hamilton's equations, equivalently, preserves Poisson brackets.

Theorem 16.6.6 (Generating functions). *Canonical transformations can be generated by functions F of mixed variables:*

- $F_1(q, Q, t)$: $p = \frac{\partial F_1}{\partial q}$, $P = -\frac{\partial F_1}{\partial Q}$.
- $F_2(q, P, t)$: $p = \frac{\partial F_2}{\partial q}$, $Q = \frac{\partial F_2}{\partial P}$.

Theorem 16.6.7 (Hamilton-Jacobi equation). *Finding $S(q, P, t)$ such that:*

$$H\left(q, \frac{\partial S}{\partial q}, t\right) + \frac{\partial S}{\partial t} = 0$$

transforms to $H' = 0$, making Q, P constants.

16.7 Variational Principles in Optics

16.7.1 Fermat's Principle

Theorem 16.7.1 (Fermat's principle). *Light travels between two points along the path of stationary optical path length:*

$$\delta \int_A^B n(\mathbf{r})\, ds = 0$$

where n is the refractive index.

Corollary 16.7.2 (Snell's law). *At an interface: $n_1 \sin\theta_1 = n_2 \sin\theta_2$.*

Example 16.7.1 (Mirage). In a medium with $n = n(y)$ decreasing upward, light curves upward, creating mirages.

16.7.2 Eikonal Equation

Theorem 16.7.3 (Eikonal equation). *In the geometric optics limit, the phase $\phi = k_0 S(\mathbf{r})$ satisfies:*

$$|\nabla S|^2 = n^2(\mathbf{r}).$$

Rays are orthogonal to surfaces of constant S.

Proposition 16.7.4 (Connection to mechanics). *The eikonal equation is the Hamilton-Jacobi equation for a particle with $H = \frac{1}{2}|\mathbf{p}|^2$ in a medium with $n^2 \leftrightarrow 2(E - V)$.*

16.8 Variational Methods in Quantum Mechanics

16.8.1 Variational Principle for Ground State

Theorem 16.8.1 (Rayleigh-Ritz variational principle). *For Hamiltonian H with ground state energy E_0:*

$$E_0 = \min_{\psi \neq 0} \frac{\langle \psi | H | \psi \rangle}{\langle \psi | \psi \rangle} = \min_{\|\psi\|=1} \langle \psi | H | \psi \rangle.$$

For any trial $|\psi\rangle$: $\langle \psi | H | \psi \rangle \geq E_0$.

Theorem 16.8.2 (Variational method). *Parametrize trial function $|\psi(\alpha_1, \ldots, \alpha_n)\rangle$. Minimize:*

$$E(\alpha_1, \ldots, \alpha_n) = \frac{\langle \psi | H | \psi \rangle}{\langle \psi | \psi \rangle}$$

over parameters. Result is an upper bound on E_0.

Example 16.8.1 (Helium atom ground state). Trial: $\psi = e^{-Z_{\text{eff}}(r_1 + r_2)/a_0}$ with variational parameter Z_{eff}.

Minimizing gives $Z_{\text{eff}} = Z - 5/16 \approx 1.69$ and $E \approx -77.5$ eV (experiment: -79.0 eV).

Example 16.8.2 (Hydrogen molecule ion H_2^+). Trial: $\psi = c_1 \phi_A + c_2 \phi_B$ (LCAO). Leads to bonding and antibonding orbitals.

16.8.2 Excited States

Theorem 16.8.3 (Hylleraas-Undheim theorem). *If $|\psi_1\rangle, \ldots, |\psi_n\rangle$ are orthonormal trial functions, the eigenvalues of the matrix $H_{ij} = \langle \psi_i | H | \psi_j \rangle$ provide upper bounds to the first n energy levels.*

Theorem 16.8.4 (Orthogonality constraint). *To estimate the first excited state, minimize $\langle \psi | H | \psi \rangle$ subject to $\langle \psi | \psi_0 \rangle = 0$ where $|\psi_0\rangle$ is the ground state.*

16.8.3 Time-Dependent Variational Principle

Theorem 16.8.5 (Dirac-Frenkel principle). *For time-dependent trial state $|\psi(t)\rangle$, require:*

$$\langle \delta\psi | (i\hbar \partial_t - H) | \psi \rangle = 0$$

for all admissible variations $|\delta\psi\rangle$.

16.9 Field Theory and Variational Principles

16.9.1 Lagrangian Field Theory

Definition 16.9.1 (Field Lagrangian). For field $\phi(\mathbf{x}, t)$, the action is:

$$S[\phi] = \int \mathcal{L}(\phi, \partial_\mu \phi, x) \, d^4x$$

where $\mathcal{L}$ is the **Lagrangian density**.

Theorem 16.9.1 (Field equations). *Euler-Lagrange equations for fields:*

$$\frac{\partial \mathcal{L}}{\partial \phi} - \partial_\mu \frac{\partial \mathcal{L}}{\partial(\partial_\mu \phi)} = 0.$$

Example 16.9.1 (Klein-Gordon field). $\mathcal{L} = \frac{1}{2}\partial_\mu \phi \partial^\mu \phi - \frac{1}{2}m^2\phi^2$.
Field equation: $(\Box + m^2)\phi = 0$.

Example 16.9.2 (Electromagnetic field). $\mathcal{L} = -\frac{1}{4}F_{\mu\nu}F^{\mu\nu} - J^\mu A_\mu$ where $F_{\mu\nu} = \partial_\mu A_\nu - \partial_\nu A_\mu$.
Field equations: $\partial_\mu F^{\mu\nu} = J^\nu$ (Maxwell's equations).

Example 16.9.3 (Dirac field). $\mathcal{L} = \bar{\psi}(i\gamma^\mu \partial_\mu - m)\psi$.
Field equation: $(i\gamma^\mu \partial_\mu - m)\psi = 0$ (Dirac equation).

16.9.2 Noether's Theorem for Fields

Theorem 16.9.2 (Noether's theorem for fields). *If $\mathcal{L}$ is invariant under $\phi \to \phi + \epsilon\delta\phi$, there is a conserved current:*

$$j^\mu = \frac{\partial \mathcal{L}}{\partial(\partial_\mu \phi)}\delta\phi, \quad \partial_\mu j^\mu = 0.$$

Example 16.9.4 (Energy-momentum tensor). Translation invariance gives conserved $T^{\mu\nu}$:

$$T^{\mu\nu} = \frac{\partial \mathcal{L}}{\partial(\partial_\mu \phi)}\partial^\nu \phi - \eta^{\mu\nu}\mathcal{L}.$$

T^{00} is energy density, T^{0i} momentum density.

Example 16.9.5 (Charge conservation). For complex ϕ with $U(1)$ symmetry ($\phi \to e^{i\alpha}\phi$):

$$j^\mu = i(\phi^* \partial^\mu \phi - \phi \partial^\mu \phi^*), \quad Q = \int j^0 \, d^3x.$$

16.9.3 Gauge Theories

Definition 16.9.2 (Gauge invariance). A theory is **gauge invariant** if $\mathcal{L}$ is unchanged under local transformations $\phi(x) \to e^{i\alpha(x)}\phi(x)$.

Theorem 16.9.3 (Minimal coupling). *Gauge invariance requires replacing $\partial_\mu \to D_\mu = \partial_\mu - ieA_\mu$ and adding kinetic term for A_μ.*

Example 16.9.6 (QED Lagrangian). $\mathcal{L} = \bar{\psi}(i\gamma^\mu D_\mu - m)\psi - \frac{1}{4}F_{\mu\nu}F^{\mu\nu}$.

16.10 Numerical Variational Methods

16.10.1 Finite Element Method

Theorem 16.10.1 (Variational formulation). *Many PDEs have variational form: $-\nabla^2 u = f$ minimizes $J[u] = \frac{1}{2}\int |\nabla u|^2 - \int fu$.*

Theorem 16.10.2 (Finite element approximation). *Approximate $u \approx u_h = \sum_i c_i\phi_i$ where ϕ_i are piecewise polynomial basis functions on a mesh. Minimize over c_i.*

Proposition 16.10.3 (Galerkin equations). *$\sum_j K_{ij}c_j = F_i$ where $K_{ij} = \int \nabla\phi_i \cdot \nabla\phi_j$ and $F_i = \int f\phi_i$.*

16.10.2 Variational Monte Carlo

Theorem 16.10.4 (VMC method). *For many-body quantum systems, evaluate $E[\psi] = \frac{\langle\psi|H|\psi\rangle}{\langle\psi|\psi\rangle}$ by Monte Carlo sampling:*

$$E = \int |\psi(\mathbf{R})|^2 E_L(\mathbf{R})\, d\mathbf{R} \Big/ \int |\psi(\mathbf{R})|^2\, d\mathbf{R}$$

where $E_L = \psi^{-1}H\psi$ is the local energy.

16.10.3 Density Functional Theory

Theorem 16.10.5 (Hohenberg-Kohn theorem). *The ground state energy is a functional of the electron density $\rho(\mathbf{r})$:*

$$E[\rho] = T[\rho] + V_{ee}[\rho] + \int v_{ext}(\mathbf{r})\rho(\mathbf{r})\, d^3r.$$

The exact ground state minimizes $E[\rho]$ over densities with correct particle number.

Theorem 16.10.6 (Kohn-Sham equations). *The density is represented by orbitals:* $\rho = \sum_i |\phi_i|^2$. *Self-consistent equations:*

$$\left[-\frac{\hbar^2}{2m}\nabla^2 + v_{\text{eff}}[\rho](\mathbf{r})\right]\phi_i = \epsilon_i\phi_i.$$

16.11 Summary

Variational principles unify theoretical physics:

Calculus of variations: Euler-Lagrange equation $\frac{\partial F}{\partial y} - \frac{d}{dx}\frac{\partial F}{\partial y'} = 0$ for extremals of $\int F\,dx$.

Constraints: Lagrange multipliers for isoperimetric and holonomic constraints.

Lagrangian mechanics: Action principle $\delta S = 0$ gives equations of motion; Noether's theorem connects symmetries to conservation laws.

Hamiltonian mechanics: Legendre transform to (q, p); Poisson brackets; canonical transformations; Hamilton-Jacobi theory.

Optics: Fermat's principle; eikonal equation analogous to Hamilton-Jacobi.

Quantum mechanics: Rayleigh-Ritz principle gives upper bounds; variational methods for ground and excited states.

Field theory: Lagrangian density; field equations from variation; Noether currents; gauge invariance.

Numerical methods: Finite elements, VMC, DFT exploit variational structure.

Exercises

Exercise 16.11.1. Derive the Euler-Lagrange equation for $J[y] = \int_a^b F(x, y, y')\,dx$.

Exercise 16.11.2. Find the curve of shortest length connecting $(0, 0)$ to $(1, 1)$.

Exercise 16.11.3. Use the Beltrami identity to solve the brachistochrone problem.

Exercise 16.11.4. Derive the Euler-Lagrange equation for $J[u] = \int\int F(x, y, u, u_x, u_y)\,dx$

Exercise 16.11.5. Find the shape of a hanging chain (catenary) using Lagrange multipliers.

Exercise 16.11.6. Show that the Legendre condition $\frac{\partial^2 F}{\partial y'^2} > 0$ is necessary for a minimum.

Exercise 16.11.7. Derive Hamilton's equations from the variational principle $\delta \int (p\dot{q} - H)\, dt = 0$.

Exercise 16.11.8. Show that $\{L_i, L_j\} = \epsilon_{ijk} L_k$ for angular momentum components.

Exercise 16.11.9. Use the variational method with $\psi = e^{-\alpha r}$ to estimate the ground state energy of hydrogen.

Exercise 16.11.10. Derive the Klein-Gordon equation from $\mathcal{L} = \frac{1}{2}\partial_\mu \phi \partial^\mu \phi - \frac{1}{2}m^2\phi^2$.

Exercise 16.11.11. Show that Noether's theorem for spacetime translations gives the energy-momentum tensor.

Exercise 16.11.12. Prove that the finite element method for $-u'' = f$ with Dirichlet BC is equivalent to minimizing $\frac{1}{2}\int u'^2 - \int fu$.

Part VI

Probability and Statistics

Probability and statistics provide essential tools for modern physics, from the interpretation of quantum measurements to the foundations of statistical mechanics and the analysis of experimental data. Unlike deterministic classical mechanics, much of modern physics deals inherently with uncertainty—whether fundamental (as in quantum mechanics) or practical (as in systems with many degrees of freedom).

This part develops probability theory rigorously, introduces stochastic processes crucial for physics, and covers statistical methods needed for data analysis. Chapter 17 establishes the mathematical foundations of probability, including random variables, distributions, and limit theorems. Chapter 18 treats stochastic processes, including Markov chains, Brownian motion, and the Langevin equation. Chapter 19 covers statistical inference, estimation theory, and hypothesis testing as applied to physical measurements.

Chapter 17

Probability Theory

From the probabilistic interpretation of the wave function to the statistical ensembles of thermodynamics, probability permeates modern physics. This chapter develops probability theory with emphasis on concepts essential for theoretical physics: random variables, characteristic functions, limit theorems, and the mathematical structures underlying quantum measurement and statistical mechanics.

17.1 Foundations of Probability

17.1.1 Probability Spaces

The modern axiomatic approach to probability, due to Kolmogorov (1933), provides the rigorous foundation for all probabilistic reasoning.

Definition 17.1.1 (Sample space). The **sample space** Ω is the set of all possible outcomes of a random experiment.

Example 17.1.1. For a single die roll, $\Omega = \{1, 2, 3, 4, 5, 6\}$. For a particle's position, $\Omega = \mathbb{R}^3$. For a quantum spin-$\frac{1}{2}$ measurement along z, $\Omega = \{\uparrow, \downarrow\}$.

Definition 17.1.2 (σ-algebra). A **σ-algebra** (or σ-field) $\mathcal{F}$ on Ω is a collection of subsets of Ω satisfying:

(i) $\Omega \in \mathcal{F}$.

(ii) If $A \in \mathcal{F}$, then $A^c \in \mathcal{F}$ (closed under complements).

(iii) If $A_1, A_2, \ldots \in \mathcal{F}$, then $\bigcup_{i=1}^{\infty} A_i \in \mathcal{F}$ (closed under countable unions).

Elements of $\mathcal{F}$ are called **events**.

The σ-algebra structure is essential: it ensures that whenever we can assign probabilities to certain events, we can also assign probabilities to their complements, unions, and intersections.

Example 17.1.2 (Borel σ-algebra)**.** On $\mathbb{R}$, the **Borel σ-algebra** $\mathcal{B}(\mathbb{R})$ is the smallest σ-algebra containing all open intervals. It includes all intervals, all open sets, all closed sets, and much more.

Definition 17.1.3 (Probability measure)**.** A **probability measure** P on $(\Omega, \mathcal{F})$ is a function $P : \mathcal{F} \to [0, 1]$ satisfying:

(i) $P(A) \geq 0$ for all $A \in \mathcal{F}$ (non-negativity).

(ii) $P(\Omega) = 1$ (normalization).

(iii) For mutually disjoint events $A_1, A_2, \ldots \in \mathcal{F}$:

$$P\left(\bigcup_{i=1}^{\infty} A_i\right) = \sum_{i=1}^{\infty} P(A_i) \quad \text{(countable additivity).}$$

Definition 17.1.4 (Probability space)**.** A **probability space** is a triple $(\Omega, \mathcal{F}, P)$ consisting of a sample space, a σ-algebra, and a probability measure.

Proposition 17.1.1 (Basic properties of probability)**.** *Let $(\Omega, \mathcal{F}, P)$ be a probability space. Then:*

(i) $P(\varnothing) = 0$.

(ii) $P(A^c) = 1 - P(A)$.

(iii) If $A \subseteq B$, then $P(A) \leq P(B)$ (monotonicity).

(iv) $P(A \cup B) = P(A) + P(B) - P(A \cap B)$ *(inclusion-exclusion).*

(v) $P\left(\bigcup_{i=1}^{n} A_i\right) \leq \sum_{i=1}^{n} P(A_i)$ *(union bound, Boole's inequality).*

(vi) If $A_n \uparrow A$ (increasing sequence), then $P(A_n) \to P(A)$ (continuity from below).

(vii) If $A_n \downarrow A$ (decreasing sequence), then $P(A_n) \to P(A)$ (continuity from above).

17.1.2 Conditional Probability and Independence

Definition 17.1.5 (Conditional probability). For events A, B with $P(B) > 0$, the **conditional probability** of A given B is:

$$P(A|B) := \frac{P(A \cap B)}{P(B)}.$$

Conditional probability can be interpreted as updating our probability assignments given new information: if we learn that B has occurred, we restrict attention to outcomes in B and renormalize.

Theorem 17.1.2 (Multiplication rule). *For any events $A_1, \dots, A_n$ with $P(A_1 \cap \dots \cap A_{n-1}) > 0$:*

$$P(A_1 \cap \dots \cap A_n) = P(A_1)P(A_2|A_1)P(A_3|A_1 \cap A_2) \cdots P(A_n|A_1 \cap \dots \cap A_{n-1}).$$

Theorem 17.1.3 (Law of total probability). *If $\{B_i\}_{i \in I}$ is a partition of Ω (mutually disjoint with $\bigcup_i B_i = \Omega$) with $P(B_i) > 0$ for all i, then:*

$$P(A) = \sum_i P(A|B_i)P(B_i).$$

Theorem 17.1.4 (Bayes' theorem). *If $P(A), P(B) > 0$:*

$$P(A|B) = \frac{P(B|A)P(A)}{P(B)}.$$

More generally, if $\{A_i\}$ partitions Ω:

$$P(A_i|B) = \frac{P(B|A_i)P(A_i)}{\sum_j P(B|A_j)P(A_j)}.$$

Bayes' theorem is fundamental to statistical inference: $P(A_i)$ represents prior probabilities, $P(B|A_i)$ the likelihood of observing B given hypothesis A_i, and $P(A_i|B)$ the posterior probability after observing B.

Example 17.1.3 (Medical testing). A disease affects 1% of the population. A test is 99% sensitive (detects disease when present) and 95% specific (negative when healthy). Given a positive test, what is the probability of having the disease?

Let D = disease, T^+ = positive test. We have $P(D) = 0.01$, $P(T^+|D) = 0.99$, $P(T^+|D^c) = 0.05$. By Bayes' theorem:

$$P(D|T^+) = \frac{0.99 \times 0.01}{0.99 \times 0.01 + 0.05 \times 0.99} = \frac{0.0099}{0.0594} \approx 0.167.$$

Despite the seemingly excellent test, only about 17% of positive tests indicate disease—a consequence of the low base rate.

Definition 17.1.6 (Independence)**.** Events A and B are **independent**, written $A \perp B$, if:

$$P(A \cap B) = P(A)P(B).$$

Equivalently, if $P(B) > 0$: $P(A|B) = P(A)$.

Definition 17.1.7 (Mutual independence)**.** Events $\{A_i\}_{i \in I}$ are **mutually independent** if for every finite subset $J \subseteq I$:

$$P\left(\bigcap_{j \in J} A_j\right) = \prod_{j \in J} P(A_j).$$

Remark 17.1.1. Pairwise independence does not imply mutual independence. Consider two fair coin flips with A = "first heads," B = "second heads," C = "both same." Any two are independent, but $P(A \cap B \cap C) = \frac{1}{4} \neq \frac{1}{8} = P(A)P(B)P(C)$.

17.2 Random Variables

17.2.1 Definitions and Distributions

Definition 17.2.1 (Random variable)**.** A **random variable** is a measurable function $X : \Omega \to \mathbb{R}$, meaning that for all $x \in \mathbb{R}$:

$$\{X \leq x\} := \{\omega \in \Omega : X(\omega) \leq x\} \in \mathcal{F}.$$

More generally, an $\mathbb{R}^n$**-valued random variable** (or **random vector**) is a measurable function $\mathbf{X} : \Omega \to \mathbb{R}^n$.

The measurability condition ensures we can assign probabilities to events like "X takes values in some interval."

Definition 17.2.2 (Cumulative distribution function)**.** The **cumulative distribution function** (CDF) of X is:

$$F_X(x) := P(X \leq x) = P(\{\omega : X(\omega) \leq x\}).$$

Proposition 17.2.1 (Properties of the CDF)**.** *Any CDF F satisfies:*

(i) F is non-decreasing.

(ii) $\lim_{x \to -\infty} F(x) = 0$ *and* $\lim_{x \to \infty} F(x) = 1$.

(iii) F is right-continuous: $\lim_{y \to x^+} F(y) = F(x)$.

(iv) $P(a < X \leq b) = F(b) - F(a)$.

(v) $P(X = x) = F(x) - F(x^-)$ *where* $F(x^-) = \lim_{y \to x^-} F(y)$.

Conversely, any function satisfying (i)–(iii) is the CDF of some random variable.

17.2.2 Discrete Random Variables

Definition 17.2.3 (Discrete random variable). X is **discrete** if it takes values in a countable set $\{x_1, x_2, \ldots\}$.

Definition 17.2.4 (Probability mass function). The **probability mass function** (PMF) of a discrete random variable is:

$$p_X(x) := P(X = x).$$

It satisfies $p_X(x) \geq 0$ and $\sum_x p_X(x) = 1$.

Example 17.2.1 (Important discrete distributions).

1. **Bernoulli**(p): $X \in \{0, 1\}$ with $P(X = 1) = p$.
2. **Binomial**(n, p): Number of successes in n independent Bernoulli(p) trials.
$$P(X = k) = \binom{n}{k} p^k (1-p)^{n-k}, \quad k = 0, 1, \ldots, n.$$
3. **Poisson**(λ): Models rare events; limit of Binomial($n, \lambda/n$) as $n \to \infty$.
$$P(X = k) = \frac{\lambda^k e^{-\lambda}}{k!}, \quad k = 0, 1, 2, \ldots$$
4. **Geometric**(p): Number of trials until first success.
$$P(X = k) = (1-p)^{k-1} p, \quad k = 1, 2, \ldots$$
5. **Negative Binomial**(r, p): Number of trials until r successes.
$$P(X = k) = \binom{k-1}{r-1} p^r (1-p)^{k-r}, \quad k = r, r+1, \ldots$$

17.2.3 Continuous Random Variables

Definition 17.2.5 (Continuous random variable). X is **(absolutely) continuous** if there exists a non-negative function f_X such that:

$$F_X(x) = \int_{-\infty}^{x} f_X(t)\, dt.$$

The function f_X is called the **probability density function** (PDF).

Proposition 17.2.2 (Properties of the PDF).

(i) $f_X(x) \geq 0$ *for all* x.

(ii) $\int_{-\infty}^{\infty} f_X(x)\, dx = 1$.

(iii) $P(a \leq X \leq b) = \int_a^b f_X(x)\, dx$.

(iv) $f_X(x) = F_X'(x)$ *wherever* F_X *is differentiable.*

(v) $P(X = x) = 0$ *for any specific value* x.

Remark 17.2.1. The PDF $f_X(x)$ is *not* a probability—it can exceed 1. Rather, $f_X(x)\, dx$ represents the infinitesimal probability of X falling in $[x, x + dx)$.

Example 17.2.2 (Important continuous distributions).

1. **Uniform**(a, b):
$$f(x) = \frac{1}{b-a}\mathbf{1}_{[a,b]}(x).$$

2. **Exponential**(λ): Models waiting times; memoryless.
$$f(x) = \lambda e^{-\lambda x}\mathbf{1}_{[0,\infty)}(x).$$

3. **Gaussian (Normal)** $N(\mu, \sigma^2)$:
$$f(x) = \frac{1}{\sqrt{2\pi\sigma^2}} \exp\left(-\frac{(x-\mu)^2}{2\sigma^2}\right).$$
The **standard normal** $N(0, 1)$ has $\mu = 0$, $\sigma = 1$.

4. **Gamma**(α, β): Generalizes exponential.
$$f(x) = \frac{\beta^\alpha}{\Gamma(\alpha)} x^{\alpha-1} e^{-\beta x}\mathbf{1}_{[0,\infty)}(x).$$

5. **Chi-squared** χ_k^2: Sum of k squared standard normals; $\chi_k^2 = \text{Gamma}(k/2, 1/2)$.

6. **Beta**(α, β): Supported on $[0, 1]$.
$$f(x) = \frac{\Gamma(\alpha+\beta)}{\Gamma(\alpha)\Gamma(\beta)} x^{\alpha-1}(1-x)^{\beta-1}\mathbf{1}_{[0,1]}(x).$$

7. **Cauchy**: Heavy-tailed; no finite mean.
$$f(x) = \frac{1}{\pi(1+x^2)}.$$

17.2.4 Transformations of Random Variables

Theorem 17.2.3 (Change of variables). *If X has PDF f_X and $Y = g(X)$ where g is strictly monotonic and differentiable:*

$$f_Y(y) = f_X(g^{-1}(y)) \left| \frac{d}{dy} g^{-1}(y) \right| = \left. \frac{f_X(x)}{|g'(x)|} \right|_{x=g^{-1}(y)}.$$

Example 17.2.3. If $X \sim N(0,1)$ and $Y = X^2$, then for $y > 0$:

$$f_Y(y) = \frac{1}{\sqrt{2\pi y}} e^{-y/2},$$

which is the χ_1^2 distribution.

For multivariate transformations $\mathbf{Y} = \mathbf{g}(\mathbf{X})$ where $\mathbf{g} : \mathbb{R}^n \to \mathbb{R}^n$ is a diffeomorphism:

$$f_{\mathbf{Y}}(\mathbf{y}) = f_{\mathbf{X}}(\mathbf{g}^{-1}(\mathbf{y})) \left| \det J_{\mathbf{g}^{-1}}(\mathbf{y}) \right|,$$

where $J_{\mathbf{g}^{-1}}$ is the Jacobian matrix of the inverse transformation.

17.3 Expectation and Moments

17.3.1 Expected Value

Definition 17.3.1 (Expected value). The **expected value** (or **mean**) of a random variable X is:

$$\mathbb{E}[X] := \begin{cases} \sum_x x \cdot p_X(x) & \text{(discrete)} \\ \int_{-\infty}^{\infty} x \cdot f_X(x)\, dx & \text{(continuous)} \end{cases}$$

when the sum or integral converges absolutely.

The expected value can be understood as the "center of mass" of the probability distribution, or as the long-run average of repeated independent observations.

Theorem 17.3.1 (Law of the unconscious statistician). *For $Y = g(X)$:*

$$\mathbb{E}[g(X)] = \begin{cases} \sum_x g(x) \cdot p_X(x) & \textit{(discrete)} \\ \int_{-\infty}^{\infty} g(x) \cdot f_X(x)\, dx & \textit{(continuous)} \end{cases}$$

This avoids computing the distribution of Y explicitly.

Proposition 17.3.2 (Properties of expectation). *(i)* ***Linearity:*** $\mathbb{E}[aX + bY] = a\mathbb{E}[X] + b\mathbb{E}[Y]$ *for any constants* a, b.

(ii) ***Monotonicity:*** *If* $X \leq Y$ *a.s., then* $\mathbb{E}[X] \leq \mathbb{E}[Y]$.

(iii) ***Independence:*** *If* $X \perp Y$*, then* $\mathbb{E}[XY] = \mathbb{E}[X]\mathbb{E}[Y]$.

(iv) ***Triangle inequality:*** $|\mathbb{E}[X]| \leq \mathbb{E}[|X|]$.

17.3.2 Variance and Higher Moments

Definition 17.3.2 (Moments). The n**-th moment** of X is $\mathbb{E}[X^n]$. The n**-th central moment** is $\mathbb{E}[(X - \mathbb{E}[X])^n]$.

Definition 17.3.3 (Variance). The **variance** of X is the second central moment:

$$\operatorname{Var}(X) := \mathbb{E}[(X - \mathbb{E}[X])^2] = \mathbb{E}[X^2] - (\mathbb{E}[X])^2.$$

The **standard deviation** is $\sigma_X := \sqrt{\operatorname{Var}(X)}$.

Proposition 17.3.3 (Properties of variance).

(i) $\operatorname{Var}(X) \geq 0$*, with equality iff* X *is constant a.s.*

(ii) $\operatorname{Var}(aX + b) = a^2 \operatorname{Var}(X)$.

(iii) $\operatorname{Var}(X + Y) = \operatorname{Var}(X) + \operatorname{Var}(Y) + 2\operatorname{Cov}(X, Y)$.

(iv) *If* $X \perp Y$*:* $\operatorname{Var}(X + Y) = \operatorname{Var}(X) + \operatorname{Var}(Y)$.

Definition 17.3.4 (Skewness and kurtosis). The **skewness** measures asymmetry:

$$\gamma_1 := \mathbb{E}\left[\left(\frac{X - \mu}{\sigma}\right)^3\right] = \frac{\mathbb{E}[(X - \mu)^3]}{\sigma^3}.$$

The **kurtosis** measures tail heaviness:

$$\gamma_2 := \mathbb{E}\left[\left(\frac{X - \mu}{\sigma}\right)^4\right] - 3 = \frac{\mathbb{E}[(X - \mu)^4]}{\sigma^4} - 3.$$

The subtraction of 3 makes the Gaussian distribution have $\gamma_2 = 0$ ("mesokurtic").

Example 17.3.1 (Moments of common distributions).

Distribution	Mean	Variance
Bernoulli(p)	p	$p(1-p)$
Binomial(n,p)	np	$np(1-p)$
Poisson(λ)	λ	λ
Geometric(p)	$1/p$	$(1-p)/p^2$
Uniform(a,b)	$(a+b)/2$	$(b-a)^2/12$
Exponential(λ)	$1/\lambda$	$1/\lambda^2$
Gaussian(μ,σ^2)	μ	σ^2
Gamma(α,β)	α/β	α/β^2

17.4 Generating Functions and Characteristic Functions

Generating functions encode the entire distribution in a single function and provide powerful tools for computing moments and proving limit theorems.

17.4.1 Moment Generating Function

Definition 17.4.1 (Moment generating function). The **moment generating function** (MGF) of X is:

$$M_X(t) := \mathbb{E}[e^{tX}] = \begin{cases} \sum_x e^{tx} p_X(x) & \text{(discrete)} \\ \int_{-\infty}^{\infty} e^{tx} f_X(x)\, dx & \text{(continuous)} \end{cases}$$

defined for t in some neighborhood of 0.

Theorem 17.4.1 (Properties of the MGF).

(i) $M_X(0) = 1$.

(ii) $\mathbb{E}[X^n] = M_X^{(n)}(0)$ *(moments from derivatives).*

(iii) If $M_X(t) = M_Y(t)$ *in a neighborhood of 0, then* X *and* Y *have the same distribution.*

(iv) If $X \perp Y$*:* $M_{X+Y}(t) = M_X(t)M_Y(t)$.

(v) $M_{aX+b}(t) = e^{bt}M_X(at)$.

Example 17.4.1 (MGFs of common distributions).

1. **Poisson(λ)**: $M(t) = \exp(\lambda(e^t - 1))$.

2. **Exponential**(λ): $M(t) = \frac{\lambda}{\lambda - t}$ for $t < \lambda$.

3. **Gaussian**(μ, σ^2): $M(t) = \exp\left(\mu t + \frac{\sigma^2 t^2}{2}\right)$.

17.4.2 Characteristic Function

Definition 17.4.2 (Characteristic function). The **characteristic function** of X is:

$$\varphi_X(t) := \mathbb{E}[e^{itX}] = \int_{-\infty}^{\infty} e^{itx}\, dF_X(x),$$

where $i = \sqrt{-1}$.

The characteristic function is the Fourier transform of the distribution and always exists (unlike the MGF).

Theorem 17.4.2 (Properties of characteristic functions).

(i) $\varphi_X(0) = 1$ *and* $|\varphi_X(t)| \leq 1$ *for all* t.

(ii) $\varphi_X(-t) = \overline{\varphi_X(t)}$ *(conjugate symmetry).*

(iii) φ_X *is uniformly continuous on* $\mathbb{R}$.

(iv) If $\mathbb{E}[|X|^n] < \infty$, *then* φ_X *is* n *times differentiable and* $\mathbb{E}[X^n] = i^{-n}\varphi_X^{(n)}(0)$.

(v) φ_X *uniquely determines the distribution (Lévy's inversion theorem).*

(vi) If $X \perp Y$*:* $\varphi_{X+Y}(t) = \varphi_X(t)\varphi_Y(t)$.

Theorem 17.4.3 (Lévy inversion formula). *If* $\varphi_X \in L^1(\mathbb{R})$, *then* X *has a continuous density:*

$$f_X(x) = \frac{1}{2\pi} \int_{-\infty}^{\infty} e^{-itx} \varphi_X(t)\, dt.$$

Example 17.4.2 (Characteristic functions).

1. **Gaussian**(μ, σ^2): $\varphi(t) = \exp\left(i\mu t - \frac{\sigma^2 t^2}{2}\right)$.

2. **Cauchy**: $\varphi(t) = e^{-|t|}$.

3. **Poisson**(λ): $\varphi(t) = \exp(\lambda(e^{it} - 1))$.

17.4.3 Cumulants

Definition 17.4.3 (Cumulant generating function). The **cumulant generating function** is:

$$K_X(t) := \log M_X(t) = \log \mathbb{E}[e^{tX}].$$

The n-**th cumulant** is $\kappa_n := K_X^{(n)}(0)$.

Proposition 17.4.4 (Cumulants vs. moments).

(i) $\kappa_1 = \mathbb{E}[X]$ *(mean).*

(ii) $\kappa_2 = \mathrm{Var}(X)$ *(variance).*

(iii) $\kappa_3 = \mathbb{E}[(X-\mu)^3]$ *(third central moment).*

(iv) $\kappa_4 = \mathbb{E}[(X-\mu)^4] - 3\sigma^4$ *(excess kurtosis* $\times\sigma^4$*).*

Theorem 17.4.5 (Cumulants of sums). *For independent* $X_1, \ldots, X_n$*:*

$$\kappa_m(X_1 + \cdots + X_n) = \kappa_m(X_1) + \cdots + \kappa_m(X_n).$$

This additivity makes cumulants natural for analyzing sums.

Remark 17.4.1 (Gaussian uniqueness). The Gaussian distribution is the only distribution for which $\kappa_n = 0$ for all $n \geq 3$.

17.5 Multiple Random Variables

17.5.1 Joint Distributions

Definition 17.5.1 (Joint distribution). The **joint CDF** of random variables X, Y is:

$$F_{X,Y}(x, y) := P(X \leq x, Y \leq y).$$

For discrete variables, the **joint PMF** is $p_{X,Y}(x, y) := P(X = x, Y = y)$. For continuous variables with joint density $f_{X,Y}$:

$$P((X, Y) \in A) = \iint_A f_{X,Y}(x, y)\, dx\, dy.$$

Definition 17.5.2 (Marginal distributions). The **marginal** distributions are obtained by "integrating out" the other variable:

$$f_X(x) = \int_{-\infty}^{\infty} f_{X,Y}(x, y)\, dy, \qquad f_Y(y) = \int_{-\infty}^{\infty} f_{X,Y}(x, y)\, dx.$$

Definition 17.5.3 (Independence of random variables). Random variables X, Y are **independent** if:

$$F_{X,Y}(x,y) = F_X(x)F_Y(y) \quad \text{for all } x, y,$$

equivalently, $f_{X,Y}(x,y) = f_X(x)f_Y(y)$ (or $p_{X,Y} = p_X p_Y$ for discrete).

17.5.2 Conditional Distributions

Definition 17.5.4 (Conditional distribution). For continuous X, Y with joint density $f_{X,Y}$ and $f_Y(y) > 0$:

$$f_{X|Y}(x|y) := \frac{f_{X,Y}(x,y)}{f_Y(y)}.$$

For discrete variables: $p_{X|Y}(x|y) = p_{X,Y}(x,y)/p_Y(y)$.

Definition 17.5.5 (Conditional expectation).

$$\mathbb{E}[X|Y = y] := \int_{-\infty}^{\infty} x\, f_{X|Y}(x|y)\, dx.$$

As a random variable: $\mathbb{E}[X|Y]$ is the function of Y that takes value $\mathbb{E}[X|Y = y]$ when $Y = y$.

Theorem 17.5.1 (Law of iterated expectation).

$$\mathbb{E}[X] = \mathbb{E}[\mathbb{E}[X|Y]].$$

More generally: $\mathbb{E}[\mathbb{E}[X|Y, Z] \mid Y] = \mathbb{E}[X|Y]$ *(tower property).*

Theorem 17.5.2 (Law of total variance).

$$\mathrm{Var}(X) = \mathbb{E}[\mathrm{Var}(X|Y)] + \mathrm{Var}(\mathbb{E}[X|Y]).$$

The first term is the "unexplained" variance within groups; the second is the variance "explained" by Y.

17.5.3 Covariance and Correlation

Definition 17.5.6 (Covariance).

$$\mathrm{Cov}(X, Y) := \mathbb{E}[(X - \mathbb{E}[X])(Y - \mathbb{E}[Y])] = \mathbb{E}[XY] - \mathbb{E}[X]\mathbb{E}[Y].$$

Proposition 17.5.3 (Properties of covariance).

(i) $\mathrm{Cov}(X, X) = \mathrm{Var}(X)$.

(ii) $\text{Cov}(X, Y) = \text{Cov}(Y, X)$ *(symmetry).*

(iii) $\text{Cov}(aX + b, cY + d) = ac \cdot \text{Cov}(X, Y)$ *(bilinearity).*

(iv) $\text{Var}\left(\sum_{i=1}^n X_i\right) = \sum_{i=1}^n \text{Var}(X_i) + 2\sum_{i<j} \text{Cov}(X_i, X_j)$.

(v) If $X \perp Y$*, then* $\text{Cov}(X, Y) = 0$ *(converse is false in general).*

Definition 17.5.7 (Correlation coefficient)**.** The **(Pearson) correlation** between X and Y is:

$$\rho_{X,Y} := \frac{\text{Cov}(X, Y)}{\sigma_X \sigma_Y} = \frac{\mathbb{E}[XY] - \mathbb{E}[X]\mathbb{E}[Y]}{\sqrt{\text{Var}(X)\,\text{Var}(Y)}}.$$

Proposition 17.5.4 (Properties of correlation)**.**

(i) $-1 \leq \rho_{X,Y} \leq 1$ *(Cauchy-Schwarz inequality).*

(ii) $|\rho_{X,Y}| = 1$ *iff* $Y = aX + b$ *for some* $a \neq 0$ *(perfect linear relationship).*

(iii) $\rho_{X,Y} = 0$ *means* X *and* Y *are* ***uncorrelated*** *(no linear relationship).*

(iv) Independence implies uncorrelated; converse fails except for jointly Gaussian.

17.5.4 The Multivariate Gaussian Distribution

Definition 17.5.8 (Multivariate Gaussian)**.** A random vector $\mathbf{X} = (X_1, \ldots, X_n)^T$ has a **multivariate Gaussian** (or normal) distribution $N(\boldsymbol{\mu}, \boldsymbol{\Sigma})$ if its density is:

$$f(\mathbf{x}) = \frac{1}{(2\pi)^{n/2}|\boldsymbol{\Sigma}|^{1/2}} \exp\left(-\frac{1}{2}(\mathbf{x} - \boldsymbol{\mu})^T \boldsymbol{\Sigma}^{-1}(\mathbf{x} - \boldsymbol{\mu})\right),$$

where $\boldsymbol{\mu} = \mathbb{E}[\mathbf{X}]$ is the mean vector and $\boldsymbol{\Sigma} = \mathbb{E}[(\mathbf{X} - \boldsymbol{\mu})(\mathbf{X} - \boldsymbol{\mu})^T]$ is the covariance matrix.

Theorem 17.5.5 (Properties of multivariate Gaussian)**.** *Let* $\mathbf{X} \sim N(\boldsymbol{\mu}, \boldsymbol{\Sigma})$*. Then:*

(i) ***Linear transformation:*** $A\mathbf{X} + \mathbf{b} \sim N(A\boldsymbol{\mu} + \mathbf{b}, A\boldsymbol{\Sigma}A^T)$.

(ii) ***Marginals:*** *Any subset of components is multivariate Gaussian.*

(iii) ***Conditionals:*** $\mathbf{X}_1 | \mathbf{X}_2 = \mathbf{x}_2$ *is multivariate Gaussian.*

(iv) ***Independence:*** *Components are independent iff they are uncorrelated* ($\boldsymbol{\Sigma}$ *diagonal).*

(v) ***Characteristic function:*** $\varphi(\mathbf{t}) = \exp\left(i\mathbf{t}^T\boldsymbol{\mu} - \frac{1}{2}\mathbf{t}^T\boldsymbol{\Sigma}\mathbf{t}\right)$.

Theorem 17.5.6 (Conditional Gaussian). *If* $\mathbf{X} = (\mathbf{X}_1, \mathbf{X}_2)^T \sim N(\boldsymbol{\mu}, \boldsymbol{\Sigma})$ *with block structure:*

$$\boldsymbol{\mu} = \begin{pmatrix} \boldsymbol{\mu}_1 \\ \boldsymbol{\mu}_2 \end{pmatrix}, \quad \boldsymbol{\Sigma} = \begin{pmatrix} \boldsymbol{\Sigma}_{11} & \boldsymbol{\Sigma}_{12} \\ \boldsymbol{\Sigma}_{21} & \boldsymbol{\Sigma}_{22} \end{pmatrix},$$

then:

$$\mathbf{X}_1 | \mathbf{X}_2 = \mathbf{x}_2 \sim N(\boldsymbol{\mu}_{1|2}, \boldsymbol{\Sigma}_{1|2}),$$

where:

$$\boldsymbol{\mu}_{1|2} = \boldsymbol{\mu}_1 + \boldsymbol{\Sigma}_{12}\boldsymbol{\Sigma}_{22}^{-1}(\mathbf{x}_2 - \boldsymbol{\mu}_2),$$
$$\boldsymbol{\Sigma}_{1|2} = \boldsymbol{\Sigma}_{11} - \boldsymbol{\Sigma}_{12}\boldsymbol{\Sigma}_{22}^{-1}\boldsymbol{\Sigma}_{21}.$$

17.6 Limit Theorems

Limit theorems describe the behavior of sums of random variables and underpin much of statistical inference.

17.6.1 Modes of Convergence

Definition 17.6.1 (Convergence of random variables). Let X_n and X be random variables. We say:

1. $X_n \xrightarrow{\text{a.s.}} X$ (**almost sure** convergence) if $P(\lim_{n\to\infty} X_n = X) = 1$.

2. $X_n \xrightarrow{P} X$ (**convergence in probability**) if for all $\epsilon > 0$:
$$\lim_{n\to\infty} P(|X_n - X| > \epsilon) = 0.$$

3. $X_n \xrightarrow{L^p} X$ (L^p **convergence**) if $\lim_{n\to\infty} \mathbb{E}[|X_n - X|^p] = 0$.

4. $X_n \xrightarrow{d} X$ (**convergence in distribution**) if $F_{X_n}(x) \to F_X(x)$ at all continuity points of F_X.

Theorem 17.6.1 (Relationships between convergence modes).

$$a.s. \Rightarrow P \Rightarrow d, \qquad L^p \Rightarrow P \Rightarrow d.$$

None of the reverse implications hold in general. However:

(i) $X_n \xrightarrow{d} c$ *(constant) implies* $X_n \xrightarrow{P} c$.

(ii) $X_n \xrightarrow{P} X$ *implies a subsequence* $X_{n_k} \xrightarrow{a.s.} X$.

Theorem 17.6.2 (Lévy's continuity theorem). $X_n \xrightarrow{d} X$ *if and only if* $\varphi_{X_n}(t) \to \varphi_X(t)$ *for all* t.

17.6.2 Laws of Large Numbers

Let $X_1, X_2, \ldots$ be i.i.d. random variables with mean μ and define $\bar{X}_n := \frac{1}{n}\sum_{i=1}^n X_i$.

Theorem 17.6.3 (Weak law of large numbers). *If* $\mathbb{E}[|X_1|] < \infty$, *then:*

$$\bar{X}_n \xrightarrow{P} \mu.$$

Theorem 17.6.4 (Strong law of large numbers). *If* $\mathbb{E}[|X_1|] < \infty$, *then:*

$$\bar{X}_n \xrightarrow{a.s.} \mu.$$

Remark 17.6.1 (Physical interpretation). The law of large numbers justifies the ergodic hypothesis in statistical mechanics: time averages equal ensemble averages for systems exploring phase space sufficiently.

17.6.3 Central Limit Theorem

Theorem 17.6.5 (Central Limit Theorem (CLT)). *Let* $X_1, X_2, \ldots$ *be i.i.d. with mean* μ *and variance* $\sigma^2 < \infty$. *Then:*

$$\frac{\bar{X}_n - \mu}{\sigma/\sqrt{n}} = \frac{\sum_{i=1}^n X_i - n\mu}{\sigma\sqrt{n}} \xrightarrow{d} N(0,1).$$

Equivalently: $\sqrt{n}(\bar{X}_n - \mu) \xrightarrow{d} N(0, \sigma^2)$.

Theorem 17.6.6 (Berry-Esseen theorem). *Under CLT conditions, if* $\mathbb{E}[|X_1|^3] = \rho < \infty$*:*

$$\sup_x \left| P\left(\frac{\bar{X}_n - \mu}{\sigma/\sqrt{n}} \leq x\right) - \Phi(x) \right| \leq \frac{C\rho}{\sigma^3\sqrt{n}},$$

where Φ *is the standard normal CDF and* $C \leq 0.4748$.

Theorem 17.6.7 (Multivariate CLT). *Let* $\mathbf{X}_1, \mathbf{X}_2, \ldots$ *be i.i.d. random vectors in* $\mathbb{R}^d$ *with mean* $\boldsymbol{\mu}$ *and covariance matrix* $\boldsymbol{\Sigma}$. *Then:*

$$\sqrt{n}(\bar{\mathbf{X}}_n - \boldsymbol{\mu}) \xrightarrow{d} N(\mathbf{0}, \boldsymbol{\Sigma}).$$

Example 17.6.1 (CLT approximation). In 100 fair coin flips, what is the probability of at least 60 heads? Here $X_i \sim \text{Bernoulli}(1/2)$, so $S_{100} = \sum X_i \sim \text{Binomial}(100, 1/2)$ with $\mu = 50$, $\sigma = 5$. By CLT with continuity correction:

$$P(S_{100} \geq 60) \approx P\left(Z \geq \frac{59.5 - 50}{5}\right) = P(Z \geq 1.9) \approx 0.029.$$

17.7 Concentration Inequalities

Concentration inequalities bound the probability that a random variable deviates from its expected value, providing finite-sample guarantees beyond the asymptotic CLT.

17.7.1 Basic Inequalities

Theorem 17.7.1 (Markov's inequality). *For non-negative X and $a > 0$:*

$$P(X \geq a) \leq \frac{\mathbb{E}[X]}{a}.$$

Theorem 17.7.2 (Chebyshev's inequality). *For any X with finite variance and $k > 0$:*

$$P(|X - \mathbb{E}[X]| \geq k\sigma) \leq \frac{1}{k^2}.$$

Equivalently: $P(|X - \mu| \geq \epsilon) \leq \sigma^2/\epsilon^2$.

Theorem 17.7.3 (Chernoff bound). *For any X and $t > 0$:*

$$P(X \geq a) \leq \inf_{t>0} e^{-ta} M_X(t) = \inf_{t>0} e^{-ta} \mathbb{E}[e^{tX}].$$

17.7.2 Sub-Gaussian Random Variables

Definition 17.7.1 (Sub-Gaussian). A mean-zero random variable X is σ-**sub-Gaussian** if:

$$\mathbb{E}[e^{tX}] \leq e^{\sigma^2 t^2/2} \quad \text{for all } t \in \mathbb{R}.$$

Proposition 17.7.4 (Sub-Gaussian tail bound). *If X is σ-sub-Gaussian:*

$$P(X \geq \epsilon) \leq e^{-\epsilon^2/(2\sigma^2)}.$$

Example 17.7.1. Bounded random variables are sub-Gaussian: if $a \leq X \leq b$, then X is $(b-a)/2$-sub-Gaussian (Hoeffding's lemma).

Theorem 17.7.5 (Hoeffding's inequality). *Let $X_1, \ldots, X_n$ be independent with $a_i \leq X_i \leq b_i$. Then:*

$$P\left(\sum_{i=1}^{n} (X_i - \mathbb{E}[X_i]) \geq t\right) \leq \exp\left(-\frac{2t^2}{\sum_{i=1}^{n}(b_i - a_i)^2}\right).$$

Corollary 17.7.6. *For i.i.d. $X_i \in [0,1]$:*

$$P(|\bar{X}_n - \mu| \geq \epsilon) \leq 2e^{-2n\epsilon^2}.$$

17.7.3 Tail Bounds for Sums

Theorem 17.7.7 (Bernstein's inequality). *Let $X_1, \ldots, X_n$ be independent mean-zero with $|X_i| \leq M$ and $\sum \mathbb{E}[X_i^2] = \sigma^2$. Then:*

$$P\left(\sum_{i=1}^{n} X_i \geq t\right) \leq \exp\left(-\frac{t^2/2}{\sigma^2 + Mt/3}\right).$$

This interpolates between sub-Gaussian behavior (small t) and sub-exponential (large t).

17.8 Applications to Physics

17.8.1 Quantum Measurement

In quantum mechanics, measurement outcomes are inherently probabilistic. For an observable $\hat{A}$ with spectral decomposition $\hat{A} = \sum_a a|a\rangle\langle a|$, the probability of obtaining eigenvalue a when measuring state $|\psi\rangle$ is:

$$P(a) = |\langle a|\psi\rangle|^2.$$

The expectation value is:

$$\langle \hat{A} \rangle = \langle \psi|\hat{A}|\psi\rangle = \sum_a a \cdot P(a).$$

The variance $(\Delta A)^2 = \langle \hat{A}^2 \rangle - \langle \hat{A} \rangle^2$ quantifies measurement uncertainty. The Heisenberg uncertainty principle:

$$\Delta A \cdot \Delta B \geq \frac{1}{2}|\langle [\hat{A}, \hat{B}] \rangle|$$

is fundamentally a statement about variances in the probability distributions of non-commuting observables.

17.8.2 Statistical Mechanics

The Boltzmann distribution gives the probability of a microstate with energy E in thermal equilibrium:

$$P(E) = \frac{e^{-E/k_B T}}{Z}, \quad Z = \sum_{\text{states}} e^{-E/k_B T}.$$

For the canonical ensemble, the partition function Z is essentially a moment generating function:

$$Z(\beta) = \sum_n e^{-\beta E_n} = \mathbb{E}[e^{-\beta E}], \quad \beta = \frac{1}{k_B T}.$$

Thermodynamic quantities follow from derivatives:

$$\langle E \rangle = -\frac{\partial \ln Z}{\partial \beta}, \quad \mathrm{Var}(E) = \frac{\partial^2 \ln Z}{\partial \beta^2} = k_B T^2 C_V.$$

Example 17.8.1 (Maxwell-Boltzmann distribution)**.** For an ideal gas molecule with kinetic energy $E = \frac{1}{2}m|\mathbf{v}|^2$, the velocity distribution is:

$$f(\mathbf{v}) = \left(\frac{m}{2\pi k_B T}\right)^{3/2} \exp\left(-\frac{m|\mathbf{v}|^2}{2k_B T}\right),$$

a 3D Gaussian. The speed distribution is:

$$f(v) = 4\pi v^2 \left(\frac{m}{2\pi k_B T}\right)^{3/2} \exp\left(-\frac{mv^2}{2k_B T}\right).$$

17.8.3 Fluctuations and Response

The fluctuation-dissipation theorem connects equilibrium fluctuations to response functions:

$$\langle \delta A \cdot \delta B \rangle = k_B T \chi_{AB},$$

where χ_{AB} is a generalized susceptibility. This profound connection between probability (fluctuations) and physics (response) underlies noise analysis and the Kubo formula.

17.8.4 Central Limit Theorem in Physics

The CLT explains why many physical quantities are normally distributed:

- Thermal fluctuations (sum of many molecular collisions)
- Measurement errors (sum of many small independent errors)
- Random walks and diffusion
- Shot noise in electronic devices

The law of large numbers justifies the thermodynamic limit: extensive quantities scale with N, while fluctuations scale as $\sqrt{N}$, becoming negligible for macroscopic systems.

Exercises

Exercise 17.8.1. Prove that $P(A \cup B) = P(A) + P(B) - P(A \cap B)$ using the axioms of probability.

Exercise 17.8.2. A test for a disease has 95% sensitivity and 90% specificity. If 2% of the population has the disease, find the probability that a person has the disease given a positive test result.

Exercise 17.8.3. Compute $\mathbb{E}[X]$ and $\mathrm{Var}(X)$ for $X \sim \mathrm{Poisson}(\lambda)$ using the probability generating function $G(s) = \mathbb{E}[s^X]$.

Exercise 17.8.4. Show that $\mathrm{Var}(X) = \mathbb{E}[X^2] - (\mathbb{E}[X])^2$ and use this to derive $\mathrm{Var}(aX + b) = a^2 \mathrm{Var}(X)$.

Exercise 17.8.5. Derive the MGF of the Gaussian distribution $N(\mu, \sigma^2)$ and use it to verify that the sum of independent Gaussians is Gaussian.

Exercise 17.8.6. For jointly Gaussian random variables (X, Y), prove that $\mathrm{Cov}(X, Y) = 0$ implies $X \perp Y$. Show by counterexample that this fails for non-Gaussian variables.

Exercise 17.8.7. Prove the law of iterated expectation: $\mathbb{E}[X] = \mathbb{E}[\mathbb{E}[X|Y]]$.

Exercise 17.8.8. Use the CLT (with continuity correction) to approximate $P(S_{100} \geq 55)$ where S_{100} is the number of heads in 100 fair coin flips. Compare with the exact value.

Exercise 17.8.9. Derive Chebyshev's inequality from Markov's inequality. Then derive the one-sided Cantelli inequality: $P(X - \mu \geq t) \leq \sigma^2/(\sigma^2 + t^2)$.

Exercise 17.8.10. Show that the sum of independent Poisson random variables is Poisson: if $X \sim \mathrm{Poisson}(\lambda)$ and $Y \sim \mathrm{Poisson}(\mu)$ are independent, then $X + Y \sim \mathrm{Poisson}(\lambda + \mu)$.

Exercise 17.8.11. Verify that the characteristic function of the Cauchy distribution is $\varphi(t) = e^{-|t|}$. Use this to explain why the Cauchy distribution has no finite mean.

Exercise 17.8.12. Derive the Maxwell-Boltzmann speed distribution from the Boltzmann factor $e^{-E/k_B T}$ with $E = \frac{1}{2}mv^2$, carefully handling the transformation from velocity to speed.

Exercise 17.8.13. For a quantum harmonic oscillator with energy levels $E_n = \hbar\omega(n + \frac{1}{2})$, compute the partition function $Z(\beta)$ and derive $\langle E \rangle$ and $\mathrm{Var}(E)$.

Exercise 17.8.14. Prove that for any random variable X with $\mathbb{E}[X^2] < \infty$ and any constant c:

$$\mathbb{E}[(X - c)^2] = \mathrm{Var}(X) + (\mathbb{E}[X] - c)^2,$$

showing that the mean minimizes the expected squared deviation.

Exercise 17.8.15. Let $X_1, \dots, X_n$ be i.i.d. Exponential(λ). Find the distribution of $\bar{X}_n = \frac{1}{n} \sum X_i$ and show directly (without the CLT) that $\bar{X}_n \xrightarrow{P} 1/\lambda$.

Chapter 18

Stochastic Processes

Stochastic processes describe systems that evolve randomly in time. They are fundamental to statistical mechanics, quantum field theory, financial physics, and the study of fluctuations in physical systems. This chapter develops the theory of stochastic processes with emphasis on Markov processes, Brownian motion, and the stochastic differential equations that model physical phenomena.

18.1 Basic Concepts

18.1.1 Definition and Classification

Definition 18.1.1 (Stochastic process). A **stochastic process** is a collection of random variables $\{X_t\}_{t \in T}$ defined on a probability space $(\Omega, \mathcal{F}, P)$, indexed by a parameter set T (usually time).

For each fixed $\omega \in \Omega$, the function $t \mapsto X_t(\omega)$ is called a **sample path**, **realization**, or **trajectory**.

Definition 18.1.2 (State space and index set). The **state space** S is the set of possible values of X_t. The **index set** T represents time:

- **Discrete-time**: $T = \{0, 1, 2, \dots\}$ or $T = \mathbb{Z}$
- **Continuous-time**: $T = [0, \infty)$ or $T = \mathbb{R}$

Definition 18.1.3 (Finite-dimensional distributions). The **finite-dimensional distributions** of $\{X_t\}$ are the joint distributions:

$$F_{t_1,\dots,t_n}(x_1, \dots, x_n) = P(X_{t_1} \le x_1, \dots, X_{t_n} \le x_n)$$

for all finite collections $t_1, \dots, t_n \in T$.

Theorem 18.1.1 (Kolmogorov extension theorem). *A consistent family of finite-dimensional distributions uniquely determines a stochastic process (on a suitable probability space).*

18.1.2 Important Properties

Definition 18.1.4 (Stationary process). $\{X_t\}$ is **(strictly) stationary** if for all n, all $t_1, \ldots, t_n$, and all h:

$$(X_{t_1}, \ldots, X_{t_n}) \stackrel{d}{=} (X_{t_1+h}, \ldots, X_{t_n+h}).$$

It is **wide-sense stationary** (WSS) if $\mathbb{E}[X_t] = \mu$ (constant) and $\text{Cov}(X_s, X_t) = C(t-s)$ depends only on the time difference.

Definition 18.1.5 (Independent increments). $\{X_t\}$ has **independent increments** if for $t_0 < t_1 < \cdots < t_n$, the increments $X_{t_1} - X_{t_0}, X_{t_2} - X_{t_1}, \ldots, X_{t_n} - X_{t_{n-1}}$ are independent.

Definition 18.1.6 (Stationary increments). $\{X_t\}$ has **stationary increments** if the distribution of $X_{t+h} - X_t$ depends only on h, not on t.

Definition 18.1.7 (Autocorrelation and autocovariance). The **autocovariance function** is:

$$C(s,t) = \text{Cov}(X_s, X_t) = \mathbb{E}[(X_s - \mu_s)(X_t - \mu_t)].$$

The **autocorrelation function** is:

$$R(s,t) = \mathbb{E}[X_s X_t].$$

For WSS processes: $C(\tau) = C(0, \tau)$ and $R(\tau) = R(0, \tau)$ where $\tau = t - s$.

Theorem 18.1.2 (Wiener-Khinchin theorem). *For a WSS process, the* ***power spectral density*** *$S(\omega)$ and autocovariance $C(\tau)$ are Fourier transform pairs:*

$$S(\omega) = \int_{-\infty}^{\infty} C(\tau) e^{-i\omega\tau}\, d\tau, \quad C(\tau) = \frac{1}{2\pi} \int_{-\infty}^{\infty} S(\omega) e^{i\omega\tau}\, d\omega.$$

18.2 Markov Chains

18.2.1 Discrete-Time Markov Chains

Definition 18.2.1 (Markov property). A stochastic process $\{X_n\}_{n\geq 0}$ with discrete state space S satisfies the **Markov property** if:

$$P(X_{n+1} = j | X_n = i, X_{n-1} = i_{n-1}, \ldots, X_0 = i_0) = P(X_{n+1} = j | X_n = i).$$

The process is called a **Markov chain**.

The Markov property states that the future depends on the past only through the present—the process is "memoryless."

Definition 18.2.2 (Transition probabilities). The **one-step transition probability** is:

$$p_{ij}(n) = P(X_{n+1} = j | X_n = i).$$

The chain is **time-homogeneous** if $p_{ij}(n) = p_{ij}$ is independent of n.

Definition 18.2.3 (Transition matrix). For a time-homogeneous chain, the **transition matrix** $P = (p_{ij})$ satisfies:

(i) $p_{ij} \geq 0$ for all i, j

(ii) $\sum_j p_{ij} = 1$ for all i (rows sum to 1)

Such a matrix is called **stochastic**.

Theorem 18.2.1 (Chapman-Kolmogorov equations). *The n-step transition probabilities $p_{ij}^{(n)} = P(X_n = j | X_0 = i)$ satisfy:*

$$p_{ij}^{(n+m)} = \sum_k p_{ik}^{(n)} p_{kj}^{(m)}.$$

In matrix form: $P^{(n+m)} = P^{(n)} P^{(m)}$, hence $P^{(n)} = P^n$.

Definition 18.2.4 (Initial distribution). The **initial distribution** $\pi^{(0)} = (\pi_i^{(0)})$ where $\pi_i^{(0)} = P(X_0 = i)$.

The distribution at time n is:

$$\pi^{(n)} = \pi^{(0)} P^n.$$

18.2.2 Classification of States

Definition 18.2.5 (Communication). State j is **accessible** from i (written $i \to j$) if $p_{ij}^{(n)} > 0$ for some $n \geq 0$.

States i and j **communicate** (written $i \leftrightarrow j$) if $i \to j$ and $j \to i$.

Proposition 18.2.2. *Communication is an equivalence relation, partitioning S into **communicating classes**.*

Definition 18.2.6 (Irreducibility). A Markov chain is **irreducible** if all states communicate (single communicating class).

Definition 18.2.7 (Periodicity)**.** The **period** of state i is:

$$d_i = \gcd\{n \geq 1 : p_{ii}^{(n)} > 0\}.$$

State i is **aperiodic** if $d_i = 1$. An irreducible chain is aperiodic if any (hence all) states are aperiodic.

Definition 18.2.8 (Recurrence and transience)**.** Let $f_{ii} = P(\text{ever return to } i | X_0 = i)$.

- State i is **recurrent** if $f_{ii} = 1$ (certain return).
- State i is **transient** if $f_{ii} < 1$ (may never return).

Theorem 18.2.3 (Recurrence criterion)**.** *State i is recurrent iff $\sum_{n=0}^{\infty} p_{ii}^{(n)} = \infty$.*

Definition 18.2.9 (Positive and null recurrence)**.** For a recurrent state i, let $\mu_i = \mathbb{E}[\text{return time to } i | X_0 = i]$.

- i is **positive recurrent** if $\mu_i < \infty$.
- i is **null recurrent** if $\mu_i = \infty$.

18.2.3 Stationary Distributions and Ergodicity

Definition 18.2.10 (Stationary distribution)**.** A distribution $\pi = (\pi_j)$ is **stationary** (or **invariant**) if:

$$\pi P = \pi, \quad \text{i.e.,} \quad \pi_j = \sum_i \pi_i p_{ij} \text{ for all } j.$$

Theorem 18.2.4 (Existence and uniqueness)**.** *For an irreducible Markov chain:*

(i) A stationary distribution exists iff the chain is positive recurrent.

(ii) If it exists, it is unique and $\pi_i = 1/\mu_i$.

Theorem 18.2.5 (Ergodic theorem for Markov chains)**.** *For an irreducible, aperiodic, positive recurrent Markov chain with stationary distribution π:*

$$\lim_{n\to\infty} p_{ij}^{(n)} = \pi_j \quad \textit{for all } i, j.$$

Moreover, for any function f with $\sum_j |f(j)|\pi_j < \infty$:

$$\frac{1}{n}\sum_{k=0}^{n-1} f(X_k) \xrightarrow{a.s.} \sum_j f(j)\pi_j = \mathbb{E}_\pi[f].$$

Definition 18.2.11 (Detailed balance). A distribution π satisfies **detailed balance** with respect to P if:

$$\pi_i p_{ij} = \pi_j p_{ji} \quad \text{for all } i, j.$$

Such a chain is called **reversible**.

Proposition 18.2.6. *If π satisfies detailed balance, then π is stationary.*

Example 18.2.1 (Random walk on a graph). For a random walk on an undirected graph where transitions occur to neighbors with equal probability, the stationary distribution is $\pi_i = d_i/(2|E|)$ where d_i is the degree of vertex i and $|E|$ is the number of edges.

18.3 Continuous-Time Markov Processes

18.3.1 Continuous-Time Markov Chains

Definition 18.3.1 (Continuous-time Markov chain). A continuous-time stochastic process $\{X_t\}_{t\geq 0}$ with countable state space S is a **continuous-time Markov chain** (CTMC) if:

$$P(X_{t+s} = j | X_s = i, \{X_u\}_{u<s}) = P(X_{t+s} = j | X_s = i)$$

for all $s, t \geq 0$.

Definition 18.3.2 (Transition rates). For a time-homogeneous CTMC, the **transition rate** from i to j is:

$$q_{ij} = \lim_{h\to 0^+} \frac{P(X_{t+h} = j | X_t = i)}{h} \quad (i \neq j).$$

The **holding rate** at i is $q_i = \sum_{j\neq i} q_{ij}$.

Definition 18.3.3 (Generator matrix). The **generator** (or **Q-matrix**) $Q = (q_{ij})$ has:

- Off-diagonal: $q_{ij} \geq 0$ for $i \neq j$
- Diagonal: $q_{ii} = -q_i = -\sum_{j\neq i} q_{ij}$

Rows sum to zero: $\sum_j q_{ij} = 0$.

Theorem 18.3.1 (Kolmogorov equations). *The transition probabilities $p_{ij}(t) = P(X_t = j | X_0 = i)$ satisfy:*
Forward equation: $\frac{d}{dt}P(t) = P(t)Q$
Backward equation: $\frac{d}{dt}P(t) = QP(t)$
Solution: $P(t) = e^{Qt}$ (matrix exponential).

Theorem 18.3.2 (Holding times). *In state i, the process remains for an exponentially distributed time with rate q_i:*

$$P(\textit{holding time} > t | X_0 = i) = e^{-q_i t}.$$

Upon leaving i, the process jumps to $j \neq i$ with probability q_{ij}/q_i.

Definition 18.3.4 (Stationary distribution for CTMC). π is stationary if $\pi Q = 0$, i.e., $\sum_i \pi_i q_{ij} = 0$ for all j.

Example 18.3.1 (Poisson process). The **Poisson process** N_t counts events occurring at rate λ:

- $N_0 = 0$
- Independent increments
- $N_{t+s} - N_s \sim \text{Poisson}(\lambda t)$

Generator: $q_{n,n+1} = \lambda$, $q_{nn} = -\lambda$, all other $q_{ij} = 0$.

Example 18.3.2 (Birth-death process). States $S = \{0, 1, 2, \ldots\}$ with birth rates λ_n and death rates μ_n:

$$q_{n,n+1} = \lambda_n, \quad q_{n,n-1} = \mu_n, \quad q_{nn} = -(\lambda_n + \mu_n).$$

Stationary distribution (if it exists): $\pi_n = \pi_0 \prod_{k=0}^{n-1} \frac{\lambda_k}{\mu_{k+1}}$.

18.4 Random Walks

18.4.1 Simple Random Walk

Definition 18.4.1 (Simple random walk). The **simple random walk** on $\mathbb{Z}$ is $S_n = \sum_{k=1}^{n} X_k$ where X_k are i.i.d. with:

$$P(X_k = +1) = p, \quad P(X_k = -1) = 1 - p = q.$$

The walk is **symmetric** if $p = 1/2$.

Theorem 18.4.1 (Properties of simple random walk). *For the simple random walk starting at $S_0 = 0$:*

(i) $\mathbb{E}[S_n] = n(p - q)$

(ii) $\text{Var}(S_n) = 4npq$

(iii) $P(S_n = k) = \binom{n}{(n+k)/2} p^{(n+k)/2} q^{(n-k)/2}$ *for* $k \equiv n \pmod 2$

Theorem 18.4.2 (Recurrence of random walk). *The simple random walk on $\mathbb{Z}$ is:*

- *Recurrent if $p = 1/2$ (symmetric)*
- *Transient if $p \neq 1/2$*

More generally, the symmetric random walk on $\mathbb{Z}^d$ is recurrent iff $d \leq 2$ (Pólya's theorem).

Theorem 18.4.3 (Reflection principle). *For a symmetric random walk:*

$$P(\max_{0 \leq k \leq n} S_k \geq a) = 2P(S_n \geq a) - P(S_n = a).$$

18.4.2 Random Walk and Diffusion

Theorem 18.4.4 (Diffusion limit). *Consider a random walk with step size Δx and time step Δt. In the limit $\Delta x, \Delta t \to 0$ with $(\Delta x)^2/(2\Delta t) = D$ (diffusion coefficient), the rescaled walk converges to Brownian motion with diffusion coefficient D.*

Theorem 18.4.5 (Central limit theorem for random walks). *For a random walk with $\mathbb{E}[X_k] = \mu$ and $\mathrm{Var}(X_k) = \sigma^2$:*

$$\frac{S_n - n\mu}{\sigma\sqrt{n}} \xrightarrow{d} N(0, 1).$$

18.5 Brownian Motion

18.5.1 Definition and Basic Properties

Definition 18.5.1 (Brownian motion / Wiener process). A stochastic process $\{B_t\}_{t \geq 0}$ is a **standard Brownian motion** (or **Wiener process**) if:

(i) $B_0 = 0$ (starts at origin)

(ii) Independent increments: $B_{t_1}, B_{t_2} - B_{t_1}, \ldots, B_{t_n} - B_{t_{n-1}}$ are independent for $t_1 < \cdots < t_n$

(iii) Stationary Gaussian increments: $B_{t+s} - B_s \sim N(0, t)$ for all $s, t \geq 0$

(iv) Continuous sample paths: $t \mapsto B_t(\omega)$ is continuous a.s.

Theorem 18.5.1 (Existence). *Brownian motion exists and is unique in distribution.*

Proposition 18.5.2 (Basic properties of Brownian motion).

(i) $\mathbb{E}[B_t] = 0$

(ii) $\text{Var}(B_t) = t$

(iii) $\text{Cov}(B_s, B_t) = \min(s, t)$

(iv) $\{B_t\}$ *is a Gaussian process: all finite-dimensional distributions are multivariate Gaussian*

(v) $\{B_t\}$ *is a martingale:* $\mathbb{E}[B_t|B_s] = B_s$ *for* $s \leq t$

Theorem 18.5.3 (Scaling properties). *If* $\{B_t\}$ *is a standard Brownian motion, then so are:*

(i) $\{-B_t\}$ *(reflection)*

(ii) $\{cB_{t/c^2}\}$ *for* $c > 0$ *(scaling)*

(iii) $\{tB_{1/t}\}$ *for* $t > 0$*, with* $0 \cdot B_\infty := 0$ *(time inversion)*

(iv) $\{B_{t+s} - B_s\}$ *for fixed* s *(time translation)*

Theorem 18.5.4 (Non-differentiability). *Brownian motion paths are almost surely:*

(i) Continuous everywhere

(ii) Differentiable nowhere

(iii) Of unbounded variation on any interval

18.5.2 Quadratic Variation

Definition 18.5.2 (Quadratic variation). The **quadratic variation** of $\{B_t\}$ on $[0, t]$ is:

$$[B]_t := \lim_{n\to\infty} \sum_{k=0}^{n-1} (B_{t_{k+1}} - B_{t_k})^2$$

where the limit is over partitions with mesh going to zero.

Theorem 18.5.5 (Quadratic variation of Brownian motion). $[B]_t = t$ *almost surely. Informally:* $(dB_t)^2 = dt$*.*

This result is fundamental to stochastic calculus and explains why Brownian motion requires special integration techniques.

18.5.3 Brownian Motion with Drift

Definition 18.5.3 (Brownian motion with drift). A **Brownian motion with drift** μ and volatility σ is:

$$X_t = \mu t + \sigma B_t.$$

Then $X_t \sim N(\mu t, \sigma^2 t)$.

Definition 18.5.4 (Geometric Brownian motion). **Geometric Brownian motion** is:

$$S_t = S_0 \exp\left(\left(\mu - \frac{\sigma^2}{2}\right)t + \sigma B_t\right).$$

This is the standard model for stock prices in mathematical finance.

18.5.4 Brownian Bridge and Other Variants

Definition 18.5.5 (Brownian bridge). A **Brownian bridge** from 0 to 0 on $[0, 1]$ is:

$$B_t^0 = B_t - tB_1,$$

or equivalently, Brownian motion conditioned on $B_1 = 0$.

Proposition 18.5.6 (Properties of Brownian bridge).

(i) $B_0^0 = B_1^0 = 0$

(ii) $\mathbb{E}[B_t^0] = 0$

(iii) $\mathrm{Cov}(B_s^0, B_t^0) = \min(s,t) - st = s(1-t)$ *for* $s \leq t$

18.6 Stochastic Calculus

18.6.1 The Itô Integral

Standard Riemann-Stieltjes integration fails for Brownian motion due to its unbounded variation. The Itô integral provides the appropriate framework.

Definition 18.6.1 (Itô integral for simple processes). For a **simple process** $f_t = \sum_{k=0}^{n-1} f_{t_k} \mathbf{1}_{(t_k, t_{k+1}]}(t)$ (piecewise constant, adapted):

$$\int_0^T f_t \, dB_t := \sum_{k=0}^{n-1} f_{t_k}(B_{t_{k+1}} - B_{t_k}).$$

Note: f is evaluated at the *left* endpoint (non-anticipating).

Theorem 18.6.1 (Properties of the Itô integral). *For adapted processes f, g with $\mathbb{E}[\int_0^T f_t^2\,dt] < \infty$:*

(i) ***Linearity:*** $\int (af + bg)\,dB = a\int f\,dB + b\int g\,dB$

(ii) ***Itô isometry:*** $\mathbb{E}\left[\left(\int_0^T f_t\,dB_t\right)^2\right] = \mathbb{E}\left[\int_0^T f_t^2\,dt\right]$

(iii) ***Martingale:*** $M_t = \int_0^t f_s\,dB_s$ *is a martingale*

(iv) ***Zero mean:*** $\mathbb{E}\left[\int_0^T f_t\,dB_t\right] = 0$

Example 18.6.1 (Fundamental Itô integrals). 1. $\int_0^t dB_s = B_t$

2. $\int_0^t B_s\,dB_s = \frac{1}{2}(B_t^2 - t)$ (note: not $\frac{1}{2}B_t^2$!)

3. $\int_0^t s\,dB_s = tB_t - \int_0^t B_s\,ds$ (integration by parts)

18.6.2 Itô's Formula

Theorem 18.6.2 (Itô's formula (one dimension)). *Let $f(t,x) \in C^{1,2}$ and $X_t = X_0 + \int_0^t \mu_s\,ds + \int_0^t \sigma_s\,dB_s$. Then:*

$$df(t, X_t) = \frac{\partial f}{\partial t}\,dt + \frac{\partial f}{\partial x}\,dX_t + \frac{1}{2}\frac{\partial^2 f}{\partial x^2}\,(dX_t)^2$$

where $(dX_t)^2 = \sigma_t^2\,dt$ (using $dB_t \cdot dB_t = dt$, $dt \cdot dB_t = 0$, $dt \cdot dt = 0$).
Explicitly:

$$f(t, X_t) = f(0, X_0) + \int_0^t \left(\frac{\partial f}{\partial s} + \mu_s\frac{\partial f}{\partial x} + \frac{\sigma_s^2}{2}\frac{\partial^2 f}{\partial x^2}\right) ds + \int_0^t \sigma_s\frac{\partial f}{\partial x}\,dB_s.$$

The key difference from ordinary calculus is the second-order term $\frac{1}{2}\sigma^2 f_{xx}\,dt$, arising from the quadratic variation of Brownian motion.

Example 18.6.2 (Deriving $\int B_s\,dB_s$). Apply Itô's formula to $f(x) = x^2/2$ with $X_t = B_t$:

$$d(B_t^2/2) = B_t\,dB_t + \frac{1}{2}\,dt.$$

Integrating: $\frac{1}{2}B_t^2 = \int_0^t B_s\,dB_s + \frac{t}{2}$, giving the result.

Theorem 18.6.3 (Itô's formula (multidimensional)). *For $\mathbf{X}_t = (X_t^1, \ldots, X_t^n)$ with $dX_t^i = \mu_t^i\,dt + \sum_j \sigma_t^{ij}\,dB_t^j$ and $f \in C^{1,2}$:*

$$df = \frac{\partial f}{\partial t}\,dt + \sum_i \frac{\partial f}{\partial x^i}\,dX_t^i + \frac{1}{2}\sum_{i,j}\frac{\partial^2 f}{\partial x^i \partial x^j}\,dX_t^i\,dX_t^j$$

where $dB_t^i\,dB_t^j = \delta_{ij}\,dt$.

18.6.3 Stratonovich Integral

Definition 18.6.2 (Stratonovich integral). The **Stratonovich integral** uses midpoint evaluation:

$$\int_0^T f_t \circ dB_t := \lim_{n\to\infty} \sum_{k=0}^{n-1} f_{(t_k+t_{k+1})/2}(B_{t_{k+1}} - B_{t_k}).$$

Theorem 18.6.4 (Itô-Stratonovich conversion).

$$\int_0^T f(B_t) \circ dB_t = \int_0^T f(B_t)\, dB_t + \frac{1}{2}\int_0^T f'(B_t)\, dt.$$

Remark 18.6.1. Stratonovich calculus obeys ordinary chain rules, making it natural for physics where equations have physical meaning independent of the noise interpretation. Itô calculus is preferred in mathematics and finance for its martingale properties.

18.7 Stochastic Differential Equations

18.7.1 Definition and Existence

Definition 18.7.1 (Stochastic differential equation). An **Itô SDE** has the form:

$$dX_t = \mu(t, X_t)\, dt + \sigma(t, X_t)\, dB_t$$

with initial condition $X_0 = x_0$. Here μ is the **drift** and σ is the **diffusion coefficient** (or **volatility**).

Integral form:

$$X_t = x_0 + \int_0^t \mu(s, X_s)\, ds + \int_0^t \sigma(s, X_s)\, dB_s.$$

Theorem 18.7.1 (Existence and uniqueness). *If μ and σ satisfy:*

(i) ***Lipschitz:*** $|\mu(t,x) - \mu(t,y)| + |\sigma(t,x) - \sigma(t,y)| \le K|x-y|$

(ii) ***Linear growth:*** $|\mu(t,x)|^2 + |\sigma(t,x)|^2 \le K^2(1+|x|^2)$

then there exists a unique strong solution.

18.7.2 Important SDEs

Example 18.7.1 (Ornstein-Uhlenbeck process).

$$dX_t = -\theta X_t \, dt + \sigma \, dB_t, \quad \theta > 0.$$

Solution: $X_t = X_0 e^{-\theta t} + \sigma \int_0^t e^{-\theta(t-s)} \, dB_s$.

Stationary distribution: $N(0, \sigma^2/(2\theta))$.

This models velocity in the Langevin equation, mean-reverting processes, and harmonic oscillators with noise.

Example 18.7.2 (Geometric Brownian motion SDE).

$$dS_t = \mu S_t \, dt + \sigma S_t \, dB_t.$$

Solution: $S_t = S_0 \exp\big((\mu - \sigma^2/2)t + \sigma B_t\big)$.

Example 18.7.3 (Cox-Ingersoll-Ross process).

$$dX_t = \kappa(\theta - X_t) \, dt + \sigma \sqrt{X_t} \, dB_t.$$

Used for interest rates and volatility; remains non-negative if $2\kappa\theta \geq \sigma^2$.

18.7.3 Connection to PDEs: Feynman-Kac Formula

Theorem 18.7.2 (Feynman-Kac formula). *Consider the SDE $dX_t = \mu(X_t) \, dt + \sigma(X_t) \, dB_t$ and define:*

$$u(t,x) = \mathbb{E}\left[e^{-\int_t^T r(X_s) \, ds} g(X_T) \,\middle|\, X_t = x \right].$$

Then u solves the PDE:

$$\frac{\partial u}{\partial t} + \mu(x) \frac{\partial u}{\partial x} + \frac{\sigma(x)^2}{2} \frac{\partial^2 u}{\partial x^2} - r(x) u = 0$$

with terminal condition $u(T,x) = g(x)$.

This connects stochastic processes to parabolic PDEs and is the basis for Monte Carlo methods for solving PDEs.

18.8 The Fokker-Planck Equation

18.8.1 Derivation and Interpretation

Theorem 18.8.1 (Fokker-Planck equation). *If X_t satisfies the SDE $dX_t = \mu(x,t) \, dt + \sigma(x,t) \, dB_t$, then the probability density $p(x,t)$ of X_t evolves according to:*

$$\frac{\partial p}{\partial t} = -\frac{\partial}{\partial x}[\mu(x,t)p] + \frac{1}{2} \frac{\partial^2}{\partial x^2}[\sigma(x,t)^2 p].$$

*This is also called the **forward Kolmogorov equation**.*

Remark 18.8.1. The Fokker-Planck equation can be written as a continuity equation:

$$\frac{\partial p}{\partial t} + \frac{\partial J}{\partial x} = 0$$

with probability current $J = \mu p - \frac{1}{2}\frac{\partial}{\partial x}(\sigma^2 p)$.

Theorem 18.8.2 (Fokker-Planck in multiple dimensions). *For $d\mathbf{X}_t = \boldsymbol{\mu}(\mathbf{X}_t)\,dt + \boldsymbol{\sigma}(\mathbf{X}_t)\,d\mathbf{B}_t$:*

$$\frac{\partial p}{\partial t} = -\sum_i \frac{\partial}{\partial x_i}[\mu_i p] + \frac{1}{2}\sum_{i,j} \frac{\partial^2}{\partial x_i \partial x_j}[D_{ij} p]$$

where $D_{ij} = \sum_k \sigma_{ik}\sigma_{jk}$ is the diffusion tensor.

18.8.2 Stationary Solutions

Theorem 18.8.3 (Stationary Fokker-Planck solution). *For time-independent $\mu(x)$ and $\sigma(x)$, a stationary solution $p_s(x)$ satisfies:*

$$\frac{d}{dx}\left[\mu p_s - \frac{1}{2}\frac{d}{dx}(\sigma^2 p_s)\right] = 0.$$

For constant σ and $\mu(x) = -V'(x)$ (gradient flow):

$$p_s(x) \propto \exp\left(-\frac{2V(x)}{\sigma^2}\right).$$

This connects to statistical mechanics: p_s is the Boltzmann distribution with $\sigma^2/2 = k_B T$.

18.9 The Langevin Equation

18.9.1 Physical Motivation

The Langevin equation models a particle subject to friction and random forces (e.g., Brownian particle in a fluid).

Definition 18.9.1 (Langevin equation). The **Langevin equation** for a particle of mass m is:

$$m\frac{dv}{dt} = -\gamma v + F(x) + \sqrt{2\gamma k_B T}\,\xi(t)$$

where:

- γ = friction coefficient

- $F(x)$ = external force
- $\xi(t)$ = Gaussian white noise: $\langle \xi(t) \rangle = 0$, $\langle \xi(t)\xi(t') \rangle = \delta(t - t')$

Remark 18.9.1. The noise strength $\sqrt{2\gamma k_B T}$ is determined by the **fluctuation-dissipation theorem**: the same environment that causes friction also causes fluctuations, with strength related by temperature.

18.9.2 Overdamped Limit

Definition 18.9.2 (Overdamped Langevin equation). In the **overdamped** (high friction) limit $m \to 0$:

$$\gamma \frac{dx}{dt} = F(x) + \sqrt{2\gamma k_B T}\, \xi(t)$$

or equivalently:

$$dx = \frac{F(x)}{\gamma}\, dt + \sqrt{\frac{2k_B T}{\gamma}}\, dB_t.$$

The corresponding Fokker-Planck equation is:

$$\frac{\partial p}{\partial t} = -\frac{1}{\gamma}\frac{\partial}{\partial x}[F(x)p] + \frac{k_B T}{\gamma}\frac{\partial^2 p}{\partial x^2}.$$

For $F(x) = -V'(x)$, the stationary distribution is:

$$p_s(x) = \frac{1}{Z} e^{-V(x)/(k_B T)}$$

— the Boltzmann distribution.

18.9.3 Kramers Problem

Theorem 18.9.1 (Kramers escape rate). *For a particle in a potential well with barrier height ΔV, the escape rate in the overdamped limit is:*

$$r \approx \frac{\omega_0 \omega_b}{2\pi\gamma} e^{-\Delta V/(k_B T)}$$

where ω_0 and ω_b are the frequencies at the well minimum and barrier top.

This is fundamental to chemical reaction rates, protein folding, and many activation processes.

18.10 Applications to Physics

18.10.1 Path Integrals

Theorem 18.10.1 (Feynman-Kac and path integrals). *The Fokker-Planck equation connects to quantum mechanics via the Feynman-Kac formula. The propagator for the diffusion equation:*

$$\frac{\partial p}{\partial t} = D\frac{\partial^2 p}{\partial x^2} - V(x)p$$

can be written as a path integral:

$$p(x,t|x_0,0) = \int_{x(0)=x_0}^{x(t)=x} \mathcal{D}[x(\tau)] \exp\left(-\frac{1}{4D}\int_0^t \left[\dot{x}^2 + V(x)\right] d\tau\right).$$

Remark 18.10.1. Under the substitution $t \to it$ (Wick rotation), this becomes the quantum mechanical path integral with $D = \hbar/(2m)$.

18.10.2 Statistical Mechanics

Example 18.10.1 (Fluctuations in equilibrium). For a system in thermal equilibrium with energy E, fluctuations satisfy:

$$\langle(\Delta E)^2\rangle = k_B T^2 C_V$$

where C_V is the heat capacity. This follows from the connection between the Boltzmann distribution and Fokker-Planck stationary solutions.

Example 18.10.2 (Diffusion). The diffusion equation $\partial_t p = D\nabla^2 p$ is the Fokker-Planck equation for free Brownian motion. Einstein's relation:

$$D = \frac{k_B T}{\gamma}$$

connects diffusion to friction and temperature.

18.10.3 Quantum Field Theory Connection

Remark 18.10.2 (Euclidean field theory). In Euclidean quantum field theory, the path integral measure:

$$\mathcal{D}\phi\, e^{-S[\phi]}$$

where S is the Euclidean action, defines a stochastic process. This connects QFT correlation functions to expectations in stochastic processes, enabling lattice Monte Carlo simulations.

18.11 Summary

Stochastic processes provide the mathematical framework for systems with randomness:

Markov chains: Memoryless processes; transition matrices; stationary distributions; ergodic theorems.

Continuous-time processes: Generator matrices; Kolmogorov equations; Poisson and birth-death processes.

Random walks: Discrete precursors to diffusion; recurrence in $d \leq 2$.

Brownian motion: Continuous paths, independent Gaussian increments, quadratic variation $[B]_t = t$.

Stochastic calculus: Itô integral (non-anticipating); Itô's formula with extra $\frac{1}{2}\sigma^2 f_{xx}\, dt$ term.

SDEs: $dX_t = \mu\, dt + \sigma\, dB_t$; existence under Lipschitz conditions; Feynman-Kac connection to PDEs.

Fokker-Planck: Evolution of probability densities; stationary solutions give Boltzmann distribution.

Langevin equation: Physical model of friction plus noise; fluctuation-dissipation theorem; Kramers escape rates.

Physics applications: Path integrals, statistical mechanics, diffusion, reaction rates.

Exercises

Exercise 18.11.1. For a two-state Markov chain with transition matrix $P = \begin{pmatrix} 1-\alpha & \alpha \\ \beta & 1-\beta \end{pmatrix}$, find the stationary distribution and prove convergence to it.

Exercise 18.11.2. Show that the simple symmetric random walk on $\mathbb{Z}$ is recurrent by computing $\sum_{n=0}^{\infty} p_{00}^{(2n)}$ using Stirling's approximation.

Exercise 18.11.3. Prove that $\text{Cov}(B_s, B_t) = \min(s, t)$ for standard Brownian motion using the definition and properties of independent increments.

Exercise 18.11.4. Use Itô's formula to find $d(e^{B_t})$ and hence compute $\mathbb{E}[e^{B_t}]$.

Exercise 18.11.5. Verify that $\int_0^t B_s\, dB_s = \frac{1}{2}(B_t^2 - t)$ by applying Itô's formula to $f(x) = x^2$.

Exercise 18.11.6. Solve the Ornstein-Uhlenbeck SDE $dX_t = -\theta X_t\, dt + \sigma\, dB_t$ and find the stationary distribution.

Exercise 18.11.7. Derive the Fokker-Planck equation for the geometric Brownian motion $dS_t = \mu S_t\,dt + \sigma S_t\,dB_t$.

Exercise 18.11.8. For a particle in a harmonic potential $V(x) = \frac{1}{2}kx^2$ with overdamped Langevin dynamics, find the stationary distribution and compute $\langle x^2 \rangle$.

Exercise 18.11.9. Show that if X_t is an Itô process with $dX_t = \mu\,dt + \sigma\,dB_t$, then $Y_t = e^{-rt}X_t$ satisfies an SDE. Use this to solve SDEs of the form $dX_t = rX_t\,dt + \sigma\,dB_t$.

Exercise 18.11.10. Prove the Itô isometry: $\mathbb{E}\left[\left(\int_0^T f_t\,dB_t\right)^2\right] = \mathbb{E}\left[\int_0^T f_t^2\,dt\right]$ for simple processes.

Exercise 18.11.11. For a birth-death process with $\lambda_n = \lambda$ and $\mu_n = \mu$ (M/M/1 queue), find the stationary distribution and the condition for its existence.

Exercise 18.11.12. Show that the generator of Brownian motion is $\mathcal{L} = \frac{1}{2}\frac{d^2}{dx^2}$ by computing $\lim_{t\to 0} \frac{\mathbb{E}[f(B_t)]-f(0)}{t}$.

Exercise 18.11.13. Derive the Einstein relation $D = k_B T/\gamma$ by requiring that the overdamped Langevin equation has the Boltzmann distribution as its stationary solution.

Exercise 18.11.14. For the Langevin equation $m\ddot{x} = -\gamma\dot{x} + \sqrt{2\gamma k_B T}\,\xi(t)$, show that $\langle v^2 \rangle = k_B T/m$ in equilibrium (equipartition theorem).

Exercise 18.11.15. Use the Feynman-Kac formula to show that $u(x,t) = \mathbb{E}[g(B_T)|B_t = x]$ solves the backward heat equation $\frac{\partial u}{\partial t} + \frac{1}{2}\frac{\partial^2 u}{\partial x^2} = 0$ with $u(x,T) = g(x)$.

Part VII

Advanced Tools

This final part collects advanced mathematical techniques that appear throughout theoretical physics but require the foundations developed earlier. These tools—integral transforms, asymptotic methods, tensor calculus, and the geometry of gauge theories—represent the cutting edge of mathematical physics.

Chapter 19 develops Fourier analysis and integral transforms, the workhorses of signal processing, quantum mechanics, and field theory. Chapter 20 presents complex-variable methods including saddle-point approximations and dispersion relations. Chapter 21 provides a systematic treatment of tensor calculus and index notation essential for relativity and continuum mechanics. Chapter 22 introduces the geometric framework of fiber bundles and gauge theories underlying the Standard Model of particle physics.

Chapter 19

Fourier and Integral Transforms

Integral transforms convert functions between different representations, often simplifying differential equations to algebraic ones. The Fourier transform connects position and momentum in quantum mechanics, time and frequency in signal analysis, and configuration and reciprocal space in solid-state physics. This chapter develops Fourier analysis rigorously and introduces related transforms essential for physics.

19.1 Fourier Series

19.1.1 Periodic Functions and Trigonometric Series

Definition 19.1.1 (Periodic function)**.** A function $f : \mathbb{R} \to \mathbb{C}$ is **periodic with period** L if $f(x + L) = f(x)$ for all x. The fundamental frequency is $\omega_0 = 2\pi/L$.

Definition 19.1.2 (Fourier series)**.** The **Fourier series** of a periodic function f with period L is:

$$f(x) = \sum_{n=-\infty}^{\infty} c_n e^{2\pi i n x/L} = \sum_{n=-\infty}^{\infty} c_n e^{i n \omega_0 x}$$

where the **Fourier coefficients** are:

$$c_n = \frac{1}{L}\int_0^L f(x) e^{-2\pi i n x/L}\, dx = \frac{1}{L}\int_0^L f(x) e^{-i n \omega_0 x}\, dx.$$

Definition 19.1.3 (Real Fourier series)**.** Equivalently, using real trigonometric functions:

$$f(x) = \frac{a_0}{2} + \sum_{n=1}^{\infty}\left(a_n \cos\frac{2\pi n x}{L} + b_n \sin\frac{2\pi n x}{L}\right)$$

where:

$$a_n = \frac{2}{L}\int_0^L f(x)\cos\frac{2\pi nx}{L}\,dx, \quad b_n = \frac{2}{L}\int_0^L f(x)\sin\frac{2\pi nx}{L}\,dx.$$

Relation to complex coefficients: $c_n = (a_n - ib_n)/2$ for $n > 0$, $c_{-n} = \overline{c_n}$, $c_0 = a_0/2$.

19.1.2 Convergence of Fourier Series

Definition 19.1.4 (Partial sums). The N-th **partial sum** is:

$$S_N(x) = \sum_{n=-N}^{N} c_n e^{in\omega_0 x}.$$

Theorem 19.1.1 (Dirichlet conditions). *If f is periodic with period L and:*

(i) f has a finite number of discontinuities in each period

(ii) f has a finite number of extrema in each period

(iii) $\int_0^L |f(x)|\,dx < \infty$

then the Fourier series converges to $f(x)$ at points of continuity and to $\frac{1}{2}[f(x^+) + f(x^-)]$ at discontinuities.

Theorem 19.1.2 (Pointwise convergence). *If f is piecewise smooth, then $S_N(x) \to f(x)$ at continuity points.*

Theorem 19.1.3 (L^2 convergence). *For $f \in L^2([0, L])$:*

$$\lim_{N\to\infty} \int_0^L |f(x) - S_N(x)|^2\,dx = 0.$$

Theorem 19.1.4 (Uniform convergence). *If f is continuous and piecewise smooth with $f(0) = f(L)$, then $S_N \to f$ uniformly.*

Theorem 19.1.5 (Gibbs phenomenon). *At a jump discontinuity, the partial sums overshoot the function value by approximately 9% of the jump height, regardless of N. This overshoot does not decrease as $N \to \infty$, though it becomes increasingly localized near the discontinuity.*

19.1.3 Properties and Parseval's Theorem

Theorem 19.1.6 (Properties of Fourier coefficients). *Let f have Fourier coefficients $\{c_n\}$ and g have coefficients $\{d_n\}$. Then:*

(i) ***Linearity:*** *$\alpha f + \beta g$ has coefficients $\alpha c_n + \beta d_n$*

(ii) ***Translation:*** *$f(x-a)$ has coefficients $c_n e^{-in\omega_0 a}$*

(iii) ***Differentiation:*** *$f'(x)$ has coefficients $in\omega_0 c_n$*

(iv) ***Conjugation:*** *$\overline{f(x)}$ has coefficients $\overline{c_{-n}}$*

(v) ***Real*** *f: $c_{-n} = \overline{c_n}$*

(vi) ***Even*** *f: $c_{-n} = c_n$ (only cosines)*

(vii) ***Odd*** *f: $c_{-n} = -c_n$ (only sines)*

Theorem 19.1.7 (Parseval's theorem). *For $f \in L^2([0, L])$:*

$$\frac{1}{L}\int_0^L |f(x)|^2\,dx = \sum_{n=-\infty}^{\infty} |c_n|^2 = \frac{a_0^2}{4} + \frac{1}{2}\sum_{n=1}^{\infty}(a_n^2 + b_n^2).$$

This is the statement that energy (power) in the signal equals the sum of energies in each frequency component.

Theorem 19.1.8 (Convolution theorem for Fourier series). *If f and g have period L with coefficients $\{c_n\}$ and $\{d_n\}$, then:*

$$(f * g)(x) := \frac{1}{L}\int_0^L f(y)g(x-y)\,dy$$

has Fourier coefficients $c_n d_n$.

Example 19.1.1 (Square wave). The square wave $f(x) = \operatorname{sgn}(\sin x)$ on $[-\pi, \pi]$ has Fourier series:

$$f(x) = \frac{4}{\pi}\sum_{n=0}^{\infty} \frac{\sin((2n+1)x)}{2n+1} = \frac{4}{\pi}\left(\sin x + \frac{\sin 3x}{3} + \frac{\sin 5x}{5} + \cdots\right).$$

Example 19.1.2 (Sawtooth wave). The sawtooth $f(x) = x$ on $(-\pi, \pi)$, extended periodically:

$$f(x) = 2\sum_{n=1}^{\infty} \frac{(-1)^{n+1}}{n}\sin(nx).$$

At $x = \pi/2$: $\frac{\pi}{4} = 1 - \frac{1}{3} + \frac{1}{5} - \frac{1}{7} + \cdots$ (Leibniz formula).

19.2 The Fourier Transform

19.2.1 Definition and Basic Properties

Definition 19.2.1 (Fourier transform). The **Fourier transform** of $f : \mathbb{R} \to \mathbb{C}$ is:

$$\hat{f}(k) = \mathcal{F}[f](k) = \int_{-\infty}^{\infty} f(x)e^{-ikx}\,dx.$$

The **inverse Fourier transform** is:

$$f(x) = \mathcal{F}^{-1}[\hat{f}](x) = \frac{1}{2\pi}\int_{-\infty}^{\infty} \hat{f}(k)e^{ikx}\,dk.$$

Remark 19.2.1 (Convention variations). Different conventions exist. Physics often uses:

$$\hat{f}(\omega) = \int f(t)e^{i\omega t}\,dt, \quad f(t) = \frac{1}{2\pi}\int \hat{f}(\omega)e^{-i\omega t}\,d\omega.$$

A symmetric convention uses $(2\pi)^{-1/2}$ in both transforms. We use the convention standard in mathematical physics.

Theorem 19.2.1 (Fourier inversion). *If f and $\hat{f}$ are both in $L^1(\mathbb{R})$, then $\mathcal{F}^{-1}[\mathcal{F}[f]] = f$ almost everywhere. If f is continuous, equality holds everywhere.*

Theorem 19.2.2 (Existence conditions). *The Fourier transform exists (as an ordinary integral) if:*

(i) $f \in L^1(\mathbb{R})$*:* $\int_{-\infty}^{\infty} |f(x)|\,dx < \infty$

It can be extended to $L^2(\mathbb{R})$ functions via the Plancherel theorem.

Theorem 19.2.3 (Properties of the Fourier transform). *Let $\mathcal{F}[f] = \hat{f}$ and $\mathcal{F}[g] = \hat{g}$. Then:*

Property	*Time domain*	*Frequency domain*
Linearity	$\alpha f + \beta g$	$\alpha\hat{f} + \beta\hat{g}$
Translation	$f(x-a)$	$e^{-ika}\hat{f}(k)$
Modulation	$e^{ik_0x}f(x)$	$\hat{f}(k-k_0)$
Scaling	$f(ax)$, $a > 0$	$\frac{1}{a}\hat{f}(k/a)$
Conjugation	$\overline{f(x)}$	$\overline{\hat{f}(-k)}$
Reflection	$f(-x)$	$\hat{f}(-k)$
Differentiation	$f'(x)$	$ik\hat{f}(k)$
Multiplication by x	$xf(x)$	$i\frac{d\hat{f}}{dk}$
Convolution	$(f * g)(x)$	$\hat{f}(k)\hat{g}(k)$
Product	$f(x)g(x)$	$\frac{1}{2\pi}(\hat{f} * \hat{g})(k)$

Theorem 19.2.4 (Riemann-Lebesgue lemma). *If $f \in L^1(\mathbb{R})$, then $\hat{f}(k) \to 0$ as $|k| \to \infty$.*

Theorem 19.2.5 (Smoothness and decay). *There is a duality between smoothness and decay:*

(i) If $f^{(n)} \in L^1$, then $\hat{f}(k) = O(|k|^{-n})$ as $|k| \to \infty$

(ii) If $x^n f(x) \in L^1$, then $\hat{f} \in C^n$

Smooth functions have rapidly decaying transforms; rapidly decaying functions have smooth transforms.

19.2.2 Important Fourier Transform Pairs

Example 19.2.1 (Gaussian).

$$\mathcal{F}[e^{-ax^2}] = \sqrt{\frac{\pi}{a}} e^{-k^2/(4a)}, \quad a > 0.$$

The Gaussian is its own Fourier transform (up to scaling)—it is an eigenfunction of $\mathcal{F}$.

Example 19.2.2 (Exponential decay).

$$\mathcal{F}[e^{-a|x|}] = \frac{2a}{a^2 + k^2}, \quad a > 0.$$

Example 19.2.3 (Rectangle function). For $\Pi(x) = \begin{cases} 1 & |x| < 1/2 \\ 0 & |x| > 1/2 \end{cases}$:

$$\mathcal{F}[\Pi(x/a)] = a\,\mathrm{sinc}(ka/2) = a\frac{\sin(ka/2)}{ka/2}.$$

Example 19.2.4 (Sinc function).

$$\mathcal{F}[\mathrm{sinc}(ax)] = \frac{\pi}{a}\Pi(k/(2a)).$$

A sinc in time gives a rectangle in frequency (ideal low-pass filter).

Example 19.2.5 (Dirac delta).

$$\mathcal{F}[\delta(x)] = 1, \quad \mathcal{F}[1] = 2\pi\delta(k).$$

Example 19.2.6 (Heaviside step function).

$$\mathcal{F}[\theta(x)] = \pi\delta(k) + \frac{1}{ik} \quad \text{(as a distribution)}.$$

Example 19.2.7 (Derivatives of delta).

$$\mathcal{F}[\delta^{(n)}(x)] = (ik)^n.$$

19.2.3 Plancherel and Parseval Theorems

Theorem 19.2.6 (Plancherel theorem). *The Fourier transform extends to a unitary operator on* $L^2(\mathbb{R})$*:*

$$\int_{-\infty}^{\infty} |f(x)|^2 \, dx = \frac{1}{2\pi} \int_{-\infty}^{\infty} |\hat{f}(k)|^2 \, dk.$$

Theorem 19.2.7 (Parseval's theorem (generalized)). *For* $f, g \in L^2(\mathbb{R})$*:*

$$\int_{-\infty}^{\infty} f(x)\overline{g(x)} \, dx = \frac{1}{2\pi} \int_{-\infty}^{\infty} \hat{f}(k)\overline{\hat{g}(k)} \, dk.$$

In quantum mechanics, Plancherel's theorem states that the probability is conserved when transforming between position and momentum representations.

19.3 Convolution and Correlation

19.3.1 Convolution

Definition 19.3.1 (Convolution). The **convolution** of f and g is:

$$(f * g)(x) = \int_{-\infty}^{\infty} f(y)g(x-y) \, dy = \int_{-\infty}^{\infty} f(x-y)g(y) \, dy.$$

Theorem 19.3.1 (Convolution theorem).

$$\mathcal{F}[f * g] = \hat{f} \cdot \hat{g}, \quad \mathcal{F}[f \cdot g] = \frac{1}{2\pi} \hat{f} * \hat{g}.$$

Convolution in one domain corresponds to multiplication in the other.

Proposition 19.3.2 (Properties of convolution).

(i) ***Commutativity:*** $f * g = g * f$

(ii) ***Associativity:*** $(f * g) * h = f * (g * h)$

(iii) ***Distributivity:*** $f * (g + h) = f * g + f * h$

(iv) ***Identity:*** $f * \delta = f$

(v) ***Differentiation:*** $(f * g)' = f' * g = f * g'$

(vi) ***Translation:*** $\tau_a f * g = \tau_a(f * g)$ *where* $(\tau_a f)(x) = f(x - a)$

Example 19.3.1 (Smoothing). Convolution with a Gaussian $g_\sigma(x) = (2\pi\sigma^2)^{-1/2}e^{-x^2/(2\sigma^2)}$ smooths f:

$$(f * g_\sigma)(x) = \int f(y)g_\sigma(x - y) \, dy$$

averages f with a Gaussian weight centered at x.

19.3.2 Correlation

Definition 19.3.2 (Cross-correlation). The **cross-correlation** of f and g is:

$$(f \star g)(x) = \int_{-\infty}^{\infty} \overline{f(y)} g(x+y)\, dy = (\bar{f}(-\cdot) * g)(x).$$

Definition 19.3.3 (Autocorrelation). The **autocorrelation** of f is:

$$(f \star f)(x) = \int_{-\infty}^{\infty} \overline{f(y)} f(x+y)\, dy.$$

Theorem 19.3.3 (Wiener-Khinchin theorem).

$$\mathcal{F}[f \star f] = |\hat{f}|^2.$$

The Fourier transform of the autocorrelation is the power spectral density.

19.4 Multidimensional Fourier Transform

Definition 19.4.1 (Fourier transform in $\mathbb{R}^n$). For $f : \mathbb{R}^n \to \mathbb{C}$:

$$\hat{f}(\mathbf{k}) = \int_{\mathbb{R}^n} f(\mathbf{x}) e^{-i\mathbf{k}\cdot\mathbf{x}}\, d^n x$$

$$f(\mathbf{x}) = \frac{1}{(2\pi)^n} \int_{\mathbb{R}^n} \hat{f}(\mathbf{k}) e^{i\mathbf{k}\cdot\mathbf{x}}\, d^n k.$$

Theorem 19.4.1 (Properties in $\mathbb{R}^n$). *All properties from Theorem 19.2.3 extend naturally:*

(i) $\mathcal{F}[\nabla f] = i\mathbf{k}\hat{f}(\mathbf{k})$

(ii) $\mathcal{F}[\nabla^2 f] = -|\mathbf{k}|^2 \hat{f}(\mathbf{k})$

(iii) $\mathcal{F}[f * g] = \hat{f}\hat{g}$

(iv) Plancherel: $\int |f|^2\, d^n x = (2\pi)^{-n} \int |\hat{f}|^2\, d^n k$

Example 19.4.1 (Gaussian in $\mathbb{R}^n$).

$$\mathcal{F}[e^{-a|\mathbf{x}|^2}] = \left(\frac{\pi}{a}\right)^{n/2} e^{-|\mathbf{k}|^2/(4a)}.$$

Example 19.4.2 (Coulomb potential in 3D).

$$\mathcal{F}\left[\frac{1}{|\mathbf{x}|}\right] = \frac{4\pi}{|\mathbf{k}|^2}.$$

This is fundamental to the Fourier analysis of electrostatics.

Example 19.4.3 (Yukawa potential).

$$\mathcal{F}\left[\frac{e^{-\mu|\mathbf{x}|}}{|\mathbf{x}|}\right] = \frac{4\pi}{|\mathbf{k}|^2 + \mu^2}.$$

19.5 Distributions and Fourier Transforms

19.5.1 Tempered Distributions

Definition 19.5.1 (Schwartz space). The **Schwartz space** $\mathcal{S}(\mathbb{R}^n)$ consists of smooth functions $\phi : \mathbb{R}^n \to \mathbb{C}$ such that ϕ and all its derivatives decay faster than any polynomial:

$$\sup_{x \in \mathbb{R}^n} |x^\alpha \partial^\beta \phi(x)| < \infty$$

for all multi-indices α, β.

Definition 19.5.2 (Tempered distributions). A **tempered distribution** is a continuous linear functional on $\mathcal{S}(\mathbb{R}^n)$. The space of tempered distributions is denoted $\mathcal{S}'(\mathbb{R}^n)$.

Theorem 19.5.1 (Fourier transform of tempered distributions). *The Fourier transform extends to a continuous bijection $\mathcal{F} : \mathcal{S}'(\mathbb{R}^n) \to \mathcal{S}'(\mathbb{R}^n)$ defined by:*

$$\langle \mathcal{F}[T], \phi \rangle = \langle T, \mathcal{F}[\phi] \rangle$$

for $T \in \mathcal{S}'$, $\phi \in \mathcal{S}$.

19.5.2 Important Distributional Transforms

Example 19.5.1 (Constant function).

$$\mathcal{F}[1] = 2\pi\delta(k).$$

A constant in time corresponds to zero frequency.

Example 19.5.2 (Complex exponential).

$$\mathcal{F}[e^{ik_0 x}] = 2\pi\delta(k - k_0).$$

A pure frequency k_0 in time gives a delta at that frequency.

Example 19.5.3 (Cosine and sine).

$$\mathcal{F}[\cos(k_0 x)] = \pi[\delta(k - k_0) + \delta(k + k_0)]$$

$$\mathcal{F}[\sin(k_0 x)] = \frac{\pi}{i}[\delta(k - k_0) - \delta(k + k_0)].$$

Example 19.5.4 (Sign function).

$$\mathcal{F}[\text{sgn}(x)] = \frac{2}{ik} \quad \text{(as distribution)}.$$

Example 19.5.5 (Principal value).

$$\mathcal{F}\left[\text{p.v.}\frac{1}{x}\right] = -i\pi\,\text{sgn}(k).$$

Theorem 19.5.2 (Poisson summation formula). *For suitable f:*

$$\sum_{n=-\infty}^{\infty} f(n) = \sum_{n=-\infty}^{\infty} \hat{f}(2\pi n).$$

More generally, for period L:

$$\sum_{n=-\infty}^{\infty} f(nL) = \frac{1}{L}\sum_{n=-\infty}^{\infty} \hat{f}\left(\frac{2\pi n}{L}\right).$$

19.6 The Laplace Transform

19.6.1 Definition and Basic Properties

Definition 19.6.1 (Laplace transform). The **Laplace transform** of $f : [0,\infty) \to \mathbb{C}$ is:

$$F(s) = \mathcal{L}[f](s) = \int_0^{\infty} f(t)e^{-st}\,dt$$

for $s \in \mathbb{C}$ in the region of convergence.

Definition 19.6.2 (Region of convergence). The Laplace transform converges absolutely for $\text{Re}(s) > \sigma_0$, where σ_0 is the **abscissa of convergence**. If $|f(t)| \le Me^{at}$, then $\sigma_0 \le a$.

Theorem 19.6.1 (Inverse Laplace transform). *If $F(s)$ is analytic for $Re(s) > \sigma_0$:*

$$f(t) = \mathcal{L}^{-1}[F](t) = \frac{1}{2\pi i}\int_{\gamma-i\infty}^{\gamma+i\infty} F(s)e^{st}\,ds$$

for any $\gamma > \sigma_0$ (Bromwich integral).

Theorem 19.6.2 (Properties of the Laplace transform). *Let $\mathcal{L}[f] = F$ and $\mathcal{L}[g] = G$. Then:*

Property	*Time domain*	*s-domain*
Linearity	$\alpha f + \beta g$	$\alpha F + \beta G$
Differentiation	$f'(t)$	$sF(s) - f(0)$
n-th derivative	$f^{(n)}(t)$	$s^n F(s) - \sum_{k=0}^{n-1} s^{n-1-k} f^{(k)}(0)$
Integration	$\int_0^t f(\tau)\, d\tau$	$F(s)/s$
Scaling	$f(at)$, $a > 0$	$\frac{1}{a} F(s/a)$
Time shift	$f(t-a)\theta(t-a)$	$e^{-as}F(s)$
s-shift	$e^{at} f(t)$	$F(s-a)$
Convolution	$(f * g)(t) = \int_0^t f(\tau) g(t-\tau)\, d\tau$	$F(s)G(s)$
Multiplication by t	$tf(t)$	$-F'(s)$
Division by t	$f(t)/t$	$\int_s^\infty F(\sigma)\, d\sigma$

Theorem 19.6.3 (Initial and final value theorems). *If the limits exist:*

$$\lim_{t \to 0^+} f(t) = \lim_{s \to \infty} sF(s) \quad \textit{(initial value)}$$

$$\lim_{t \to \infty} f(t) = \lim_{s \to 0} sF(s) \quad \textit{(final value, if stable)}$$

19.6.2 Important Laplace Transform Pairs

$f(t)$ for $t \geq 0$	$F(s)$
1	$1/s$
t^n	$n!/s^{n+1}$
e^{at}	$1/(s-a)$
$\sin(\omega t)$	$\omega/(s^2+\omega^2)$
$\cos(\omega t)$	$s/(s^2+\omega^2)$
$\sinh(\omega t)$	$\omega/(s^2-\omega^2)$
$\cosh(\omega t)$	$s/(s^2-\omega^2)$
$e^{at}\sin(\omega t)$	$\omega/((s-a)^2+\omega^2)$
$e^{at}\cos(\omega t)$	$(s-a)/((s-a)^2+\omega^2)$
$t^n e^{at}$	$n!/(s-a)^{n+1}$
$\delta(t)$	1
$\delta(t-a)$	e^{-as}
$\theta(t-a)$	e^{-as}/s

19.6.3 Applications to Differential Equations

Example 19.6.1 (Second-order ODE with initial conditions). Solve $y'' + 3y' + 2y = e^{-3t}$ with $y(0) = 0$, $y'(0) = 1$.

Taking the Laplace transform:

$$s^2Y - sy(0) - y'(0) + 3(sY - y(0)) + 2Y = \frac{1}{s+3}$$

$$(s^2 + 3s + 2)Y = 1 + \frac{1}{s+3} = \frac{s+4}{s+3}$$

$$Y(s) = \frac{s+4}{(s+1)(s+2)(s+3)}.$$

Partial fractions and inverse transform give:

$$y(t) = \frac{3}{2}e^{-t} - 2e^{-2t} + \frac{1}{2}e^{-3t}.$$

Theorem 19.6.4 (Transfer function). *For a linear time-invariant system with impulse response $h(t)$:*

$$output = h * input.$$

In the s-domain: $Y(s) = H(s)X(s)$ where $H(s) = \mathcal{L}[h]$ is the ***transfer function****.*

19.7 Other Integral Transforms

19.7.1 Fourier Sine and Cosine Transforms

Definition 19.7.1 (Fourier cosine transform). For $f : [0, \infty) \to \mathbb{R}$:

$$\hat{f}_c(k) = \sqrt{\frac{2}{\pi}} \int_0^\infty f(x) \cos(kx)\, dx$$

$$f(x) = \sqrt{\frac{2}{\pi}} \int_0^\infty \hat{f}_c(k) \cos(kx)\, dk.$$

Definition 19.7.2 (Fourier sine transform).

$$\hat{f}_s(k) = \sqrt{\frac{2}{\pi}} \int_0^\infty f(x) \sin(kx)\, dx$$

$$f(x) = \sqrt{\frac{2}{\pi}} \int_0^\infty \hat{f}_s(k) \sin(kx)\, dk.$$

These are useful for problems on $[0, \infty)$ with Dirichlet ($f(0) = 0$, use sine) or Neumann ($f'(0) = 0$, use cosine) boundary conditions.

19.7.2 Mellin Transform

Definition 19.7.3 (Mellin transform).

$$\mathcal{M}[f](s) = \int_0^\infty f(x)x^{s-1}\,dx$$

$$f(x) = \frac{1}{2\pi i}\int_{c-i\infty}^{c+i\infty} \mathcal{M}[f](s)x^{-s}\,ds.$$

Theorem 19.7.1 (Mellin convolution). *The Mellin transform converts multiplicative convolution to products:*

$$\mathcal{M}\left[\int_0^\infty f(x/y)g(y)\frac{dy}{y}\right] = \mathcal{M}[f]\cdot\mathcal{M}[g].$$

Remark 19.7.1. The Mellin transform is related to the Fourier transform by $x = e^t$. It is important in number theory (connection to the Riemann zeta function) and asymptotics.

19.7.3 Hilbert Transform

Definition 19.7.4 (Hilbert transform).

$$\mathcal{H}[f](x) = \frac{1}{\pi}\text{p.v.}\int_{-\infty}^{\infty}\frac{f(y)}{x-y}\,dy = \frac{1}{\pi x} * f(x).$$

Theorem 19.7.2 (Properties of the Hilbert transform). *(i)* $\mathcal{H}[\mathcal{H}[f]] = -f$

(ii) $\mathcal{F}[\mathcal{H}[f]](k) = -i\,sgn(k)\hat{f}(k)$

(iii) $\int|f|^2 = \int|\mathcal{H}[f]|^2$ *(isometry on L^2)*

Definition 19.7.5 (Analytic signal). The **analytic signal** associated with real f is:

$$f_a(t) = f(t) + i\mathcal{H}[f](t).$$

Its Fourier transform has support only for $k > 0$.

The Hilbert transform is fundamental in signal processing (instantaneous amplitude and phase) and in the derivation of Kramers-Kronig relations.

19.8 Applications to Physics

19.8.1 Quantum Mechanics

Theorem 19.8.1 (Position-momentum duality). *In quantum mechanics, the position and momentum representations are related by Fourier transform:*

$$\tilde{\psi}(p) = \frac{1}{\sqrt{2\pi\hbar}} \int_{-\infty}^{\infty} \psi(x) e^{-ipx/\hbar}\, dx$$

$$\psi(x) = \frac{1}{\sqrt{2\pi\hbar}} \int_{-\infty}^{\infty} \tilde{\psi}(p) e^{ipx/\hbar}\, dp.$$

Theorem 19.8.2 (Heisenberg uncertainty). *The uncertainty principle $\Delta x \cdot \Delta p \geq \hbar/2$ is a consequence of Fourier analysis: a function and its Fourier transform cannot both be arbitrarily localized.*

Example 19.8.1 (Free particle propagator). For a free particle, the propagator is:

$$K(x,t;x',0) = \sqrt{\frac{m}{2\pi i\hbar t}} \exp\left(\frac{im(x-x')^2}{2\hbar t}\right).$$

This is the inverse Fourier transform of $e^{-ip^2t/(2m\hbar)}$.

19.8.2 Solving PDEs with Transforms

Example 19.8.2 (Heat equation on $\mathbb{R}$). For $\frac{\partial u}{\partial t} = D\frac{\partial^2 u}{\partial x^2}$ with $u(x,0) = f(x)$:
Taking the Fourier transform in x:

$$\frac{\partial \hat{u}}{\partial t} = -Dk^2\hat{u} \implies \hat{u}(k,t) = \hat{f}(k)e^{-Dk^2t}.$$

Inverting:

$$u(x,t) = \frac{1}{\sqrt{4\pi Dt}} \int_{-\infty}^{\infty} f(x') \exp\left(-\frac{(x-x')^2}{4Dt}\right) dx' = f * G_t$$

where $G_t(x) = (4\pi Dt)^{-1/2}e^{-x^2/(4Dt)}$ is the heat kernel.

Example 19.8.3 (Wave equation on $\mathbb{R}$). For $\frac{\partial^2 u}{\partial t^2} = c^2\frac{\partial^2 u}{\partial x^2}$ with $u(x,0) = f(x)$, $u_t(x,0) = g(x)$:
Taking the Fourier transform:

$$\frac{\partial^2 \hat{u}}{\partial t^2} = -c^2k^2\hat{u} \implies \hat{u}(k,t) = \hat{f}(k)\cos(ckt) + \frac{\hat{g}(k)}{ck}\sin(ckt).$$

Inverting gives d'Alembert's solution:

$$u(x,t) = \frac{1}{2}[f(x+ct) + f(x-ct)] + \frac{1}{2c}\int_{x-ct}^{x+ct} g(s)\, ds.$$

Example 19.8.4 (Laplace equation in half-plane)**.** For $\nabla^2 u = 0$ in $y > 0$ with $u(x, 0) = f(x)$:

Taking Fourier transform in x:

$$-k^2\hat{u} + \frac{\partial^2 \hat{u}}{\partial y^2} = 0 \implies \hat{u}(k, y) = \hat{f}(k)e^{-|k|y}$$

(choosing the decaying solution). Inverting:

$$u(x, y) = \frac{1}{\pi}\int_{-\infty}^{\infty} \frac{y f(x')}{(x - x')^2 + y^2}\, dx'$$

— the Poisson integral formula.

19.8.3 Green's Functions via Transforms

Example 19.8.5 (Green's function for Laplacian in 3D)**.** The Green's function satisfies $\nabla^2 G = \delta^3(\mathbf{x})$. Taking 3D Fourier transform:

$$-|\mathbf{k}|^2\hat{G} = 1 \implies \hat{G}(\mathbf{k}) = -\frac{1}{|\mathbf{k}|^2}.$$

Inverting (using contour integration or known transform):

$$G(\mathbf{x}) = -\frac{1}{4\pi|\mathbf{x}|}.$$

Example 19.8.6 (Helmholtz equation)**.** For $(\nabla^2 + k_0^2)G = \delta^3(\mathbf{x})$:

$$\hat{G}(\mathbf{k}) = \frac{1}{k_0^2 - |\mathbf{k}|^2}.$$

The inverse transform requires careful treatment of poles, giving:

$$G^{\pm}(\mathbf{x}) = -\frac{e^{\pm ik_0|\mathbf{x}|}}{4\pi|\mathbf{x}|}$$

for outgoing/incoming waves.

19.8.4 Signal Processing and Spectral Analysis

Definition 19.8.1 (Power spectrum)**.** For a signal $f(t)$, the **power spectrum** is $|\hat{f}(\omega)|^2$, giving the energy distribution across frequencies.

Theorem 19.8.3 (Sampling theorem (Nyquist-Shannon)). *A band-limited signal with $\hat{f}(\omega) = 0$ for $|\omega| > \omega_{\max}$ can be perfectly reconstructed from samples at rate $f_s > 2\omega_{\max}/(2\pi)$:*

$$f(t) = \sum_{n=-\infty}^{\infty} f(nT)\, sinc\left(\frac{t - nT}{T}\right)$$

where $T = 1/f_s$ is the sampling period.

Theorem 19.8.4 (Uncertainty principle for signals). *For any signal f:*

$$\Delta t \cdot \Delta\omega \geq \frac{1}{2}$$

where Δt and $\Delta\omega$ are the RMS widths in time and frequency. Equality holds for Gaussians.

19.9 Summary

Integral transforms are powerful tools for solving differential equations and analyzing signals:

Fourier series: Periodic functions decompose into harmonics $e^{in\omega_0 x}$; convergence under Dirichlet conditions; Parseval's theorem relates L^2 norms.

Fourier transform: $\hat{f}(k) = \int f(x)e^{-ikx}\,dx$ converts convolution to multiplication; Plancherel theorem preserves L^2 norms; derivatives become algebraic.

Convolution: $(f * g)(x) = \int f(y)g(x-y)\,dy$; convolution theorem $\mathcal{F}[f * g] = \hat{f}\hat{g}$.

Distributions: Extend Fourier analysis to δ, θ, and other generalized functions.

Laplace transform: $F(s) = \int_0^\infty f(t)e^{-st}\,dt$; ideal for initial value problems and transfer functions.

Other transforms: Sine/cosine (half-line), Mellin (multiplicative), Hilbert (analytic signals).

Physics applications: Position-momentum duality in QM; solving heat, wave, Laplace equations; Green's functions; spectral analysis.

Exercises

Exercise 19.9.1. Find the Fourier series of $f(x) = |x|$ on $[-\pi, \pi]$ and use Parseval's theorem to evaluate $\sum_{n=0}^{\infty} \frac{1}{(2n+1)^4}$.

Exercise 19.9.2. Prove the convolution theorem: $\mathcal{F}[f * g] = \hat{f} \cdot \hat{g}$.

Exercise 19.9.3. Compute the Fourier transform of $f(x) = e^{-a|x|}$ for $a > 0$ and verify Plancherel's theorem.

Exercise 19.9.4. Show that $\mathcal{F}[e^{-ax^2}] = \sqrt{\pi/a}\, e^{-k^2/(4a)}$ by completing the square.

Exercise 19.9.5. Derive the Fourier transform of $\delta'(x)$ using the distributional definition.

Exercise 19.9.6. Use the Laplace transform to solve $y'' + 4y = \sin(2t)$ with $y(0) = y'(0) = 0$.

Exercise 19.9.7. Prove the initial value theorem: $\lim_{s\to\infty} sF(s) = \lim_{t\to 0^+} f(t)$ for suitable f.

Exercise 19.9.8. Find the Green's function for $\frac{d^2G}{dx^2} - k^2G = \delta(x)$ on $\mathbb{R}$ using Fourier transform.

Exercise 19.9.9. Solve the heat equation $u_t = u_{xx}$ on $\mathbb{R}$ with $u(x, 0) = e^{-|x|}$ using Fourier transform.

Exercise 19.9.10. Derive the Poisson integral formula for the Laplace equation in the upper half-plane.

Exercise 19.9.11. Show that $\mathcal{F}[\text{p.v.}(1/x)] = -i\pi\,\text{sgn}(k)$.

Exercise 19.9.12. Prove the Poisson summation formula: $\sum_n f(n) = \sum_n \hat{f}(2\pi n)$.

Exercise 19.9.13. For the damped harmonic oscillator $\ddot{x}+2\gamma\dot{x}+\omega_0^2 x = F(t)$, find the transfer function $H(s)$ and the impulse response $h(t)$.

Exercise 19.9.14. Compute the Fourier transform of the Coulomb potential $V(r) = 1/r$ in 3D.

Exercise 19.9.15. Verify the uncertainty relation $\Delta x \cdot \Delta k \geq 1/2$ for a Gaussian wave packet and show that equality holds.

Chapter 20

Complex Methods in Physics

Complex analysis provides powerful techniques for evaluating integrals, solving differential equations, and understanding analytic structure. This chapter develops advanced complex-variable methods essential for theoretical physics: asymptotic evaluation of integrals via steepest descent, the WKB approximation for differential equations, and dispersion relations connecting real and imaginary parts of response functions.

20.1 Review of Complex Analysis

20.1.1 Analytic Functions and Singularities

Definition 20.1.1 (Analytic function). A function $f : \mathbb{C} \to \mathbb{C}$ is **analytic** (or **holomorphic**) at z_0 if it is complex differentiable in a neighborhood of z_0:

$$f'(z_0) = \lim_{h\to 0} \frac{f(z_0 + h) - f(z_0)}{h}$$

exists (independent of the direction of $h \to 0$).

Theorem 20.1.1 (Cauchy-Riemann equations). *$f = u + iv$ is analytic iff u, v have continuous partial derivatives satisfying:*

$$\frac{\partial u}{\partial x} = \frac{\partial v}{\partial y}, \quad \frac{\partial u}{\partial y} = -\frac{\partial v}{\partial x}.$$

Definition 20.1.2 (Types of singularities). For f analytic except at isolated point z_0:

- **Removable singularity**: $\lim_{z\to z_0} f(z)$ exists and is finite
- **Pole of order** n: $f(z) = g(z)/(z - z_0)^n$ with g analytic and $g(z_0) \neq 0$

- **Essential singularity**: Neither removable nor a pole

Definition 20.1.3 (Branch point and branch cut). A **branch point** is a point around which f is multi-valued (e.g., $z = 0$ for $\sqrt{z}$ or $\log z$). A **branch cut** is a curve connecting branch points across which f is discontinuous, making f single-valued on the cut plane.

20.1.2 Contour Integration and Residues

Theorem 20.1.2 (Cauchy's integral theorem). *If f is analytic inside and on a simple closed contour C:*

$$\oint_C f(z)\, dz = 0.$$

Theorem 20.1.3 (Cauchy's integral formula). *If f is analytic inside and on C, and z_0 is inside C:*

$$f(z_0) = \frac{1}{2\pi i} \oint_C \frac{f(z)}{z - z_0}\, dz.$$

For derivatives:

$$f^{(n)}(z_0) = \frac{n!}{2\pi i} \oint_C \frac{f(z)}{(z - z_0)^{n+1}}\, dz.$$

Definition 20.1.4 (Residue). The **residue** of f at an isolated singularity z_0 is:

$$\text{Res}(f, z_0) = \frac{1}{2\pi i} \oint_C f(z)\, dz$$

where C is a small circle around z_0. For a pole of order n:

$$\text{Res}(f, z_0) = \frac{1}{(n-1)!} \lim_{z \to z_0} \frac{d^{n-1}}{dz^{n-1}}[(z - z_0)^n f(z)].$$

For a simple pole: $\text{Res}(f, z_0) = \lim_{z \to z_0}(z - z_0) f(z)$.

Theorem 20.1.4 (Residue theorem). *If f is analytic inside and on C except at isolated singularities $z_1, \ldots, z_n$ inside C:*

$$\oint_C f(z)\, dz = 2\pi i \sum_{k=1}^{n} Res(f, z_k).$$

Theorem 20.1.5 (Jordan's lemma). *If $f(z) \to 0$ uniformly as $|z| \to \infty$ in the upper half-plane and $a > 0$:*

$$\lim_{R \to \infty} \int_{C_R} f(z) e^{iaz}\, dz = 0$$

where C_R is the semicircle $|z| = R$, $Im(z) \geq 0$.

20.1.3 Evaluation of Real Integrals

Example 20.1.1 (Rational functions). For $\int_{-\infty}^{\infty} \frac{dx}{1+x^4}$: close in upper half-plane, poles at $e^{i\pi/4}$ and $e^{3i\pi/4}$:

$$\int_{-\infty}^{\infty} \frac{dx}{1+x^4} = 2\pi i \left[\text{Res}\left(\frac{1}{1+z^4}, e^{i\pi/4}\right) + \text{Res}\left(\frac{1}{1+z^4}, e^{3i\pi/4}\right)\right] = \frac{\pi}{\sqrt{2}}.$$

Example 20.1.2 (Fourier-type integrals). For $\int_{-\infty}^{\infty} \frac{\cos x}{1+x^2}\,dx = \text{Re}\int_{-\infty}^{\infty} \frac{e^{ix}}{1+x^2}\,dx$: Close in upper half-plane (Jordan's lemma), pole at $z = i$:

$$\int_{-\infty}^{\infty} \frac{e^{ix}}{1+x^2}\,dx = 2\pi i \cdot \frac{e^{i\cdot i}}{2i} = \frac{\pi}{e}.$$

Example 20.1.3 (Principal value integrals). For p.v. $\int_{-\infty}^{\infty} \frac{dx}{x-a}$ with a real, indent around a with semicircle C_ϵ:

$$\lim_{\epsilon\to 0} \int_{C_\epsilon} \frac{dz}{z-a} = \pm i\pi$$

(sign depends on indentation direction).

20.2 Asymptotic Expansions

20.2.1 Asymptotic Series

Definition 20.2.1 (Asymptotic expansion). The series $\sum_{n=0}^{\infty} a_n/z^n$ is an **asymptotic expansion** of $f(z)$ as $z \to \infty$, written:

$$f(z) \sim \sum_{n=0}^{\infty} \frac{a_n}{z^n}$$

if for each N:

$$f(z) - \sum_{n=0}^{N} \frac{a_n}{z^n} = O(z^{-N-1}) \quad \text{as } z \to \infty.$$

Remark 20.2.1. Asymptotic series need not converge! The error after N terms may decrease initially but eventually grows. The optimal truncation minimizes the error.

Example 20.2.1 (Exponential integral). The exponential integral $\text{Ei}(x) = \int_{-\infty}^{x} \frac{e^t}{t}\,dt$ has asymptotic expansion:

$$e^{-x}\text{Ei}(x) \sim \frac{1}{x}\left(1 + \frac{1!}{x} + \frac{2!}{x^2} + \cdots\right) \quad (x \to \infty).$$

The series diverges (factorially), but truncating at $n \approx x$ gives exponentially small error.

Definition 20.2.2 (Big-O and little-o notation).

- $f(z) = O(g(z))$ as $z \to z_0$ if $|f(z)/g(z)|$ is bounded near z_0
- $f(z) = o(g(z))$ as $z \to z_0$ if $f(z)/g(z) \to 0$ as $z \to z_0$

20.2.2 Watson's Lemma

Theorem 20.2.1 (Watson's lemma). *If* $f(t) \sim \sum_{n=0}^{\infty} a_n t^{n+\lambda-1}$ *as* $t \to 0^+$ *with* $\lambda > 0$, *and* $|f(t)| \leq Me^{bt}$ *for large* t, *then:*

$$\int_0^\infty f(t)e^{-zt}\,dt \sim \sum_{n=0}^{\infty} \frac{a_n \Gamma(n+\lambda)}{z^{n+\lambda}} \qquad (z \to \infty, |\arg z| < \pi/2).$$

Example 20.2.2 (Gamma function asymptotics). For $\Gamma(z+1) = \int_0^\infty t^z e^{-t}\,dt$, substituting $t = z(1+s)$:

$$\Gamma(z+1) = z^{z+1}e^{-z} \int_{-1}^{\infty} (1+s)^z e^{-zs}\,ds.$$

Watson's lemma (after expanding $(1+s)^z$) yields Stirling's series:

$$\Gamma(z+1) \sim \sqrt{2\pi z}\left(\frac{z}{e}\right)^z \left(1 + \frac{1}{12z} + \frac{1}{288z^2} - \cdots\right).$$

20.3 The Method of Steepest Descent

20.3.1 Laplace's Method

Theorem 20.3.1 (Laplace's method). *For real* $\phi(t)$ *with unique maximum at* $t_0 \in (a,b)$ *where* $\phi''(t_0) < 0$*:*

$$\int_a^b g(t)e^{\lambda\phi(t)}\,dt \sim g(t_0)e^{\lambda\phi(t_0)}\sqrt{\frac{2\pi}{\lambda|\phi''(t_0)|}} \qquad (\lambda \to \infty).$$

Proof sketch. Near t_0: $\phi(t) \approx \phi(t_0) + \frac{1}{2}\phi''(t_0)(t-t_0)^2$. The dominant contribution comes from $|t - t_0| = O(\lambda^{-1/2})$. Extending limits to $\pm\infty$ and evaluating the Gaussian integral gives the result. □

Example 20.3.1 (Stirling's formula). For $n! = \int_0^\infty t^n e^{-t}\,dt = \int_0^\infty e^{n\ln t - t}\,dt$:
Set $\phi(t) = \ln t - t/n$. Maximum at $t_0 = n$ where $\phi(t_0) = \ln n - 1$, $\phi''(t_0) = -1/n^2$.

$$n! \sim e^{n(\ln n - 1)}\sqrt{2\pi n} = \sqrt{2\pi n}\left(\frac{n}{e}\right)^n.$$

20.3.2 Steepest Descent (Saddle Point Method)

Definition 20.3.1 (Saddle point). For analytic $\phi(z)$, a **saddle point** z_0 satisfies $\phi'(z_0) = 0$. Near z_0:

$$\phi(z) \approx \phi(z_0) + \frac{1}{2}\phi''(z_0)(z - z_0)^2.$$

Theorem 20.3.2 (Method of steepest descent). *For the integral:*

$$I(\lambda) = \int_C g(z)e^{\lambda\phi(z)}\,dz$$

as $\lambda \to \infty$:

(i) Find saddle points where $\phi'(z_0) = 0$

(ii) Deform contour C to pass through relevant saddle points along paths of steepest descent (constant Im$[\phi(z)]$, decreasing Re$[\phi(z)]$)

(iii) Near each saddle point z_k:

$$I(\lambda) \sim \sum_k g(z_k)e^{\lambda\phi(z_k)}\sqrt{\frac{2\pi}{\lambda|\phi''(z_k)|}}e^{i\theta_k}$$

where θ_k is determined by the direction of the steepest descent path.

Definition 20.3.2 (Steepest descent contour). At a saddle point z_0, the **steepest descent** directions are where $\mathrm{Re}[\phi(z)]$ decreases most rapidly. If $\phi''(z_0) = |\phi''(z_0)|e^{i\alpha}$, these directions are $\theta = (\pi - \alpha)/2 + n\pi/2$.

Example 20.3.2 (Airy function asymptotics). The Airy function $\mathrm{Ai}(x) = \frac{1}{2\pi}\int_{-\infty}^{\infty} e^{i(t^3/3+xt)}\,dt$.

For large $|x|$, write $\mathrm{Ai}(x) = \frac{1}{2\pi i}\int_C e^{\phi(t)}\,dt$ where $\phi(t) = t^3/3 + xt$.
Saddle points: $\phi'(t) = t^2 + x = 0 \Rightarrow t = \pm\sqrt{-x}$.
For $x > 0$: saddle points at $t = \pm i\sqrt{x}$. Steepest descent gives:

$$\mathrm{Ai}(x) \sim \frac{1}{2\sqrt{\pi}x^{1/4}}e^{-\frac{2}{3}x^{3/2}} \quad (x \to +\infty).$$

For $x < 0$: saddle points at $t = \pm\sqrt{|x|}$ (real). Both contribute:

$$\mathrm{Ai}(x) \sim \frac{1}{\sqrt{\pi}|x|^{1/4}}\cos\left(\frac{2}{3}|x|^{3/2} - \frac{\pi}{4}\right) \quad (x \to -\infty).$$

20.3.3 Stokes Phenomenon

Definition 20.3.3 (Stokes lines and anti-Stokes lines). For an asymptotic expansion with multiple exponentials $e^{\lambda\phi_k(z)}$:

- **Stokes lines**: where $\text{Re}[\phi_i(z) - \phi_j(z)] = 0$ (exponentials have equal magnitude)
- **Anti-Stokes lines**: where $\text{Im}[\phi_i(z) - \phi_j(z)] = 0$ (exponentials have equal phase)

Theorem 20.3.3 (Stokes phenomenon). *As z crosses a Stokes line in the complex plane, subdominant exponentials can switch on or off, causing discontinuous changes in the coefficients of the asymptotic expansion. The full asymptotic expansion is generally different in different sectors of the complex plane.*

20.4 The WKB Approximation

20.4.1 Derivation

The WKB (Wentzel-Kramers-Brillouin) method provides approximate solutions to linear ODEs with slowly varying coefficients.

Definition 20.4.1 (Standard form). Consider the equation:

$$\epsilon^2 \frac{d^2\psi}{dx^2} = Q(x)\psi$$

where $\epsilon \ll 1$ is a small parameter (often $\epsilon = \hbar$ in quantum mechanics).

Theorem 20.4.1 (WKB ansatz). *Substitute* $\psi(x) = \exp\left(\frac{1}{\epsilon}\sum_{n=0}^{\infty} \epsilon^n S_n(x)\right)$. *To leading order:*

$$\psi(x) \approx \frac{C}{[Q(x)]^{1/4}} \exp\left(\pm\frac{1}{\epsilon}\int^x \sqrt{Q(x')}\, dx'\right).$$

Derivation. Substituting $\psi = e^{S/\epsilon}$ into the ODE:

$$\epsilon^2 \left[\frac{1}{\epsilon^2}(S')^2 + \frac{1}{\epsilon}S''\right] e^{S/\epsilon} = Qe^{S/\epsilon}.$$

Expanding $S = S_0 + \epsilon S_1 + \epsilon^2 S_2 + \cdots$:

$$O(1): \quad (S_0')^2 = Q \Rightarrow S_0 = \pm \int \sqrt{Q}\, dx$$

$$O(\epsilon): \quad 2S_0'S_1' + S_0'' = 0 \Rightarrow S_1 = -\frac{1}{4}\ln Q$$

Hence $\psi \approx Q^{-1/4} \exp(\pm\epsilon^{-1} \int \sqrt{Q}\, dx)$. □

20.4.2 Classical and Quantum Regions

For the Schrödinger equation $-\frac{\hbar^2}{2m}\psi'' + V(x)\psi = E\psi$:

$$\psi'' = \frac{2m}{\hbar^2}[V(x) - E]\psi = -k^2(x)\psi$$

where $k^2(x) = \frac{2m}{\hbar^2}[E - V(x)]$.

Definition 20.4.2 (Classical and forbidden regions)**.**

- **Classically allowed** $(E > V)$: $k^2 > 0$, k real, oscillatory solutions
- **Classically forbidden** $(E < V)$: $k^2 < 0$, $\kappa = ik$ real, exponential solutions
- **Turning points**: where $E = V(x)$, i.e., $k(x) = 0$

Theorem 20.4.2 (WKB solutions)**.** ***Classically allowed region*** *$(E > V)$:*

$$\psi(x) \approx \frac{A}{\sqrt{k(x)}} \cos\left(\int^x k(x')\,dx' + \phi\right)$$

or equivalently complex exponentials $e^{\pm i\int k\,dx}/\sqrt{k}$.

Classically forbidden region *$(E < V)$:*

$$\psi(x) \approx \frac{B}{\sqrt{\kappa(x)}} e^{-\int^x \kappa(x')\,dx'} + \frac{C}{\sqrt{\kappa(x)}} e^{+\int^x \kappa(x')\,dx'}$$

where $\kappa = \sqrt{2m(V - E)}/\hbar$.

20.4.3 Connection Formulas

The WKB approximation breaks down at turning points where $Q(x) = 0$. Connection formulas relate solutions across turning points.

Theorem 20.4.3 (Connection formulas at a turning point)**.** *Near a simple turning point $x = a$ where $V(a) = E$ and $V'(a) > 0$ (forbidden region on left):*

From allowed to forbidden *(right to left):*

$$\frac{2}{\sqrt{k}} \cos\left(\int_a^x k\,dx' - \frac{\pi}{4}\right) \longleftrightarrow \frac{1}{\sqrt{\kappa}} e^{-\int_x^a \kappa\,dx'}$$

From forbidden to allowed *(left to right):*

$$\frac{1}{\sqrt{\kappa}} e^{-\int_x^a \kappa\,dx'} \longleftrightarrow \frac{2}{\sqrt{k}} \cos\left(\int_a^x k\,dx' - \frac{\pi}{4}\right)$$

$$\frac{1}{\sqrt{\kappa}} e^{+\int_x^a \kappa\,dx'} \longleftrightarrow \frac{1}{\sqrt{k}} \sin\left(\int_a^x k\,dx' - \frac{\pi}{4}\right)$$

The phase shift $\pi/4$ arises from matching through the Airy function near the turning point.

20.4.4 Bohr-Sommerfeld Quantization

Theorem 20.4.4 (Bohr-Sommerfeld quantization condition). *For a bound state in a potential well with turning points a and b where $V(a) = V(b) = E$:*

$$\int_a^b k(x)\,dx = \int_a^b \sqrt{\frac{2m[E - V(x)]}{\hbar^2}}\,dx = \left(n + \frac{1}{2}\right)\pi, \quad n = 0, 1, 2, \ldots$$

Equivalently:

$$\oint p\,dx = 2\int_a^b p\,dx = \left(n + \frac{1}{2}\right)h$$

where $p = \hbar k = \sqrt{2m(E - V)}$ is the classical momentum.

Derivation. For a bound state, $\psi \to 0$ as $x \to \pm\infty$. Apply connection formulas at both turning points and require single-valuedness. The phase accumulated in the allowed region must satisfy:

$$\int_a^b k\,dx - \frac{\pi}{4} - \frac{\pi}{4} = n\pi \Rightarrow \int_a^b k\,dx = \left(n + \frac{1}{2}\right)\pi.$$

□

Example 20.4.1 (Harmonic oscillator). For $V(x) = \frac{1}{2}m\omega^2 x^2$, turning points at $x = \pm\sqrt{2E/(m\omega^2)}$:

$$\int_{-a}^a \sqrt{2m(E - \frac{1}{2}m\omega^2 x^2)}\,dx = \frac{\pi E}{\omega} = \left(n + \frac{1}{2}\right)\pi\hbar.$$

Hence $E_n = (n + \frac{1}{2})\hbar\omega$—exact for the harmonic oscillator!

20.4.5 Tunneling

Theorem 20.4.5 (WKB tunneling probability). *For a particle incident on a barrier from the left with $E < V_{\max}$, the transmission coefficient is:*

$$T \approx \exp\left(-\frac{2}{\hbar}\int_a^b \sqrt{2m[V(x) - E]}\,dx\right) = \exp\left(-2\int_a^b \kappa(x)\,dx\right)$$

where a and b are the classical turning points.

Example 20.4.2 (Rectangular barrier). For a barrier of height V_0 and width L with $E < V_0$:

$$T \approx e^{-2\kappa L}, \quad \kappa = \sqrt{2m(V_0 - E)}/\hbar.$$

Example 20.4.3 (Gamow's alpha decay formula). For alpha decay, the alpha particle tunnels through the Coulomb barrier. The decay rate:

$$\Gamma \propto \exp\left(-\frac{2}{\hbar}\int_R^{r_c}\sqrt{2m\left(\frac{Zze^2}{4\pi\epsilon_0 r}-E\right)}\,dr\right)$$

explains the strong dependence of half-life on decay energy.

20.5 Dispersion Relations

20.5.1 Causality and Analyticity

Definition 20.5.1 (Causal response function). A system is **causal** if its response at time t depends only on inputs at earlier times $t' \leq t$. The response function $G(t)$ satisfies:

$$G(t) = 0 \quad \text{for } t < 0.$$

The output for input $f(t)$ is $y(t) = \int_{-\infty}^{t} G(t-t')f(t')\,dt' = \int_0^\infty G(\tau)f(t-\tau)\,d\tau$.

Theorem 20.5.1 (Analyticity from causality). *If $G(t) = 0$ for $t < 0$ and $G(t)$ is sufficiently well-behaved, then its Fourier transform:*

$$\tilde{G}(\omega) = \int_0^\infty G(t)e^{i\omega t}\,dt$$

is analytic in the upper half-plane $Im(\omega) > 0$.

Proof sketch. For $\text{Im}(\omega) > 0$, $|e^{i\omega t}| = e^{-\text{Im}(\omega)t}$ decays for $t > 0$, ensuring convergence. Analyticity follows from differentiation under the integral. □

20.5.2 Kramers-Kronig Relations

Theorem 20.5.2 (Kramers-Kronig relations). *If $\tilde{G}(\omega)$ is analytic in the upper half-plane and $\tilde{G}(\omega) \to 0$ as $|\omega| \to \infty$, then the real and imaginary parts satisfy:*

$$Re\,\tilde{G}(\omega) = \frac{1}{\pi}p.v.\int_{-\infty}^{\infty}\frac{Im\,\tilde{G}(\omega')}{\omega'-\omega}\,d\omega'$$

$$Im\,\tilde{G}(\omega) = -\frac{1}{\pi}p.v.\int_{-\infty}^{\infty}\frac{Re\,\tilde{G}(\omega')}{\omega'-\omega}\,d\omega'$$

Derivation. Consider the contour integral $\oint_C \frac{\tilde{G}(\omega')}{\omega'-\omega}\,d\omega'$ around the upper half-plane with indentation around $\omega' = \omega$ on the real axis. Since $\tilde{G}$ is analytic in the upper half-plane, the integral is zero. Taking limits and separating real/imaginary parts gives the result. □

Theorem 20.5.3 (Kramers-Kronig for susceptibility). *For linear response* $\chi(\omega) = \chi'(\omega) + i\chi''(\omega)$*:*

$$\chi'(\omega) = \frac{1}{\pi} p.v. \int_{-\infty}^{\infty} \frac{\chi''(\omega')}{\omega' - \omega} d\omega'$$

$$\chi''(\omega) = -\frac{1}{\pi} p.v. \int_{-\infty}^{\infty} \frac{\chi'(\omega')}{\omega' - \omega} d\omega'$$

Remark 20.5.1 (Physical interpretation). $\chi''(\omega)$ is related to dissipation (energy absorption), while $\chi'(\omega)$ is related to dispersion (phase shift). Kramers-Kronig relations show these are not independent—they are Hilbert transforms of each other.

20.5.3 Applications

Example 20.5.1 (Dielectric function). For the complex dielectric function $\epsilon(\omega) = \epsilon'(\omega) + i\epsilon''(\omega)$:

$$\epsilon'(\omega) - 1 = \frac{2}{\pi}\text{p.v.} \int_{0}^{\infty} \frac{\omega'\epsilon''(\omega')}{\omega'^2 - \omega^2} d\omega'$$

$$\epsilon''(\omega) = -\frac{2\omega}{\pi}\text{p.v.} \int_{0}^{\infty} \frac{\epsilon'(\omega') - 1}{\omega'^2 - \omega^2} d\omega'$$

(using symmetry $\epsilon(-\omega) = \epsilon^*(\omega)$ for real response).

Example 20.5.2 (Optical sum rules). From Kramers-Kronig and physical constraints:

$$\int_{0}^{\infty} \omega \, \text{Im}\, \epsilon(\omega) \, d\omega = \frac{\pi}{2}\omega_p^2$$

where $\omega_p = \sqrt{ne^2/(m\epsilon_0)}$ is the plasma frequency. This is the **f-sum rule**.

Example 20.5.3 (Refractive index). The complex refractive index $\tilde{n}(\omega) = n(\omega) + i\kappa(\omega)$ satisfies:

$$n(\omega) - 1 = \frac{2}{\pi}\text{p.v.} \int_{0}^{\infty} \frac{\omega'\kappa(\omega')}{\omega'^2 - \omega^2} d\omega'$$

relating the real refractive index to the absorption coefficient.

20.5.4 Dispersion Relations in Scattering

Theorem 20.5.4 (Forward scattering dispersion relation). *For the forward scattering amplitude $f(E)$ in potential scattering:*

$$Re\, f(E) = \frac{1}{\pi} p.v. \int_0^\infty \frac{Im\, f(E')}{E' - E} dE' + \text{subtraction terms}$$

relating the real part (phase shift) to the imaginary part (total cross-section via optical theorem).

Theorem 20.5.5 (Optical theorem).

$$Im\, f(0) = \frac{k}{4\pi}\sigma_{tot}$$

where $f(0)$ is the forward scattering amplitude, k is the wave number, and σ_{tot} is the total cross-section.

20.6 Analytic Continuation and Resummation

20.6.1 Analytic Continuation

Definition 20.6.1 (Analytic continuation). If f_1 is analytic on domain D_1 and f_2 is analytic on D_2 with $D_1 \cap D_2 \neq \varnothing$, and $f_1 = f_2$ on $D_1 \cap D_2$, then f_2 is the **analytic continuation** of f_1 to D_2.

Theorem 20.6.1 (Uniqueness of analytic continuation). *The analytic continuation, if it exists, is unique.*

Example 20.6.1 (Riemann zeta function). $\zeta(s) = \sum_{n=1}^\infty n^{-s}$ converges for $\text{Re}(s) > 1$. Analytic continuation extends ζ to all $\mathbb{C}$ except for a simple pole at $s = 1$. Famously, $\zeta(-1) = -1/12$ (giving meaning to "1+2+3+$\cdots = -1/12$").

Example 20.6.2 (Gamma function). $\Gamma(z) = \int_0^\infty t^{z-1}e^{-t}\, dt$ converges for $\text{Re}(z) > 0$. Using $\Gamma(z+1) = z\Gamma(z)$, we analytically continue to $\text{Re}(z) > -1$, then > -2, etc., with simple poles at $z = 0, -1, -2, \ldots$

20.6.2 Borel Summation

Definition 20.6.2 (Borel transform). For a (possibly divergent) series $f(z) = \sum_{n=0}^\infty a_n z^n$, the **Borel transform** is:

$$\mathcal{B}[f](t) = \sum_{n=0}^\infty \frac{a_n}{n!} t^n.$$

Definition 20.6.3 (Borel sum). If $\mathcal{B}[f](t)$ converges for $|t| < R$ and can be analytically continued along $[0, \infty)$, the **Borel sum** is:

$$f^B(z) = \int_0^\infty \mathcal{B}[f](t)e^{-t/z}\,\frac{dt}{z}.$$

Example 20.6.3. The divergent series $\sum_{n=0}^\infty n!(-z)^n$ has Borel transform $\sum_{n=0}^\infty (-t)^n = 1/(1+t)$. The Borel sum:

$$f^B(z) = \frac{1}{z}\int_0^\infty \frac{e^{-t/z}}{1+t}\,dt$$

which equals $e^{1/z}\mathrm{Ei}(-1/z)$ and provides the analytic function that the series "represents."

Theorem 20.6.2 (Borel summability). *If $\sum a_n z^n$ is asymptotic to an analytic function $f(z)$ in a sector, and $\mathcal{B}[f]$ can be continued along $\mathbb{R}^+$, then the Borel sum equals $f(z)$.*

20.6.3 Padé Approximants

Definition 20.6.4 (Padé approximant). The $[M/N]$ **Padé approximant** to $f(z) = \sum_{n=0}^\infty c_n z^n$ is the rational function:

$$[M/N]_f(z) = \frac{P_M(z)}{Q_N(z)} = \frac{a_0 + a_1 z + \cdots + a_M z^M}{1 + b_1 z + \cdots + b_N z^N}$$

whose Taylor expansion agrees with $f(z)$ through order z^{M+N}.

Theorem 20.6.3 (Properties of Padé approximants). *(i) Padé approximants can approximate functions beyond the radius of convergence of the power series*

(ii) Poles of $[M/N]$ approximate singularities of f

(iii) Diagonal Padé approximants $[N/N]$ often converge where the series diverges

20.7 Applications to Physics

20.7.1 Perturbation Theory and Resummation

Example 20.7.1 (Anharmonic oscillator). For $H = p^2/2 + x^2/2 + gx^4$, perturbation theory gives:

$$E_0(g) = \frac{1}{2} + \frac{3g}{4} - \frac{21g^2}{8} + \frac{333g^3}{16} - \cdots$$

This series diverges! Coefficients grow as $\sim n!$. Borel summation or Padé approximants recover the correct (unique) ground state energy.

Example 20.7.2 (QED perturbation theory)**.** In quantum electrodynamics, perturbative calculations give asymptotic series in $\alpha \approx 1/137$. Despite formal divergence at large order, the first several terms give spectacularly accurate predictions.

20.7.2 Scattering Theory

Example 20.7.3 (Partial wave expansion)**.** In scattering theory, the amplitude is expanded in partial waves:

$$f(\theta) = \sum_{\ell=0}^{\infty}(2\ell+1)f_\ell P_\ell(\cos\theta)$$

Each $f_\ell = (e^{2i\delta_\ell} - 1)/(2ik)$ where δ_ℓ is the phase shift. Analyticity of $f_\ell(k)$ in the complex k-plane relates bound states (poles on imaginary axis) to scattering (real axis).

Example 20.7.4 (Regge theory)**.** Analytic continuation of partial waves to complex angular momentum ℓ reveals Regge poles whose trajectories $\ell = \alpha(E)$ connect bound states and resonances. High-energy scattering is dominated by Regge poles with largest $\text{Re}(\alpha)$.

20.7.3 Green's Functions and Spectral Representations

Theorem 20.7.1 (Lehmann representation)**.** *The retarded Green's function:*

$$G^R(\omega) = \int_{-\infty}^{\infty} \frac{\rho(\omega')}{\omega - \omega' + i\eta}\, d\omega'$$

where $\rho(\omega) = -\frac{1}{\pi} Im\, G^R(\omega)$ *is the spectral function.*

Example 20.7.5 (Matsubara formalism)**.** In finite-temperature field theory, Green's functions are defined at discrete imaginary frequencies $i\omega_n = i\pi T(2n+1)$ (fermions) or $i\omega_n = 2\pi i T n$ (bosons). Analytic continuation $i\omega_n \to \omega + i\eta$ recovers real-time Green's functions.

20.8 Summary

Complex analysis provides essential tools for theoretical physics:

Contour integration: Residue theorem evaluates real integrals; Jordan's lemma handles oscillatory integrands.

Asymptotic expansions: Divergent series can be useful; optimal truncation minimizes error; Watson's lemma for Laplace-type integrals.

Steepest descent: For $\int g(z)e^{\lambda\phi(z)}dz$, deform contour through saddle points along paths of steepest descent; yields asymptotic expansions for large λ.

WKB approximation: $\psi \approx Q^{-1/4}e^{\pm\int\sqrt{Q}dx}$ for $\epsilon^2\psi'' = Q\psi$; connection formulas at turning points; Bohr-Sommerfeld quantization; tunneling rates.

Dispersion relations: Causality implies analyticity; Kramers-Kronig relates real and imaginary parts of response functions; sum rules constrain physical quantities.

Resummation: Borel summation assigns values to divergent series; Padé approximants extend beyond convergence radius; essential for quantum field theory.

Exercises

Exercise 20.8.1. Evaluate $\int_0^\infty \frac{x^2}{(x^2+1)(x^2+4)}\,dx$ using contour integration.

Exercise 20.8.2. Compute $\int_0^\infty \frac{\sin x}{x}\,dx$ by considering $\int_C \frac{e^{iz}}{z}\,dz$ with appropriate contour.

Exercise 20.8.3. Use Laplace's method to find the leading asymptotic behavior of $\int_0^1 x^n(1-x)^n\,dx$ as $n \to \infty$.

Exercise 20.8.4. Apply the method of steepest descent to find the asymptotic expansion of $\int_0^\infty t^{z-1}e^{-t}\,dt$ for large real z, recovering Stirling's formula.

Exercise 20.8.5. Find the WKB approximation for the bound states of a particle in the potential $V(x) = V_0|x|$ and use Bohr-Sommerfeld quantization to find the energy levels.

Exercise 20.8.6. Derive the connection formula at a turning point by matching WKB solutions to the Airy function.

Exercise 20.8.7. Calculate the WKB tunneling probability through a parabolic barrier $V(x) = V_0(1 - x^2/a^2)$ for $|x| < a$.

Exercise 20.8.8. Derive the Kramers-Kronig relations starting from the analyticity of the response function in the upper half-plane.

Exercise 20.8.9. For a damped harmonic oscillator $\ddot{x} + \gamma\dot{x} + \omega_0^2 x = F(t)/m$, find the susceptibility $\chi(\omega)$ and verify the Kramers-Kronig relations.

Exercise 20.8.10. Show that the Drude model dielectric function $\epsilon(\omega) = 1 - \omega_p^2/(\omega^2 + i\gamma\omega)$ satisfies the Kramers-Kronig relations.

Exercise 20.8.11. Compute the Borel sum of $\sum_{n=0}^{\infty}(-1)^n n! z^{n+1}$ and identify the resulting function.

Exercise 20.8.12. Find the [1/1] and [2/2] Padé approximants to e^z and compare their accuracy for $z = 1$ and $z = 2$.

Exercise 20.8.13. Use complex analysis to prove the optical theorem: $\operatorname{Im} f(0) = \frac{k}{4\pi}\sigma_{\text{tot}}$.

Exercise 20.8.14. For the anharmonic oscillator $H = p^2/2 + x^2/2 + gx^4$, show that perturbation theory coefficients grow factorially and explain why Borel summation is necessary.

Exercise 20.8.15. Derive the spectral representation $G^R(\omega) = \int \frac{\rho(\omega')}{\omega - \omega' + i\eta}\, d\omega'$ and show how poles correspond to bound states.

Chapter 21

Tensor Calculus and Index Notation

Tensor calculus provides the natural language for physics in curved spacetime, continuum mechanics, and any situation where coordinate independence is essential. This chapter develops the machinery of index notation, covariant differentiation, and curvature tensors, with emphasis on applications to general relativity and field theory.

21.1 Index Notation and Einstein Convention

21.1.1 The Einstein Summation Convention

Definition 21.1.1 (Einstein summation convention). When an index appears twice in a term—once as a superscript (upper) and once as a subscript (lower)—summation over all values of that index is implied:

$$A^\mu B_\mu \equiv \sum_{\mu=0}^{n-1} A^\mu B_\mu.$$

The repeated index is called a **dummy index** or **contraction index**.

Definition 21.1.2 (Free and dummy indices).

- A **free index** appears exactly once on each side of an equation
- A **dummy index** appears exactly twice in a term and is summed over
- Free indices must match on both sides; dummy indices can be renamed

Example 21.1.1 (Index notation examples).

$$\begin{aligned} A_i B^i &= A_1 B^1 + A_2 B^2 + A_3 B^3 \quad \text{(scalar)} \\ T^i{}_j V^j &= T^i{}_1 V^1 + T^i{}_2 V^2 + T^i{}_3 V^3 \quad \text{(vector with free index } i\text{)} \\ A^{ij} B_{jk} &= \sum_j A^{ij} B_{jk} \quad \text{(tensor with free indices } i, k\text{)} \end{aligned}$$

Remark 21.1.1 (Notation conventions). Common conventions:

- Greek indices $(\mu, \nu, \rho, \ldots)$: spacetime indices, running $0, 1, 2, 3$
- Latin indices $(i, j, k, \ldots)$: spatial indices, running $1, 2, 3$
- Early alphabet $(a, b, c, \ldots)$: sometimes used for tangent space or abstract indices

21.1.2 Basic Operations

Definition 21.1.3 (Kronecker delta). The **Kronecker delta** is:

$$\delta^\mu_\nu = \delta^\mu_\nu = \begin{cases} 1 & \mu = \nu \\ 0 & \mu \neq \nu \end{cases}$$

It acts as the identity: $\delta^\mu_\nu A^\nu = A^\mu$ and $\delta^\mu_\nu B_\mu = B_\nu$.

Definition 21.1.4 (Levi-Civita symbol). In n dimensions, the **Levi-Civita symbol** (or permutation symbol) is:

$$\epsilon_{\mu_1 \mu_2 \cdots \mu_n} = \begin{cases} +1 & \text{even permutation of } (0, 1, \ldots, n-1) \\ -1 & \text{odd permutation} \\ 0 & \text{any repeated index} \end{cases}$$

In 3D: ϵ_{ijk} with $\epsilon_{123} = +1$. In 4D: $\epsilon_{\mu\nu\rho\sigma}$ with $\epsilon_{0123} = +1$ (conventions vary).

Proposition 21.1.1 (Levi-Civita identities). *In 3D:*

$$\begin{aligned} \epsilon_{ijk}\epsilon_{lmn} &= \det \begin{pmatrix} \delta_{il} & \delta_{im} & \delta_{in} \\ \delta_{jl} & \delta_{jm} & \delta_{jn} \\ \delta_{kl} & \delta_{km} & \delta_{kn} \end{pmatrix} \\ \epsilon_{ijk}\epsilon_{lmk} &= \delta_{il}\delta_{jm} - \delta_{im}\delta_{jl} \\ \epsilon_{ijk}\epsilon_{ljk} &= 2\delta_{il} \\ \epsilon_{ijk}\epsilon_{ijk} &= 6 \end{aligned}$$

Example 21.1.2 (Vector cross product). The cross product $(\mathbf{A} \times \mathbf{B})_i = \epsilon_{ijk} A_j B_k$. The identity $\mathbf{A} \times (\mathbf{B} \times \mathbf{C}) = \mathbf{B}(\mathbf{A} \cdot \mathbf{C}) - \mathbf{C}(\mathbf{A} \cdot \mathbf{B})$ becomes:

$$\epsilon_{ijk} A_j \epsilon_{klm} B_l C_m = (\delta_{il}\delta_{jm} - \delta_{im}\delta_{jl}) A_j B_l C_m = B_i A_j C_j - C_i A_j B_j.$$

21.2 Vectors and Covectors

21.2.1 Contravariant and Covariant Vectors

Definition 21.2.1 (Contravariant vector). A **contravariant vector** (or simply **vector**) V^μ transforms under coordinate change $x^\mu \to x'^\mu$ as:

$$V'^\mu = \frac{\partial x'^\mu}{\partial x^\nu} V^\nu.$$

Components carry upper indices. Examples: displacement dx^μ, velocity $\frac{dx^\mu}{d\tau}$.

Definition 21.2.2 (Covariant vector). A **covariant vector** (or **covector**, **one-form**) ω_μ transforms as:

$$\omega'_\mu = \frac{\partial x^\nu}{\partial x'^\mu} \omega_\nu.$$

Components carry lower indices. Example: gradient $\partial_\mu \phi = \frac{\partial \phi}{\partial x^\mu}$.

Theorem 21.2.1 (Transformation matrices are inverses). *Let* $\Lambda^\mu{}_\nu = \frac{\partial x'^\mu}{\partial x^\nu}$ *and* $(\Lambda^{-1})^\mu{}_\nu = \frac{\partial x^\mu}{\partial x'^\nu}$. *Then:*

$$\Lambda^\mu{}_\rho (\Lambda^{-1})^\rho{}_\nu = \delta^\mu_\nu.$$

Theorem 21.2.2 (Scalar invariance). *The contraction of a vector with a covector is a scalar (invariant):*

$$V'^\mu \omega'_\mu = V^\mu \omega_\mu.$$

21.2.2 Coordinate Bases

Definition 21.2.3 (Coordinate basis vectors). At each point, the **coordinate basis vectors** are:

$$\mathbf{e}_\mu = \frac{\partial}{\partial x^\mu}$$

(tangent to coordinate lines). A vector $\mathbf{V} = V^\mu \mathbf{e}_\mu$.

Definition 21.2.4 (Dual basis). The **dual basis** (or **cobasis**) consists of one-forms:

$$\mathbf{e}^\mu = dx^\mu$$

satisfying $\mathbf{e}^\mu(\mathbf{e}_\nu) = \delta^\mu_\nu$. A covector $\boldsymbol{\omega} = \omega_\mu \mathbf{e}^\mu$.

21.3 Tensors

21.3.1 Definition and Transformation

Definition 21.3.1 (Tensor). A **tensor of type** (p, q) (or **rank** (p, q)) has p contravariant and q covariant indices: $T^{\mu_1 \cdots \mu_p}{}_{\nu_1 \cdots \nu_q}$. It transforms as:

$$T'^{\mu_1 \cdots \mu_p}{}_{\nu_1 \cdots \nu_q} = \frac{\partial x'^{\mu_1}}{\partial x^{\alpha_1}} \cdots \frac{\partial x'^{\mu_p}}{\partial x^{\alpha_p}} \frac{\partial x^{\beta_1}}{\partial x'^{\nu_1}} \cdots \frac{\partial x^{\beta_q}}{\partial x'^{\nu_q}} T^{\alpha_1 \cdots \alpha_p}{}_{\beta_1 \cdots \beta_q}.$$

Example 21.3.1 (Tensor types).

- $(0, 0)$: scalar
- $(1, 0)$: contravariant vector
- $(0, 1)$: covariant vector (covector)
- $(1, 1)$: mixed tensor, e.g., $T^\mu{}_\nu$
- $(0, 2)$: covariant 2-tensor, e.g., metric $g_{\mu\nu}$
- $(2, 0)$: contravariant 2-tensor, e.g., inverse metric $g^{\mu\nu}$
- $(1, 3)$: e.g., Riemann tensor $R^\rho{}_{\sigma\mu\nu}$

21.3.2 Tensor Operations

Definition 21.3.2 (Tensor operations).

1. **Addition**: $(A + B)^{\mu \cdots}{}_{\nu \cdots} = A^{\mu \cdots}{}_{\nu \cdots} + B^{\mu \cdots}{}_{\nu \cdots}$ (same type)
2. **Scalar multiplication**: $(\lambda T)^{\mu \cdots}{}_{\nu \cdots} = \lambda T^{\mu \cdots}{}_{\nu \cdots}$
3. **Tensor product**: $(A \otimes B)^{\mu\nu \cdots}{}_{\rho\sigma \cdots} = A^{\mu \cdots}{}_{\rho \cdots} B^{\nu \cdots}{}_{\sigma \cdots}$
4. **Contraction**: Sum over one upper and one lower index:

   $$T^{\mu \cdots}{}_{\mu \cdots} = \sum_\mu T^{\mu \cdots}{}_{\mu \cdots}$$

   Reduces rank by $(1, 1)$.
5. **Symmetrization**: $T_{(\mu\nu)} = \frac{1}{2}(T_{\mu\nu} + T_{\nu\mu})$
6. **Antisymmetrization**: $T_{[\mu\nu]} = \frac{1}{2}(T_{\mu\nu} - T_{\nu\mu})$

Definition 21.3.3 (Symmetric and antisymmetric tensors). A tensor is **symmetric** in indices μ, ν if $T_{\cdots\mu\cdots\nu\cdots} = T_{\cdots\nu\cdots\mu\cdots}$.

A tensor is **antisymmetric** if $T_{\cdots\mu\cdots\nu\cdots} = -T_{\cdots\nu\cdots\mu\cdots}$.

Proposition 21.3.1 (Decomposition). *Any tensor can be decomposed into symmetric and antisymmetric parts:*

$$T_{\mu\nu} = T_{(\mu\nu)} + T_{[\mu\nu]}.$$

21.4 The Metric Tensor

21.4.1 Definition and Properties

Definition 21.4.1 (Metric tensor). The **metric tensor** $g_{\mu\nu}$ is a symmetric $(0, 2)$ tensor that defines the inner product and line element:

$$ds^2 = g_{\mu\nu}\, dx^\mu dx^\nu.$$

For vectors: $\mathbf{U} \cdot \mathbf{V} = g_{\mu\nu} U^\mu V^\nu$.

Definition 21.4.2 (Inverse metric). The **inverse metric** $g^{\mu\nu}$ satisfies:

$$g^{\mu\rho} g_{\rho\nu} = \delta^\mu_\nu.$$

Example 21.4.1 (Common metrics). **Euclidean 3D** (Cartesian): $g_{ij} = \delta_{ij}$, $ds^2 = dx^2 + dy^2 + dz^2$.

Minkowski spacetime: $\eta_{\mu\nu} = \text{diag}(-1, +1, +1, +1)$ (or $(+, -, -, -)$):

$$ds^2 = -c^2 dt^2 + dx^2 + dy^2 + dz^2.$$

Spherical coordinates: $ds^2 = dr^2 + r^2 d\theta^2 + r^2 \sin^2\theta\, d\phi^2$:

$$g_{ij} = \text{diag}(1, r^2, r^2 \sin^2\theta).$$

Schwarzschild:

$$ds^2 = -\left(1 - \frac{r_s}{r}\right) c^2 dt^2 + \left(1 - \frac{r_s}{r}\right)^{-1} dr^2 + r^2 (d\theta^2 + \sin^2\theta\, d\phi^2).$$

21.4.2 Raising and Lowering Indices

Definition 21.4.3 (Index raising and lowering). The metric and inverse metric convert between contravariant and covariant components:

$$V_\mu = g_{\mu\nu} V^\nu \quad \text{(lowering)}$$

$$V^\mu = g^{\mu\nu} V_\nu \quad \text{(raising)}$$

For tensors: $T_{\mu\nu} = g_{\mu\rho} g_{\nu\sigma} T^{\rho\sigma}$, etc.

Remark 21.4.1. With the metric, contravariant and covariant vectors are different representations of the same geometric object. The distinction becomes crucial in curved spacetime or non-Cartesian coordinates.

Definition 21.4.4 (Metric determinant). The **determinant** of the metric is $g = \det(g_{\mu\nu})$. Under coordinate transformations:

$$g' = \left(\det \frac{\partial x}{\partial x'}\right)^2 g.$$

The quantity $\sqrt{|g|}$ transforms as a scalar density and appears in invariant volume elements:

$$d^n V = \sqrt{|g|}\, d^n x.$$

21.5 Covariant Differentiation

21.5.1 The Problem with Partial Derivatives

Proposition 21.5.1 (Partial derivatives of vectors are not tensors). *If V^μ is a vector, $\partial_\nu V^\mu$ does not transform as a tensor:*

$$\partial'_\nu V'^\mu = \frac{\partial x^\rho}{\partial x'^\nu}\frac{\partial x'^\mu}{\partial x^\sigma}\partial_\rho V^\sigma + \frac{\partial^2 x'^\mu}{\partial x^\rho \partial x^\sigma}\frac{\partial x^\rho}{\partial x'^\nu} V^\sigma.$$

The second term spoils tensor transformation.

21.5.2 The Covariant Derivative

Definition 21.5.1 (Covariant derivative). The **covariant derivative** ∇_μ (or $;_\mu$) is defined to produce tensors from tensors. For vectors:

$$\nabla_\mu V^\nu = \partial_\mu V^\nu + \Gamma^\nu_{\mu\rho} V^\rho$$

$$\nabla_\mu \omega_\nu = \partial_\mu \omega_\nu - \Gamma^\rho_{\mu\nu}\omega_\rho$$

where $\Gamma^\rho_{\mu\nu}$ are the **connection coefficients**.

Definition 21.5.2 (Christoffel symbols). For the **Levi-Civita connection** (metric-compatible, torsion-free), the connection coefficients are the **Christoffel symbols**:

$$\Gamma^\rho_{\mu\nu} = \frac{1}{2} g^{\rho\sigma}\left(\partial_\mu g_{\nu\sigma} + \partial_\nu g_{\mu\sigma} - \partial_\sigma g_{\mu\nu}\right).$$

Theorem 21.5.2 (Properties of Christoffel symbols).

(i) Symmetric in lower indices: $\Gamma^\rho_{\mu\nu} = \Gamma^\rho_{\nu\mu}$

(ii) Not a tensor (transforms inhomogeneously)

(iii) $\Gamma^{\rho}_{\mu\nu} = 0$ *in Cartesian coordinates in flat space*

Theorem 21.5.3 (Covariant derivative of tensors). *For a general tensor:*

$$\begin{aligned}\nabla_\sigma T^{\mu_1\cdots\mu_p}{}_{\nu_1\cdots\nu_q} &= \partial_\sigma T^{\mu_1\cdots\mu_p}{}_{\nu_1\cdots\nu_q} \\ &+ \Gamma^{\mu_1}_{\sigma\rho} T^{\rho\mu_2\cdots\mu_p}{}_{\nu_1\cdots\nu_q} + \cdots + \Gamma^{\mu_p}_{\sigma\rho} T^{\mu_1\cdots\mu_{p-1}\rho}{}_{\nu_1\cdots\nu_q} \\ &- \Gamma^{\rho}_{\sigma\nu_1} T^{\mu_1\cdots\mu_p}{}_{\rho\nu_2\cdots\nu_q} - \cdots - \Gamma^{\rho}_{\sigma\nu_q} T^{\mu_1\cdots\mu_p}{}_{\nu_1\cdots\nu_{q-1}\rho}.\end{aligned}$$

One $+\Gamma$ *term per upper index, one* $-\Gamma$ *term per lower index.*

Theorem 21.5.4 (Metric compatibility). *For the Levi-Civita connection:*

$$\nabla_\rho g_{\mu\nu} = 0, \quad \nabla_\rho g^{\mu\nu} = 0.$$

This allows raising/lowering indices to commute with covariant differentiation.

Theorem 21.5.5 (Covariant derivative of scalars). *For a scalar* ϕ*:* $\nabla_\mu \phi = \partial_\mu \phi$.

21.5.3 Useful Formulas

Proposition 21.5.6 (Christoffel symbol identities).

(i) $\Gamma^{\mu}_{\mu\nu} = \frac{1}{\sqrt{|g|}} \partial_\nu \sqrt{|g|} = \partial_\nu \ln \sqrt{|g|}$

(ii) $\Gamma^{\rho}_{\mu\nu} = \frac{1}{2} g^{\rho\sigma} (\partial_\mu g_{\nu\sigma} + \partial_\nu g_{\mu\sigma} - \partial_\sigma g_{\mu\nu})$

Theorem 21.5.7 (Divergence formula). *For a vector field:*

$$\nabla_\mu V^\mu = \frac{1}{\sqrt{|g|}} \partial_\mu (\sqrt{|g|} V^\mu).$$

For an antisymmetric tensor:

$$\nabla_\mu F^{\mu\nu} = \frac{1}{\sqrt{|g|}} \partial_\mu (\sqrt{|g|} F^{\mu\nu}).$$

Theorem 21.5.8 (Laplacian). *The covariant Laplacian of a scalar:*

$$\nabla^2 \phi = g^{\mu\nu} \nabla_\mu \nabla_\nu \phi = \frac{1}{\sqrt{|g|}} \partial_\mu \left(\sqrt{|g|} g^{\mu\nu} \partial_\nu \phi \right).$$

21.6 Geodesics and Parallel Transport

21.6.1 Geodesic Equation

Definition 21.6.1 (Geodesic). A **geodesic** is a curve that parallel transports its own tangent vector. It extremizes the proper length/time between two points.

Theorem 21.6.1 (Geodesic equation). *A geodesic $x^\mu(\lambda)$ satisfies:*

$$\frac{d^2x^\mu}{d\lambda^2} + \Gamma^\mu_{\rho\sigma}\frac{dx^\rho}{d\lambda}\frac{dx^\sigma}{d\lambda} = 0.$$

Equivalently, with tangent vector $U^\mu = dx^\mu/d\lambda$:

$$U^\nu \nabla_\nu U^\mu = 0.$$

Remark 21.6.1. For timelike geodesics (massive particles), λ can be proper time τ. For null geodesics (light), λ is an affine parameter.

Example 21.6.1 (Geodesics in Schwarzschild spacetime). For radial infall with $\theta = \pi/2$, $d\phi/d\tau = 0$:

$$\frac{d^2r}{d\tau^2} = -\frac{GM}{r^2}\left(1 - \frac{2GM}{c^2 r}\right)^{-1}$$

reducing to Newtonian gravity for $r \gg r_s = 2GM/c^2$.

21.6.2 Parallel Transport

Definition 21.6.2 (Parallel transport). A vector V^μ is **parallel transported** along a curve $x^\mu(\lambda)$ if:

$$\frac{DV^\mu}{D\lambda} \equiv \frac{dx^\nu}{d\lambda}\nabla_\nu V^\mu = \frac{dV^\mu}{d\lambda} + \Gamma^\mu_{\nu\rho}\frac{dx^\nu}{d\lambda}V^\rho = 0.$$

Theorem 21.6.2 (Parallel transport preserves inner products). *If U^μ and V^μ are parallel transported along a curve:*

$$\frac{d}{d\lambda}(g_{\mu\nu}U^\mu V^\nu) = 0.$$

Lengths and angles are preserved.

Remark 21.6.2 (Holonomy). In curved spacetime, parallel transport around a closed loop generally does not return a vector to itself. The resulting rotation is called **holonomy** and is a manifestation of curvature.

21.7 Curvature Tensors

21.7.1 The Riemann Curvature Tensor

Definition 21.7.1 (Riemann tensor). The **Riemann curvature tensor** $R^{\rho}{}_{\sigma\mu\nu}$ is defined by the failure of covariant derivatives to commute:

$$[\nabla_\mu, \nabla_\nu]V^\rho = R^\rho{}_{\sigma\mu\nu}V^\sigma$$

or for covectors:

$$[\nabla_\mu, \nabla_\nu]\omega_\rho = -R^\sigma{}_{\rho\mu\nu}\omega_\sigma.$$

Theorem 21.7.1 (Riemann tensor in terms of Christoffel symbols).

$$R^\rho{}_{\sigma\mu\nu} = \partial_\mu\Gamma^\rho_{\nu\sigma} - \partial_\nu\Gamma^\rho_{\mu\sigma} + \Gamma^\rho_{\mu\lambda}\Gamma^\lambda_{\nu\sigma} - \Gamma^\rho_{\nu\lambda}\Gamma^\lambda_{\mu\sigma}.$$

Theorem 21.7.2 (Symmetries of the Riemann tensor). *With all indices lowered, $R_{\rho\sigma\mu\nu} = g_{\rho\lambda}R^\lambda{}_{\sigma\mu\nu}$:*

(i) Antisymmetric in last two indices: $R_{\rho\sigma\mu\nu} = -R_{\rho\sigma\nu\mu}$

(ii) Antisymmetric in first two indices: $R_{\rho\sigma\mu\nu} = -R_{\sigma\rho\mu\nu}$

(iii) Symmetric under pair exchange: $R_{\rho\sigma\mu\nu} = R_{\mu\nu\rho\sigma}$

(iv) First Bianchi identity: $R_{\rho\sigma\mu\nu} + R_{\rho\mu\nu\sigma} + R_{\rho\nu\sigma\mu} = 0$

Theorem 21.7.3 (Independent components). *In n dimensions, the Riemann tensor has $\frac{n^2(n^2-1)}{12}$ independent components:*

- $n = 2$: *1 component*
- $n = 3$: *6 components*
- $n = 4$: *20 components*

Theorem 21.7.4 (Second Bianchi identity).

$$\nabla_\lambda R^\rho{}_{\sigma\mu\nu} + \nabla_\mu R^\rho{}_{\sigma\nu\lambda} + \nabla_\nu R^\rho{}_{\sigma\lambda\mu} = 0.$$

21.7.2 Ricci Tensor and Scalar

Definition 21.7.2 (Ricci tensor). The **Ricci tensor** is the contraction:

$$R_{\mu\nu} = R^\rho{}_{\mu\rho\nu} = g^{\rho\sigma}R_{\rho\mu\sigma\nu}.$$

It is symmetric: $R_{\mu\nu} = R_{\nu\mu}$.

Definition 21.7.3 (Ricci scalar). The **Ricci scalar** (or **scalar curvature**) is:

$$R = g^{\mu\nu} R_{\mu\nu} = R^{\mu}{}_{\mu}.$$

Example 21.7.1 (Curvature of a 2-sphere). For a sphere of radius a with $ds^2 = a^2(d\theta^2 + \sin^2\theta\, d\phi^2)$:

$$R = \frac{2}{a^2} \quad \text{(constant positive curvature).}$$

21.7.3 Einstein Tensor

Definition 21.7.4 (Einstein tensor). The **Einstein tensor** is:

$$G_{\mu\nu} = R_{\mu\nu} - \frac{1}{2} g_{\mu\nu} R.$$

Theorem 21.7.5 (Contracted Bianchi identity). *The Einstein tensor is divergence-free:*

$$\nabla^{\mu} G_{\mu\nu} = 0.$$

This follows from the second Bianchi identity.

Remark 21.7.1. The divergence-free property of $G_{\mu\nu}$ is essential for the consistency of Einstein's field equations, ensuring energy-momentum conservation.

21.7.4 Weyl Tensor

Definition 21.7.5 (Weyl tensor). The **Weyl tensor** (conformal curvature tensor) is the trace-free part of the Riemann tensor:

$$C_{\rho\sigma\mu\nu} = R_{\rho\sigma\mu\nu} - \frac{2}{n-2}\left(g_{\rho[\mu} R_{\nu]\sigma} - g_{\sigma[\mu} R_{\nu]\rho}\right) + \frac{2}{(n-1)(n-2)} R\, g_{\rho[\mu} g_{\nu]\sigma}.$$

Theorem 21.7.6 (Properties of the Weyl tensor).

(i) Same symmetries as Riemann tensor

(ii) Trace-free: all contractions vanish

(iii) Invariant under conformal transformations $g_{\mu\nu} \to \Omega^2 g_{\mu\nu}$

(iv) In $n \leq 3$: $C_{\rho\sigma\mu\nu} = 0$ identically

(v) Describes tidal forces and gravitational waves (vacuum curvature)

21.8 Applications to General Relativity

21.8.1 Einstein's Field Equations

Theorem 21.8.1 (Einstein field equations).

$$G_{\mu\nu} + \Lambda g_{\mu\nu} = \frac{8\pi G}{c^4} T_{\mu\nu}$$

or equivalently:

$$R_{\mu\nu} - \frac{1}{2} g_{\mu\nu} R + \Lambda g_{\mu\nu} = \frac{8\pi G}{c^4} T_{\mu\nu}$$

where:

- $G_{\mu\nu}$ = *Einstein tensor (geometry)*
- $T_{\mu\nu}$ = *stress-energy tensor (matter/energy)*
- Λ = *cosmological constant*
- G = *Newton's gravitational constant*

Theorem 21.8.2 (Energy-momentum conservation). *From* $\nabla^\mu G_{\mu\nu} = 0$*:*

$$\nabla^\mu T_{\mu\nu} = 0$$

— the covariant conservation of energy and momentum.

Theorem 21.8.3 (Vacuum equations). *In vacuum* ($T_{\mu\nu} = 0$, $\Lambda = 0$)*:*

$$R_{\mu\nu} = 0.$$

Note: $R_{\mu\nu} = 0$ *does not imply* $R_{\rho\sigma\mu\nu} = 0$*; spacetime can still be curved (gravitational waves, black holes).*

21.8.2 Stress-Energy Tensor

Definition 21.8.1 (Stress-energy tensor). The **stress-energy tensor** $T_{\mu\nu}$ encodes energy density, momentum density, and stress:

- T_{00}: energy density
- T_{0i}: momentum density / energy flux
- T_{ij}: stress (momentum flux)

Example 21.8.1 (Perfect fluid).

$$T_{\mu\nu} = (\rho + p/c^2)u_\mu u_\nu + p\, g_{\mu\nu}$$

where ρ is energy density, p is pressure, and u^μ is 4-velocity.

Example 21.8.2 (Electromagnetic field).

$$T_{\mu\nu} = \frac{1}{\mu_0}\left(F_{\mu\rho}F_\nu{}^\rho - \frac{1}{4}g_{\mu\nu}F_{\rho\sigma}F^{\rho\sigma}\right)$$

where $F_{\mu\nu}$ is the electromagnetic field tensor.

21.8.3 Geodesic Deviation

Theorem 21.8.4 (Geodesic deviation equation). *For two nearby geodesics with separation vector ξ^μ:*

$$\frac{D^2\xi^\mu}{D\tau^2} = R^\mu{}_{\nu\rho\sigma}u^\nu u^\rho \xi^\sigma$$

where $u^\mu = dx^\mu/d\tau$ is the 4-velocity. This describes tidal forces.

Remark 21.8.1. The geodesic deviation equation shows that curvature causes initially parallel geodesics to converge or diverge—the geometric manifestation of gravity.

21.9 Summary

Tensor calculus provides the geometric framework for physics:

Index notation: Einstein convention sums repeated upper-lower indices; Kronecker delta δ^μ_ν; Levi-Civita symbol $\epsilon_{\mu\nu\rho\sigma}$.

Vectors and tensors: Contravariant V^μ (upper indices), covariant ω_μ (lower indices); transformation laws ensure coordinate independence.

Metric tensor: $g_{\mu\nu}$ defines inner products and line element; raises/lowers indices; $g = \det(g_{\mu\nu})$ for volume elements.

Covariant derivative: $\nabla_\mu V^\nu = \partial_\mu V^\nu + \Gamma^\nu_{\mu\rho}V^\rho$; Christoffel symbols from metric; $\nabla_\rho g_{\mu\nu} = 0$.

Geodesics: $\ddot{x}^\mu + \Gamma^\mu_{\rho\sigma}\dot{x}^\rho\dot{x}^\sigma = 0$; extremize proper time/length.

Curvature: Riemann tensor $R^\rho{}_{\sigma\mu\nu}$ from commutator of ∇; Ricci tensor $R_{\mu\nu}$; scalar curvature R; Einstein tensor $G_{\mu\nu} = R_{\mu\nu} - \frac{1}{2}Rg_{\mu\nu}$.

General relativity: $G_{\mu\nu} = 8\pi G T_{\mu\nu}/c^4$; geometry equals matter; $\nabla^\mu T_{\mu\nu} = 0$ (conservation).

Exercises

Exercise 21.9.1. Use index notation to prove $\nabla\times(\nabla\times\mathbf{A}) = \nabla(\nabla\cdot\mathbf{A}) - \nabla^2\mathbf{A}$.

Exercise 21.9.2. Show that $\epsilon_{ijk}\epsilon_{lmk} = \delta_{il}\delta_{jm} - \delta_{im}\delta_{jl}$ by explicit computation.

Exercise 21.9.3. Compute the Christoffel symbols for spherical coordinates (r, θ, ϕ) with $ds^2 = dr^2 + r^2 d\theta^2 + r^2 \sin^2\theta\, d\phi^2$.

Exercise 21.9.4. Derive the geodesic equations for the 2-sphere and show that great circles are geodesics.

Exercise 21.9.5. Prove that $\Gamma^{\mu}_{\mu\nu} = \partial_\nu \ln\sqrt{|g|}$.

Exercise 21.9.6. Show that the covariant divergence formula $\nabla_\mu V^\mu = \frac{1}{\sqrt{|g|}}\partial_\mu(\sqrt{|g|}V^\mu)$ follows from the Christoffel symbol identity.

Exercise 21.9.7. Compute the Riemann tensor, Ricci tensor, and scalar curvature for the 2-sphere of radius a.

Exercise 21.9.8. Verify the first Bianchi identity $R_{\rho\sigma\mu\nu} + R_{\rho\mu\nu\sigma} + R_{\rho\nu\sigma\mu} = 0$.

Exercise 21.9.9. Derive the contracted Bianchi identity $\nabla^\mu G_{\mu\nu} = 0$ from the second Bianchi identity.

Exercise 21.9.10. For the Schwarzschild metric, compute $g^{\mu\nu}$ and verify that $g^{\mu\rho}g_{\rho\nu} = \delta^\mu_\nu$.

Exercise 21.9.11. Show that parallel transport around a closed loop on a sphere rotates vectors by an angle proportional to the enclosed area.

Exercise 21.9.12. Derive the geodesic deviation equation from the definition of the Riemann tensor.

Exercise 21.9.13. Compute the stress-energy tensor for a perfect fluid and verify $\nabla^\mu T_{\mu\nu} = 0$ implies the relativistic Euler equation.

Exercise 21.9.14. Show that the Weyl tensor vanishes identically in 3 dimensions.

Exercise 21.9.15. For weak gravitational fields $g_{\mu\nu} = \eta_{\mu\nu} + h_{\mu\nu}$ with $|h_{\mu\nu}| \ll 1$, linearize the Einstein equations and identify the Newtonian limit.

Chapter 22

Fiber Bundles and Gauge Theory

Fiber bundles provide the geometric framework for gauge theories, which underlie the Standard Model of particle physics. The electromagnetic field is a connection on a $U(1)$ bundle; the weak and strong forces arise from $SU(2)$ and $SU(3)$ bundles. This chapter develops the mathematics of fiber bundles, connections, and curvature, culminating in the structure of Yang-Mills theory.

22.1 Fiber Bundles

22.1.1 Basic Definitions

Definition 22.1.1 (Fiber bundle). A **fiber bundle** consists of:

- A **total space** E
- A **base space** M (typically a manifold)
- A **fiber** F
- A **projection** $\pi : E \to M$

such that E is locally a product: for each $p \in M$, there exists a neighborhood U and a diffeomorphism $\phi : \pi^{-1}(U) \to U \times F$ with $\pi_1 \circ \phi = \pi$ (where π_1 is projection onto the first factor).

The inverse image $\pi^{-1}(p) \cong F$ is called the **fiber over** p.

Definition 22.1.2 (Trivial bundle). A bundle is **trivial** if $E \cong M \times F$ globally. A bundle that is not globally trivial is **nontrivial**.

Definition 22.1.3 (Section). A **section** of a bundle is a map $s : M \to E$ such that $\pi \circ s = \mathrm{id}_M$. That is, $s(p) \in \pi^{-1}(p)$ for all $p \in M$.

The space of sections is denoted $\Gamma(E)$ or $\Gamma(M, E)$.

Example 22.1.1 (Tangent bundle). The **tangent bundle** TM of a manifold M has fiber $T_pM \cong \mathbb{R}^n$ at each point. Sections are vector fields.

Example 22.1.2 (Cotangent bundle). The **cotangent bundle** T^*M has fiber T_p^*M. Sections are one-forms.

Example 22.1.3 (Möbius strip). The Möbius strip is a nontrivial line bundle over S^1. It has no nonvanishing global section.

22.1.2 Vector Bundles

Definition 22.1.4 (Vector bundle). A **vector bundle** is a fiber bundle where each fiber is a vector space and the local trivializations are linear on fibers. The **rank** is the dimension of the fiber.

Definition 22.1.5 (Transition functions). For overlapping charts $U_\alpha \cap U_\beta \neq \varnothing$, the **transition functions** $g_{\alpha\beta} : U_\alpha \cap U_\beta \to GL(n, \mathbb{R})$ (or $GL(n, \mathbb{C})$) relate local trivializations:

$$\phi_\alpha \circ \phi_\beta^{-1} : (U_\alpha \cap U_\beta) \times F \to (U_\alpha \cap U_\beta) \times F$$

$$(p, v) \mapsto (p, g_{\alpha\beta}(p) \cdot v).$$

Theorem 22.1.1 (Cocycle conditions). *Transition functions satisfy:*

(i) $g_{\alpha\alpha} = id$

(ii) $g_{\alpha\beta} = g_{\beta\alpha}^{-1}$

(iii) $g_{\alpha\beta} g_{\beta\gamma} g_{\gamma\alpha} = id$ *on* $U_\alpha \cap U_\beta \cap U_\gamma$

Conversely, transition functions satisfying these conditions define a bundle.

22.1.3 Principal Bundles

Definition 22.1.6 (Principal bundle). A **principal G-bundle** $P \xrightarrow{\pi} M$ is a fiber bundle where:

- The fiber is a Lie group G
- G acts freely on P from the right: $R_g : P \to P$, $R_g(p) = p \cdot g$
- The action preserves fibers: $\pi(p \cdot g) = \pi(p)$

- $M = P/G$ (orbits of G are the fibers)

Example 22.1.4 (Frame bundle). The **frame bundle** FM over an n-manifold M has fiber over p consisting of all ordered bases of T_pM. It is a principal $GL(n, \mathbb{R})$-bundle.

Example 22.1.5 (Hopf fibration). The **Hopf fibration** $S^3 \xrightarrow{\pi} S^2$ with fiber $S^1 \cong U(1)$ is a principal $U(1)$-bundle. It is nontrivial—S^3 is not $S^2 \times S^1$.

Definition 22.1.7 (Associated bundle). Given a principal G-bundle P and a representation $\rho : G \to GL(V)$, the **associated vector bundle** is:

$$E = P \times_\rho V = (P \times V)/G$$

where $(p, v) \sim (p \cdot g, \rho(g^{-1})v)$.

Remark 22.1.1. Matter fields in gauge theories are sections of associated bundles. The principal bundle carries the gauge structure; the representation determines how matter transforms.

22.2 Connections on Principal Bundles

22.2.1 The Connection One-Form

Definition 22.2.1 (Vertical subspace). At $p \in P$, the **vertical subspace** $V_pP \subset T_pP$ consists of vectors tangent to the fiber:

$$V_pP = \ker(d\pi_p) = \{X \in T_pP : d\pi(X) = 0\}.$$

Definition 22.2.2 (Connection on a principal bundle). A **connection** on a principal G-bundle P is a smooth assignment of a **horizontal subspace** $H_pP \subset T_pP$ at each $p \in P$ such that:

(i) $T_pP = V_pP \oplus H_pP$ (direct sum)

(ii) $(R_g)_*H_pP = H_{p \cdot g}P$ (equivariance under right action)

Definition 22.2.3 (Connection one-form). A connection is equivalently specified by a **connection one-form** $\omega \in \Omega^1(P, \mathfrak{g})$ (a $\mathfrak{g}$-valued one-form on P) satisfying:

(i) $\omega(X^\#) = X$ for all $X \in \mathfrak{g}$, where $X^\#$ is the fundamental vector field

(ii) $R_g^*\omega = \mathrm{Ad}_{g^{-1}}\omega$ (equivariance)

The horizontal space is $H_pP = \ker(\omega_p)$.

Definition 22.2.4 (Fundamental vector field)**.** For $X \in \mathfrak{g}$, the **fundamental vector field** $X^\#$ on P is:

$$X^\#_p = \frac{d}{dt}\bigg|_{t=0} p \cdot \exp(tX).$$

It generates the G-action and is tangent to fibers.

22.2.2 Local Connection Forms (Gauge Potentials)

Definition 22.2.5 (Local section and gauge potential)**.** Given a local section $\sigma : U \to P$ over $U \subset M$, the **local connection form** (or **gauge potential**) is:

$$A = \sigma^*\omega \in \Omega^1(U, \mathfrak{g}).$$

In physics notation, $A = A^a_\mu T_a \, dx^\mu$ where T_a are generators of $\mathfrak{g}$.

Theorem 22.2.1 (Transformation under gauge transformations)**.** *If $\sigma' = \sigma \cdot g$ for $g : U \to G$, then:*

$$A' = g^{-1}Ag + g^{-1}dg = Ad_{g^{-1}}(A) + g^{-1}dg.$$

For matrix groups: $A'_\mu = g^{-1}A_\mu g + g^{-1}\partial_\mu g$.

Example 22.2.1 (Electromagnetism: $U(1)$ bundle)**.** For $G = U(1)$, $\mathfrak{g} = i\mathbb{R}$. The gauge potential is $A = iA_\mu dx^\mu$ where A_μ is the electromagnetic 4-potential. Gauge transformations:

$$A_\mu \to A_\mu + \partial_\mu \chi$$

for $g = e^{i\chi}$.

22.2.3 Horizontal Lift and Parallel Transport

Definition 22.2.6 (Horizontal lift)**.** Given a curve $\gamma : [0,1] \to M$ and $p_0 \in \pi^{-1}(\gamma(0))$, the **horizontal lift** $\tilde{\gamma}$ is the unique curve in P with $\tilde{\gamma}(0) = p_0$, $\pi \circ \tilde{\gamma} = \gamma$, and $\dot{\tilde{\gamma}}(t) \in H_{\tilde{\gamma}(t)}P$.

Definition 22.2.7 (Parallel transport)**. Parallel transport** along γ from $\gamma(0)$ to $\gamma(1)$ defines a map between fibers. For associated bundles, this transports vectors/fields along curves.

Definition 22.2.8 (Holonomy)**.** For a closed loop γ based at $p \in M$, parallel transport defines an element $g \in G$: the **holonomy** of the connection around γ. The set of all holonomies forms the **holonomy group**.

Theorem 22.2.2 (Wilson loop). *The holonomy around a loop γ is given by the path-ordered exponential:*

$$W[\gamma] = \mathcal{P}\exp\left(-\oint_\gamma A\right) = \mathcal{P}\exp\left(-\oint_\gamma A_\mu dx^\mu\right).$$

This is gauge-covariant: $W[\gamma] \to g(\gamma(0))^{-1}W[\gamma]g(\gamma(0))$.

22.3 Curvature

22.3.1 The Curvature Two-Form

Definition 22.3.1 (Curvature of a connection). The **curvature** of a connection ω is the $\mathfrak{g}$-valued two-form:

$$\Omega = d\omega + \frac{1}{2}[\omega, \omega] = d\omega + \omega \wedge \omega$$

where the bracket is in $\mathfrak{g}$ and the wedge is in forms.

Theorem 22.3.1 (Structure equation (Cartan)).

$$\Omega = d\omega + \omega \wedge \omega.$$

This is the Maurer-Cartan equation with the connection.

Theorem 22.3.2 (Bianchi identity).

$$d\Omega + [\omega, \Omega] = D\Omega = 0$$

where D is the covariant exterior derivative.

22.3.2 Local Curvature (Field Strength)

Definition 22.3.2 (Field strength tensor). The local curvature $F = \sigma^*\Omega$ is the **field strength**:

$$F = dA + A \wedge A.$$

In components:

$$F_{\mu\nu} = \partial_\mu A_\nu - \partial_\nu A_\mu + [A_\mu, A_\nu].$$

For matrix groups: $F_{\mu\nu} = \partial_\mu A_\nu - \partial_\nu A_\mu + [A_\mu, A_\nu]$.

Theorem 22.3.3 (Transformation of field strength). *Under gauge transformation $g : U \to G$:*

$$F' = g^{-1}Fg = Ad_{g^{-1}}(F).$$

The field strength transforms covariantly (homogeneously), unlike the potential.

Example 22.3.1 (Electromagnetism). For $U(1)$: $[A_\mu, A_\nu] = 0$, so:

$$F_{\mu\nu} = \partial_\mu A_\nu - \partial_\nu A_\mu.$$

This is the electromagnetic field tensor with $F_{0i} = E_i/c$, $F_{ij} = \epsilon_{ijk} B_k$.

Example 22.3.2 (Yang-Mills: $SU(N)$ bundle). For $G = SU(N)$ with generators T^a satisfying $[T^a, T^b] = if^{abc}T^c$:

$$F^a_{\mu\nu} = \partial_\mu A^a_\nu - \partial_\nu A^a_\mu + gf^{abc}A^b_\mu A^c_\nu$$

where g is the coupling constant. The non-Abelian structure gives self-interactions.

22.3.3 Curvature and Holonomy

Theorem 22.3.4 (Ambrose-Singer theorem). *The Lie algebra of the holonomy group at p is spanned by elements $\Omega_q(X, Y)$ for all q reachable from p by horizontal paths and all horizontal $X, Y \in H_q P$.*

Theorem 22.3.5 (Infinitesimal holonomy). *For a small loop enclosing area $\delta S^{\mu\nu}$:*

$$W \approx 1 - \frac{1}{2} F_{\mu\nu} \delta S^{\mu\nu}.$$

Curvature measures infinitesimal holonomy.

22.4 Covariant Derivative

22.4.1 Covariant Derivative on Associated Bundles

Definition 22.4.1 (Covariant derivative). For a section ψ of an associated bundle $E = P \times_\rho V$, the **covariant derivative** is:

$$D_\mu \psi = \partial_\mu \psi + \rho_*(A_\mu)\psi$$

where $\rho_* : \mathfrak{g} \to \mathfrak{gl}(V)$ is the induced Lie algebra representation.

Example 22.4.1 (Fundamental representation). For ψ in the fundamental representation of $SU(N)$:

$$D_\mu \psi = \partial_\mu \psi + igA_\mu \psi = \partial_\mu \psi + igA^a_\mu T^a \psi.$$

Example 22.4.2 (Adjoint representation). For ϕ in the adjoint representation:

$$D_\mu \phi = \partial_\mu \phi + [A_\mu, \phi].$$

Theorem 22.4.1 (Covariant derivative transforms covariantly). *Under $\psi \to g^{-1}\psi$ (or $\psi \to \rho(g^{-1})\psi$):*

$$D_\mu\psi \to g^{-1}D_\mu\psi.$$

Theorem 22.4.2 (Curvature from commutator).

$$[D_\mu, D_\nu]\psi = \rho_*(F_{\mu\nu})\psi.$$

For fundamental representation: $[D_\mu, D_\nu]\psi = igF_{\mu\nu}\psi$.

22.4.2 Gauge-Covariant Objects

Definition 22.4.2 (Gauge-invariant vs gauge-covariant).

- **Gauge-invariant**: unchanged under gauge transformations (e.g., $\text{tr}(F_{\mu\nu}F^{\mu\nu})$)
- **Gauge-covariant**: transforms homogeneously (e.g., $F_{\mu\nu}$, $D_\mu\psi$)

Proposition 22.4.3 (Building gauge-invariant quantities).

(i) Traces: $tr(F_{\mu\nu}F^{\mu\nu})$

(ii) Covariant derivatives contracted with matter: $\bar{\psi}D_\mu\psi$

(iii) Wilson loops: $tr(W[\gamma])$ for closed γ

22.5 Yang-Mills Theory

22.5.1 The Yang-Mills Action

Definition 22.5.1 (Yang-Mills action). The **Yang-Mills action** for a gauge field on a principal G-bundle over spacetime M is:

$$S_{YM}[A] = -\frac{1}{2g^2}\int_M \text{tr}(F \wedge *F) = -\frac{1}{4g^2}\int_M \text{tr}(F_{\mu\nu}F^{\mu\nu})\sqrt{|g|}\, d^4x$$

where $*$ is the Hodge dual and g is the coupling constant.

Remark 22.5.1 (Normalization). For $SU(N)$ with $\text{tr}(T^aT^b) = \frac{1}{2}\delta^{ab}$:

$$S_{YM} = -\frac{1}{4}\int F^a_{\mu\nu}F^{a\mu\nu}\, d^4x.$$

Theorem 22.5.1 (Yang-Mills equations). *The Euler-Lagrange equations are:*

$$D_\mu F^{\mu\nu} = \partial_\mu F^{\mu\nu} + [A_\mu, F^{\mu\nu}] = 0$$

or with source J^ν:

$$D_\mu F^{\mu\nu} = J^\nu.$$

In components: $\partial_\mu F^{a\mu\nu} + gf^{abc}A^b_\mu F^{c\mu\nu} = J^{a\nu}$.

Theorem 22.5.2 (Bianchi identity (local form)).

$$D_\mu F_{\nu\rho} + D_\nu F_{\rho\mu} + D_\rho F_{\mu\nu} = 0$$

or $D_{[\mu}F_{\nu\rho]} = 0$, *equivalently* $D * F = 0$ *where* $*F$ *is the dual.*

Remark 22.5.2 (Comparison with electromagnetism). For $U(1)$, Yang-Mills reduces to Maxwell:

$$\partial_\mu F^{\mu\nu} = J^\nu \quad \text{(inhomogeneous Maxwell)}$$
$$\partial_{[\mu}F_{\nu\rho]} = 0 \quad \text{(homogeneous Maxwell / Bianchi)}$$

22.5.2 Matter Coupling

Definition 22.5.2 (Gauge-matter action). For a matter field ψ (fermion) in representation ρ:

$$S = S_{YM} + \int \bar{\psi}(i\gamma^\mu D_\mu - m)\psi \, d^4x$$

where $D_\mu = \partial_\mu + ig\rho_*(A_\mu)$.

Theorem 22.5.3 (Noether current). *The gauge symmetry gives conserved current:*

$$J^a_\mu = \bar{\psi}\gamma_\mu T^a \psi.$$

The Yang-Mills equation $D_\mu F^{\mu\nu} = J^\nu$ *is the non-Abelian generalization of* $\partial_\mu F^{\mu\nu} = J^\nu$.

22.5.3 Self-Duality and Instantons

Definition 22.5.3 (Self-dual and anti-self-dual). In 4D Euclidean space, a field strength is:

- **Self-dual** if $F = *F$, i.e., $F_{\mu\nu} = \frac{1}{2}\epsilon_{\mu\nu\rho\sigma}F^{\rho\sigma}$
- **Anti-self-dual** if $F = - * F$

Theorem 22.5.4 (Instantons are Yang-Mills solutions). *Self-dual (or anti-self-dual) connections automatically satisfy the Yang-Mills equations:*

$$F = \pm * F \implies D_\mu F^{\mu\nu} = 0.$$

Definition 22.5.4 (Instanton number). The **instanton number** (or **topological charge**) is:

$$k = \frac{1}{8\pi^2}\int \text{tr}(F \wedge F) = \frac{1}{16\pi^2}\int \text{tr}(F_{\mu\nu}\tilde{F}^{\mu\nu})\, d^4x$$

where $\tilde{F}^{\mu\nu} = \frac{1}{2}\epsilon^{\mu\nu\rho\sigma}F_{\rho\sigma}$. It is an integer for $SU(N)$.

Example 22.5.1 (BPST instanton). The basic $SU(2)$ instanton with $k = 1$ centered at origin with scale ρ:

$$A_\mu = \frac{\eta_{a\mu\nu} x^\nu}{x^2 + \rho^2} \sigma^a$$

where $\eta_{a\mu\nu}$ are 't Hooft symbols and σ^a are Pauli matrices.

22.6 Characteristic Classes

22.6.1 Chern Classes

Definition 22.6.1 (Chern classes). For a complex vector bundle E with curvature F, the **total Chern class** is:

$$c(E) = \det\left(I + \frac{iF}{2\pi}\right) = 1 + c_1(E) + c_2(E) + \cdots$$

where $c_k(E) \in H^{2k}(M, \mathbb{Z})$.

Theorem 22.6.1 (Chern class formulas).

$$c_1(E) = \frac{i}{2\pi} tr(F)$$
$$c_2(E) = \frac{1}{8\pi^2} \left[tr(F \wedge F) - tr(F) \wedge tr(F)\right]$$

For $SU(N)$ bundles, $c_1 = 0$ and:

$$c_2(E) = \frac{1}{8\pi^2} tr(F \wedge F).$$

Theorem 22.6.2 (Chern classes are topological). *Chern classes depend only on the bundle, not on the choice of connection. They are topological invariants.*

Example 22.6.1 (First Chern class and magnetic charge). For a $U(1)$ bundle over S^2:

$$c_1 = \frac{1}{2\pi} \int_{S^2} F = n \in \mathbb{Z}.$$

This is the magnetic monopole charge (Dirac quantization).

22.6.2 Chern-Simons Form

Definition 22.6.2 (Chern-Simons form). The **Chern-Simons 3-form** is:

$$\mathrm{CS}(A) = \mathrm{tr}\left(A \wedge dA + \frac{2}{3} A \wedge A \wedge A\right).$$

It satisfies $d(\mathrm{CS}(A)) = \mathrm{tr}(F \wedge F)$.

Theorem 22.6.3 (Chern-Simons action). *In 3D, the Chern-Simons action:*

$$S_{CS} = \frac{k}{4\pi} \int_M CS(A) = \frac{k}{4\pi} \int_M tr\left(A \wedge dA + \frac{2}{3} A \wedge A \wedge A\right)$$

is gauge-invariant up to integers when $k \in \mathbb{Z}$. It defines a topological quantum field theory.

22.6.3 Pontryagin Classes

Definition 22.6.3 (Pontryagin classes). For a real vector bundle E with curvature F, the **total Pontryagin class** is:

$$p(E) = \det\left(I + \frac{F}{2\pi}\right)^{1/2} = 1 + p_1(E) + p_2(E) + \cdots$$

where $p_k(E) \in H^{4k}(M, \mathbb{Z})$.

Theorem 22.6.4 (First Pontryagin class).

$$p_1(E) = -\frac{1}{8\pi^2} tr(F \wedge F).$$

22.7 The Standard Model Structure

22.7.1 Gauge Group and Matter Content

Definition 22.7.1 (Standard Model gauge group). The gauge group of the Standard Model is:

$$G_{SM} = SU(3)_C \times SU(2)_L \times U(1)_Y$$

where:

- $SU(3)_C$: color (strong interaction), 8 gluons
- $SU(2)_L$: weak isospin, 3 W bosons

- $U(1)_Y$: hypercharge, 1 B boson

Definition 22.7.2 (Covariant derivative in SM).

$$D_\mu = \partial_\mu + ig_s G^a_\mu T^a + igW^i_\mu \tau^i + ig' Y B_\mu$$

where T^a are $SU(3)$ generators, $\tau^i = \sigma^i/2$ are $SU(2)$ generators, and Y is hypercharge.

Theorem 22.7.1 (Electroweak unification). *After spontaneous symmetry breaking via the Higgs mechanism:*

$$SU(2)_L \times U(1)_Y \to U(1)_{EM}$$

The photon A_μ and Z boson are mixtures:

$$\begin{pmatrix} A_\mu \\ Z_\mu \end{pmatrix} = \begin{pmatrix} \cos\theta_W & \sin\theta_W \\ -\sin\theta_W & \cos\theta_W \end{pmatrix} \begin{pmatrix} B_\mu \\ W^3_\mu \end{pmatrix}$$

where θ_W is the Weinberg angle.

22.7.2 Anomalies

Definition 22.7.3 (Gauge anomaly). A **gauge anomaly** occurs when a classical gauge symmetry is broken by quantum effects. For chiral fermions:

$$\partial_\mu J^\mu_a = \frac{1}{16\pi^2}\epsilon^{\mu\nu\rho\sigma}\mathrm{tr}(T_a F_{\mu\nu} F_{\rho\sigma}).$$

Theorem 22.7.2 (Anomaly cancellation in SM). *The Standard Model is anomaly-free due to the specific hypercharge assignments of quarks and leptons. For each generation:*

$$\sum_{\text{fermions}} Y^3 = 0, \qquad \sum_{\text{fermions}} Y = 0.$$

22.8 Applications to Physics

22.8.1 Magnetic Monopoles

Theorem 22.8.1 (Dirac quantization). *If magnetic monopoles exist with charge g_m, then:*

$$\frac{eg_m}{2\pi\hbar c} = n \in \mathbb{Z}.$$

This follows from requiring the $U(1)$ bundle over S^2 surrounding the monopole to be well-defined.

Theorem 22.8.2 ('t Hooft-Polyakov monopole). *In non-Abelian gauge theories with symmetry breaking $G \to H$, magnetic monopoles exist if $\pi_2(G/H) \neq 0$. They are solitonic (finite energy) solutions.*

22.8.2 Aharonov-Bohm Effect

Theorem 22.8.3 (Aharonov-Bohm effect). *A charged particle acquires a phase:*

$$\phi = \frac{e}{\hbar} \oint A_\mu dx^\mu$$

when encircling a region of magnetic flux, even if $F_{\mu\nu} = 0$ *along its path. This demonstrates the physical reality of the gauge potential* A_μ.

Remark 22.8.1. The Aharonov-Bohm effect is a manifestation of holonomy: the connection (not just curvature) has physical meaning.

22.8.3 Topological Field Theory

Definition 22.8.1 (Topological field theory). A **topological field theory** (TFT) has an action independent of the metric. Observables are topological invariants.

Example 22.8.1 (Chern-Simons theory). 3D Chern-Simons theory is a TFT. Its observables (Wilson loops) compute knot invariants like the Jones polynomial.

22.9 Summary

Fiber bundles and gauge theory provide the geometric foundation for fundamental physics:

Fiber bundles: Total space E, base M, fiber F, projection π; sections are fields; transition functions encode topology.

Principal bundles: Fiber is Lie group G; frame bundles, Hopf fibration; associated bundles carry matter representations.

Connections: Horizontal distribution / connection one-form ω; local gauge potential $A = \sigma^*\omega$; parallel transport and holonomy.

Curvature: $\Omega = d\omega + \omega \wedge \omega$; field strength $F = dA + A \wedge A$; $F_{\mu\nu} = \partial_\mu A_\nu - \partial_\nu A_\mu + [A_\mu, A_\nu]$.

Gauge transformations: $A' = g^{-1}Ag + g^{-1}dg$; $F' = g^{-1}Fg$; covariant derivative $D_\mu\psi = \partial_\mu\psi + igA_\mu\psi$.

Yang-Mills: Action $\sim \int \operatorname{tr}(F \wedge *F)$; equations $D_\mu F^{\mu\nu} = J^\nu$; instantons are self-dual solutions.

Characteristic classes: Chern classes c_k, Pontryagin classes p_k; topological invariants; Chern-Simons form.

Standard Model: $SU(3) \times SU(2) \times U(1)$; anomaly cancellation; Higgs mechanism.

Physical applications: Dirac quantization, monopoles, Aharonov-Bohm effect, topological field theory.

Exercises

Exercise 22.9.1. Show that the Möbius strip is a nontrivial line bundle over S^1 by computing its transition function.

Exercise 22.9.2. Verify that the transition functions for a vector bundle satisfy the cocycle conditions.

Exercise 22.9.3. For a $U(1)$ bundle, show that the gauge transformation $A_\mu \to A_\mu + \partial_\mu \chi$ leaves $F_{\mu\nu}$ invariant.

Exercise 22.9.4. Derive the transformation law $F' = g^{-1}Fg$ for the field strength under gauge transformations.

Exercise 22.9.5. Show that $[D_\mu, D_\nu]\psi = igF_{\mu\nu}\psi$ for a field in the fundamental representation.

Exercise 22.9.6. Verify that $\text{tr}(F_{\mu\nu}F^{\mu\nu})$ is gauge-invariant.

Exercise 22.9.7. Derive the Yang-Mills equations from the action $S = -\frac{1}{4}\int F^a_{\mu\nu}F^{a\mu\nu}\, d^4x$.

Exercise 22.9.8. Show that self-dual configurations ($F = *F$) automatically satisfy the Yang-Mills equations.

Exercise 22.9.9. Compute the first Chern class of a $U(1)$ bundle over S^2 with uniform magnetic field and relate it to the total magnetic flux.

Exercise 22.9.10. Verify that $d(\text{CS}(A)) = \text{tr}(F \wedge F)$ where CS is the Chern-Simons 3-form.

Exercise 22.9.11. For the Hopf fibration $S^3 \to S^2$, identify the fiber, base, and structure group.

Exercise 22.9.12. Derive the Dirac quantization condition from the requirement that the wave function be single-valued around a magnetic monopole.

Exercise 22.9.13. Show that the Standard Model hypercharge assignments lead to anomaly cancellation: $\sum Y^3 = 0$.

Exercise 22.9.14. Compute the holonomy of a constant magnetic field B around a loop of area A and relate it to the Aharonov-Bohm phase.

Exercise 22.9.15. For $SU(2)$ gauge theory, express the field strength $F^a_{\mu\nu}$ in terms of potentials A^a_μ and structure constants ϵ^{abc}.

Appendix A

Mathematical Notation and Symbols

This appendix provides a comprehensive reference for the mathematical notation used throughout this book.

A.1 Sets and Logic

Symbol	Meaning
$\mathbb{N}$	Natural numbers $\{0, 1, 2, \ldots\}$ or $\{1, 2, 3, \ldots\}$
$\mathbb{Z}$	Integers $\{\ldots, -2, -1, 0, 1, 2, \ldots\}$
$\mathbb{Q}$	Rational numbers
$\mathbb{R}$	Real numbers
$\mathbb{C}$	Complex numbers
$\mathbb{R}^n$	n-dimensional Euclidean space
$\mathbb{C}^n$	n-dimensional complex space
$\varnothing$	Empty set
$\in$	Element of
$\notin$	Not an element of
$\subset, \subseteq$	Subset, subset or equal
$\supset, \supseteq$	Superset, superset or equal
$\cup$	Union
$\cap$	Intersection
$\setminus$	Set difference: $A \setminus B = \{x \in A : x \notin B\}$
A^c	Complement of A
$A \times B$	Cartesian product
$\|A\|$ or $\#A$	Cardinality of A
$\mathcal{P}(A)$	Power set of A
$\forall$	For all (universal quantifier)
$\exists$	There exists (existential quantifier)
$\implies$	Implies
$\iff$	If and only if
$\wedge$	Logical AND
$\vee$	Logical OR
$\neg$	Logical NOT

A.2 Functions and Mappings

Symbol	**Meaning**
$f : A \to B$	Function from A to B
$f(x)$	Value of f at x
$f \circ g$	Composition: $(f \circ g)(x) = f(g(x))$
f^{-1}	Inverse function
$f\|_S$	Restriction of f to subset S
$\mathrm{dom}(f)$	Domain of f
$\mathrm{ran}(f)$ or $\mathrm{im}(f)$	Range/image of f
$\ker(f)$	Kernel of f: $\{x : f(x) = 0\}$
id or id_A	Identity function on A
$\mathrm{supp}(f)$	Support of f: closure of $\{x : f(x) \neq 0\}$
f^*	Pullback
f_*	Pushforward
δ_{ij}, δ^i_j	Kronecker delta
ϵ_{ijk}, ϵ^{ijk}	Levi-Civita symbol
$\theta(x)$	Heaviside step function
$\mathrm{sgn}(x)$	Sign function

A.3 Linear Algebra

Symbol	**Meaning**
$\mathbf{v}$, $\vec{v}$, v	Vector
$\mathbf{A}$, A	Matrix
A^T	Transpose
A^*, $A^\dagger$	Conjugate transpose (adjoint)
A^{-1}	Inverse matrix
$\det(A)$ or $\|A\|$	Determinant
$\mathrm{tr}(A)$	Trace
$\mathrm{rank}(A)$	Rank
$\dim(V)$	Dimension of vector space V
$\mathrm{span}\{v_1, \ldots, v_n\}$	Linear span
$V \oplus W$	Direct sum
$V \otimes W$	Tensor product
V^*	Dual space
$\langle u, v \rangle$ or (u, v)	Inner product
$\\|v\\|$	Norm
$u \perp v$	Orthogonality: $\langle u, v \rangle = 0$
$[A, B]$	Commutator: $AB - BA$
$\{A, B\}$	Anticommutator: $AB + BA$
$\mathbf{e}_i$	Standard basis vector
I or $\mathbf{1}$	Identity matrix
$O(n)$, $SO(n)$	Orthogonal, special orthogonal group
$U(n)$, $SU(n)$	Unitary, special unitary group
$GL(n)$, $SL(n)$	General linear, special linear group

A.4 Calculus and Analysis

Symbol	Meaning
$\frac{df}{dx}$, $f'(x)$, $\dot{f}$	Derivative
$\frac{\partial f}{\partial x}$, $\partial_x f$, f_x	Partial derivative
$\frac{d^n f}{dx^n}$, $f^{(n)}$	n-th derivative
$\int f\,dx$	Indefinite integral
$\int_a^b f\,dx$	Definite integral
$\oint$	Contour/closed integral
$\iint$, $\iiint$	Double, triple integral
$\lim_{x\to a}$	Limit as x approaches a
$\sum_{n=1}^{\infty}$	Infinite sum
$\prod_{n=1}^{N}$	Product
$O(x^n)$	Big-O: bounded by $C\|x\|^n$
$o(x^n)$	Little-o: $\lim \|f\|/\|x\|^n = 0$
$f \sim g$	Asymptotically equal
∇	Gradient operator (del)
$\nabla\cdot$	Divergence
$\nabla\times$	Curl
∇^2, Δ	Laplacian
$\square$	d'Alembertian: $\partial_t^2 - c^2\nabla^2$
D_t, $\frac{D}{Dt}$	Material derivative
$\delta(x)$	Dirac delta function
p.v.	Principal value

A.5 Complex Analysis

Symbol	Meaning
i	Imaginary unit: $i^2 = -1$
$z = x + iy$	Complex number
$\bar{z}$, z^*	Complex conjugate
$\|z\|$	Modulus: $\sqrt{x^2 + y^2}$
$\arg(z)$	Argument (phase)
$\mathrm{Re}(z)$	Real part
$\mathrm{Im}(z)$	Imaginary part
$e^{i\theta}$	$\cos\theta + i\sin\theta$ (Euler's formula)
$\mathrm{Res}(f, z_0)$	Residue at z_0
$\log z$	Complex logarithm
z^a	Complex power: $e^{a\log z}$

A.6 Differential Geometry

Symbol	Meaning
M	Manifold
T_pM	Tangent space at p
T_p^*M	Cotangent space at p
TM	Tangent bundle
T^*M	Cotangent bundle
$\Omega^k(M)$	Space of k-forms on M
d	Exterior derivative
$\wedge$	Wedge product
ι_X or $X \lrcorner$	Interior product with X
$\mathcal{L}_X$	Lie derivative along X
$[X, Y]$	Lie bracket of vector fields
$g_{\mu\nu}$	Metric tensor
$g^{\mu\nu}$	Inverse metric
$\Gamma^\rho_{\mu\nu}$	Christoffel symbols
∇_μ, $;_\mu$	Covariant derivative
$R^\rho{}_{\sigma\mu\nu}$	Riemann curvature tensor
$R_{\mu\nu}$	Ricci tensor
R	Ricci scalar
$G_{\mu\nu}$	Einstein tensor
$*$	Hodge star operator

A.7 Probability and Statistics

Symbol	**Meaning**
$P(A)$	Probability of event A
$P(A\|B)$	Conditional probability
$\mathbb{E}[X]$, $\langle X \rangle$	Expectation of X
$\mathrm{Var}(X)$, σ^2	Variance
σ, $\mathrm{std}(X)$	Standard deviation
$\mathrm{Cov}(X, Y)$	Covariance
ρ_{XY}	Correlation coefficient
$f_X(x)$	Probability density function
$F_X(x)$	Cumulative distribution function
$p_X(x)$	Probability mass function
$N(\mu, \sigma^2)$	Normal distribution
$X \sim \mathrm{Dist}$	X distributed as Dist
$X \xrightarrow{d} Y$	Convergence in distribution
$X \xrightarrow{P} Y$	Convergence in probability
$X \xrightarrow{a.s.} Y$	Almost sure convergence

A.8 Physics Notation

Symbol	Meaning
$\hbar$	Reduced Planck constant: $h/(2\pi)$
c	Speed of light
k_B	Boltzmann constant
G	Gravitational constant
e	Elementary charge
m_e, m_p	Electron, proton mass
α	Fine structure constant
$\|\psi\rangle$	Ket (quantum state)
$\langle\psi\|$	Bra (dual state)
$\langle\phi\|\psi\rangle$	Inner product (bracket)
$\hat{A}$, $\mathbf{A}$	Operator
$[A, B]$	Commutator
$\{A, B\}$	Anticommutator / Poisson bracket
H	Hamiltonian
L	Lagrangian
S	Action
Z	Partition function
β	Inverse temperature: $1/(k_BT)$
$\eta_{\mu\nu}$	Minkowski metric
A_μ	Gauge potential (4-vector)
$F_{\mu\nu}$	Field strength tensor

Appendix B

Physical Constants and Units

B.1 Fundamental Constants

Constant	Symbol	Value	Units
Speed of light	c	2.998×10^{8}	m/s
Planck constant	h	6.626×10^{-34}	J·s
Reduced Planck	$\hbar$	1.055×10^{-34}	J·s
Gravitational constant	G	6.674×10^{-11}	$m^3/(kg \cdot s^2)$
Boltzmann constant	k_B	1.381×10^{-23}	J/K
Elementary charge	e	1.602×10^{-19}	C
Permittivity of vacuum	ϵ_0	8.854×10^{-12}	F/m
Permeability of vacuum	μ_0	$4\pi \times 10^{-7}$	H/m
Fine structure constant	α	7.297×10^{-3}	(dimensionless)
Electron mass	m_e	9.109×10^{-31}	kg
Proton mass	m_p	1.673×10^{-27}	kg
Avogadro number	N_A	6.022×10^{23}	mol^{-1}
Gas constant	R	8.314	J/(mol·K)
Stefan-Boltzmann	σ	5.670×10^{-8}	$W/(m^2 \cdot K^4)$
Bohr radius	a_0	5.292×10^{-11}	m
Bohr magneton	μ_B	9.274×10^{-24}	J/T

B.2 Natural Units

In **natural units**, we set $\hbar = c = k_B = 1$. Conversions:

Quantity	Conversion
Length	$1\ \text{GeV}^{-1} = 1.973 \times 10^{-16}$ m
Time	$1\ \text{GeV}^{-1} = 6.582 \times 10^{-25}$ s
Mass	$1\ \text{GeV} = 1.783 \times 10^{-27}$ kg
Temperature	$1\ \text{GeV} = 1.160 \times 10^{13}$ K

Planck units ($\hbar = c = G = 1$):

$$\ell_P = \sqrt{\frac{\hbar G}{c^3}} = 1.616 \times 10^{-35}\ \text{m}$$

$$t_P = \sqrt{\frac{\hbar G}{c^5}} = 5.391 \times 10^{-44}\ \text{s}$$

$$m_P = \sqrt{\frac{\hbar c}{G}} = 2.176 \times 10^{-8}\ \text{kg} = 1.221 \times 10^{19}\ \text{GeV}$$

Gaussian units: $4\pi\epsilon_0 = 1$, $\mu_0/(4\pi) = 1/c^2$
Heaviside-Lorentz units: $\epsilon_0 = \mu_0 = 1$, rationalized

B.3 Useful Numerical Values

Quantity	Value
π	3.14159265358979...
e (Euler's number)	2.71828182845905...
γ (Euler-Mascheroni)	0.57721566490153...
$\ln 2$	0.69314718055995...
$\ln 10$	2.30258509299405...
$\sqrt{2}$	1.41421356237310...
$\sqrt{3}$	1.73205080756888...
1 eV	1.602×10^{-19} J
1 Å	10^{-10} m
1 fm	10^{-15} m
1 barn	10^{-28} m^2

Appendix C

Useful Formulas and Identities

C.1 Algebraic Identities

Binomial theorem:

$$(a+b)^n = \sum_{k=0}^{n} \binom{n}{k} a^{n-k} b^k, \quad \binom{n}{k} = \frac{n!}{k!(n-k)!}$$

Geometric series:

$$\sum_{k=0}^{n-1} r^k = \frac{1-r^n}{1-r}, \quad \sum_{k=0}^{\infty} r^k = \frac{1}{1-r} \quad (|r| < 1)$$

Exponential series:

$$e^x = \sum_{n=0}^{\infty} \frac{x^n}{n!}$$

Logarithm series ($|x| < 1$):

$$\ln(1+x) = x - \frac{x^2}{2} + \frac{x^3}{3} - \cdots = \sum_{n=1}^{\infty} \frac{(-1)^{n+1} x^n}{n}$$

Useful sums:

$$\sum_{k=1}^{n} k = \frac{n(n+1)}{2}$$

$$\sum_{k=1}^{n} k^2 = \frac{n(n+1)(2n+1)}{6}$$

$$\sum_{k=1}^{n} k^3 = \left[\frac{n(n+1)}{2}\right]^2$$

C.2 Trigonometric Identities

Pythagorean:

$$\sin^2\theta + \cos^2\theta = 1, \quad 1 + \tan^2\theta = \sec^2\theta$$

Addition formulas:

$$\sin(\alpha \pm \beta) = \sin\alpha\cos\beta \pm \cos\alpha\sin\beta$$
$$\cos(\alpha \pm \beta) = \cos\alpha\cos\beta \mp \sin\alpha\sin\beta$$
$$\tan(\alpha \pm \beta) = \frac{\tan\alpha \pm \tan\beta}{1 \mp \tan\alpha\tan\beta}$$

Double angle:

$$\sin 2\theta = 2\sin\theta\cos\theta$$
$$\cos 2\theta = \cos^2\theta - \sin^2\theta = 2\cos^2\theta - 1 = 1 - 2\sin^2\theta$$

Half angle:

$$\sin^2\frac{\theta}{2} = \frac{1-\cos\theta}{2}, \quad \cos^2\frac{\theta}{2} = \frac{1+\cos\theta}{2}$$

Product to sum:

$$\sin\alpha\sin\beta = \frac{1}{2}[\cos(\alpha-\beta) - \cos(\alpha+\beta)]$$
$$\cos\alpha\cos\beta = \frac{1}{2}[\cos(\alpha-\beta) + \cos(\alpha+\beta)]$$
$$\sin\alpha\cos\beta = \frac{1}{2}[\sin(\alpha+\beta) + \sin(\alpha-\beta)]$$

Sum to product:

$$\sin\alpha + \sin\beta = 2\sin\frac{\alpha+\beta}{2}\cos\frac{\alpha-\beta}{2}$$
$$\cos\alpha + \cos\beta = 2\cos\frac{\alpha+\beta}{2}\cos\frac{\alpha-\beta}{2}$$

Euler's formula:

$$e^{i\theta} = \cos\theta + i\sin\theta, \quad \cos\theta = \frac{e^{i\theta}+e^{-i\theta}}{2}, \quad \sin\theta = \frac{e^{i\theta}-e^{-i\theta}}{2i}$$

Hyperbolic functions:

$$\sinh x = \frac{e^x - e^{-x}}{2}, \quad \cosh x = \frac{e^x + e^{-x}}{2}$$
$$\cosh^2 x - \sinh^2 x = 1$$
$$\sinh(x \pm y) = \sinh x\cosh y \pm \cosh x\sinh y$$
$$\cosh(x \pm y) = \cosh x\cosh y \pm \sinh x\sinh y$$

C.3 Calculus Formulas

Derivatives:

$\frac{d}{dx}x^n = nx^{n-1}$ $\quad$ $\frac{d}{dx}e^x = e^x$ $\quad$ $\frac{d}{dx}\ln x = \frac{1}{x}$ $\quad$ $\frac{d}{dx}a^x = a^x \ln a$

$\frac{d}{dx}\sin x = \cos x$ $\quad$ $\frac{d}{dx}\cos x = -\sin x$ $\quad$ $\frac{d}{dx}\tan x = \sec^2 x$ $\quad$ $\frac{d}{dx}\cot x = -\csc^2 x$

$\frac{d}{dx}\arcsin x = \frac{1}{\sqrt{1-x^2}}$ $\quad$ $\frac{d}{dx}\arccos x = \frac{-1}{\sqrt{1-x^2}}$ $\quad$ $\frac{d}{dx}\arctan x = \frac{1}{1+x^2}$

$\frac{d}{dx}\sinh x = \cosh x$ $\quad$ $\frac{d}{dx}\cosh x = \sinh x$ $\quad$ $\frac{d}{dx}\tanh x = \operatorname{sech}^2 x$

Integrals:

$$\int x^n\,dx = \frac{x^{n+1}}{n+1} + C \quad (n \neq -1) \qquad \int \frac{dx}{x} = \ln|x| + C$$

$$\int e^{ax}\,dx = \frac{e^{ax}}{a} + C \qquad \int a^x\,dx = \frac{a^x}{\ln a} + C$$

$$\int \sin x\,dx = -\cos x + C \qquad \int \cos x\,dx = \sin x + C$$

$$\int \tan x\,dx = -\ln|\cos x| + C \qquad \int \sec^2 x\,dx = \tan x + C$$

$$\int \frac{dx}{\sqrt{1-x^2}} = \arcsin x + C \qquad \int \frac{dx}{1+x^2} = \arctan x + C$$

$$\int \frac{dx}{x^2+a^2} = \frac{1}{a}\arctan\frac{x}{a} + C \qquad \int \frac{dx}{x^2-a^2} = \frac{1}{2a}\ln\left|\frac{x-a}{x+a}\right| + C$$

Integration by parts: $\int u\,dv = uv - \int v\,du$

Leibniz rule:

$$\frac{d}{dx}\int_{a(x)}^{b(x)} f(x,t)\,dt = f(x,b(x))b'(x) - f(x,a(x))a'(x) + \int_{a(x)}^{b(x)} \frac{\partial f}{\partial x}\,dt$$

C.4 Definite Integrals

Gaussian integrals:

$$\int_{-\infty}^{\infty} e^{-ax^2}\,dx = \sqrt{\frac{\pi}{a}} \quad (a > 0)$$

$$\int_{-\infty}^{\infty} x^2 e^{-ax^2}\,dx = \frac{1}{2}\sqrt{\frac{\pi}{a^3}}$$

$$\int_{-\infty}^{\infty} x^{2n} e^{-ax^2}\,dx = \frac{(2n-1)!!}{(2a)^n}\sqrt{\frac{\pi}{a}}$$

$$\int_{-\infty}^{\infty} e^{-ax^2+bx}\,dx = \sqrt{\frac{\pi}{a}}e^{b^2/(4a)}$$

Trigonometric integrals:

$$\int_0^{\pi/2} \sin^n x\,dx = \int_0^{\pi/2} \cos^n x\,dx = \begin{cases} \frac{(n-1)!!}{n!!} \cdot \frac{\pi}{2} & n \text{ even} \\ \frac{(n-1)!!}{n!!} & n \text{ odd} \end{cases}$$

$$\int_0^{\infty} \frac{\sin x}{x}\,dx = \frac{\pi}{2}$$

Gamma and Beta functions:

$$\Gamma(n) = (n-1)! \quad (n \in \mathbb{N})$$

$$\Gamma(1/2) = \sqrt{\pi}$$

$$\int_0^{\infty} x^{n-1} e^{-x}\,dx = \Gamma(n)$$

$$B(m,n) = \int_0^1 x^{m-1}(1-x)^{n-1}\,dx = \frac{\Gamma(m)\Gamma(n)}{\Gamma(m+n)}$$

Other useful integrals:

$$\int_0^{\infty} \frac{x^{n-1}}{e^x - 1}\,dx = \Gamma(n)\zeta(n) \quad (n > 1)$$

$$\int_0^{\infty} \frac{x^3}{e^x - 1}\,dx = \frac{\pi^4}{15}$$

$$\int_0^{\infty} e^{-x^2} \cos(2bx)\,dx = \frac{\sqrt{\pi}}{2} e^{-b^2}$$

C.5 Series Expansions

Taylor series:

$$e^x = 1 + x + \frac{x^2}{2!} + \frac{x^3}{3!} + \cdots$$

$$\sin x = x - \frac{x^3}{3!} + \frac{x^5}{5!} - \cdots$$

$$\cos x = 1 - \frac{x^2}{2!} + \frac{x^4}{4!} - \cdots$$

$$\ln(1+x) = x - \frac{x^2}{2} + \frac{x^3}{3} - \cdots \quad (|x| < 1)$$

$$(1+x)^{\alpha} = 1 + \alpha x + \frac{\alpha(\alpha-1)}{2!} x^2 + \cdots \quad (|x| < 1)$$

$$\frac{1}{1-x} = 1 + x + x^2 + x^3 + \cdots \quad (|x| < 1)$$

$$\arctan x = x - \frac{x^3}{3} + \frac{x^5}{5} - \cdots \quad (|x| \le 1)$$

Stirling's approximation:

$$n! \approx \sqrt{2\pi n}\left(\frac{n}{e}\right)^n, \quad \ln(n!) \approx n\ln n - n + \frac{1}{2}\ln(2\pi n)$$

Appendix D

Special Functions Reference

D.1 Gamma and Related Functions

Gamma function:

$$\Gamma(z) = \int_0^\infty t^{z-1} e^{-t}\, dt \quad (\mathrm{Re}(z) > 0)$$

- Recurrence: $\Gamma(z+1) = z\Gamma(z)$
- Reflection: $\Gamma(z)\Gamma(1-z) = \frac{\pi}{\sin(\pi z)}$
- Duplication: $\Gamma(z)\Gamma(z+\frac{1}{2}) = \frac{\sqrt{\pi}}{2^{2z-1}}\Gamma(2z)$
- Special values: $\Gamma(n) = (n-1)!$, $\Gamma(\frac{1}{2}) = \sqrt{\pi}$, $\Gamma(\frac{3}{2}) = \frac{\sqrt{\pi}}{2}$

Beta function:

$$B(a,b) = \int_0^1 t^{a-1}(1-t)^{b-1}\, dt = \frac{\Gamma(a)\Gamma(b)}{\Gamma(a+b)}$$

Digamma function: $\psi(z) = \frac{d}{dz}\ln\Gamma(z) = \frac{\Gamma'(z)}{\Gamma(z)}$
Riemann zeta function:

$$\zeta(s) = \sum_{n=1}^{\infty} \frac{1}{n^s} \quad (\mathrm{Re}(s) > 1)$$

Special values: $\zeta(2) = \frac{\pi^2}{6}$, $\zeta(4) = \frac{\pi^4}{90}$, $\zeta(3) \approx 1.202$ (Apéry's constant)

D.2 Bessel Functions

Bessel's equation: $x^2y'' + xy' + (x^2 - \nu^2)y = 0$

Bessel function of the first kind:

$$J_\nu(x) = \sum_{k=0}^{\infty} \frac{(-1)^k}{k!\Gamma(k+\nu+1)} \left(\frac{x}{2}\right)^{2k+\nu}$$

Bessel function of the second kind (Neumann function):

$$Y_\nu(x) = \frac{J_\nu(x)\cos(\nu\pi) - J_{-\nu}(x)}{\sin(\nu\pi)}$$

Modified Bessel functions: $I_\nu(x) = i^{-\nu}J_\nu(ix)$, $K_\nu(x)$

Hankel functions: $H_\nu^{(1)} = J_\nu + iY_\nu$, $H_\nu^{(2)} = J_\nu - iY_\nu$

Recurrence relations:

$$J_{\nu-1}(x) + J_{\nu+1}(x) = \frac{2\nu}{x}J_\nu(x)$$
$$J_{\nu-1}(x) - J_{\nu+1}(x) = 2J'_\nu(x)$$

Spherical Bessel functions:

$$j_n(x) = \sqrt{\frac{\pi}{2x}}J_{n+1/2}(x), \quad y_n(x) = \sqrt{\frac{\pi}{2x}}Y_{n+1/2}(x)$$

$j_0(x) = \frac{\sin x}{x}$, $j_1(x) = \frac{\sin x}{x^2} - \frac{\cos x}{x}$

D.3 Legendre Functions

Legendre's equation: $(1-x^2)y'' - 2xy' + \ell(\ell+1)y = 0$

Legendre polynomials (Rodrigues formula):

$$P_\ell(x) = \frac{1}{2^\ell \ell!}\frac{d^\ell}{dx^\ell}(x^2-1)^\ell$$

First few: $P_0 = 1$, $P_1 = x$, $P_2 = \frac{1}{2}(3x^2-1)$, $P_3 = \frac{1}{2}(5x^3 - 3x)$

Orthogonality:

$$\int_{-1}^{1} P_\ell(x)P_m(x)\,dx = \frac{2}{2\ell+1}\delta_{\ell m}$$

Generating function:

$$\frac{1}{\sqrt{1-2xt+t^2}} = \sum_{\ell=0}^{\infty} P_\ell(x)t^\ell$$

Associated Legendre functions:

$$P_\ell^m(x) = (-1)^m(1-x^2)^{m/2}\frac{d^m}{dx^m}P_\ell(x)$$

Spherical harmonics:

$$Y_\ell^m(\theta,\phi) = \sqrt{\frac{2\ell+1}{4\pi}\frac{(\ell-m)!}{(\ell+m)!}}P_\ell^m(\cos\theta)e^{im\phi}$$

Orthonormality: $\int Y_\ell^m Y_{\ell'}^{m'*}\, d\Omega = \delta_{\ell\ell'}\delta_{mm'}$

D.4 Hermite and Laguerre Functions

Hermite polynomials (physicist's convention):

$$H_n(x) = (-1)^n e^{x^2}\frac{d^n}{dx^n}e^{-x^2}$$

First few: $H_0 = 1$, $H_1 = 2x$, $H_2 = 4x^2 - 2$, $H_3 = 8x^3 - 12x$
Orthogonality: $\int_{-\infty}^{\infty} H_m(x)H_n(x)e^{-x^2}\, dx = \sqrt{\pi}2^n n!\delta_{mn}$
Laguerre polynomials:

$$L_n(x) = \frac{e^x}{n!}\frac{d^n}{dx^n}(x^n e^{-x})$$

Associated Laguerre polynomials:

$$L_n^k(x) = \frac{d^k}{dx^k}L_{n+k}(x)$$

Orthogonality: $\int_0^\infty L_n^k(x)L_m^k(x)x^k e^{-x}\, dx = \frac{(n+k)!}{n!}\delta_{mn}$

D.5 Other Special Functions

Airy functions: Solutions of $y'' - xy = 0$

$$\text{Ai}(x) = \frac{1}{\pi}\int_0^\infty \cos\left(\frac{t^3}{3} + xt\right) dt$$
$$\text{Bi}(x) = \frac{1}{\pi}\int_0^\infty \left[e^{-t^3/3+xt} + \sin\left(\frac{t^3}{3} + xt\right)\right] dt$$

Error function:

$$\text{erf}(x) = \frac{2}{\sqrt{\pi}}\int_0^x e^{-t^2}\, dt, \quad \text{erfc}(x) = 1 - \text{erf}(x)$$

Exponential integral:

$$\mathrm{Ei}(x) = -\int_{-x}^{\infty} \frac{e^{-t}}{t}\,dt = \int_{-\infty}^{x} \frac{e^{t}}{t}\,dt$$

Dilogarithm (Spence's function):

$$\mathrm{Li}_2(x) = -\int_0^x \frac{\ln(1-t)}{t}\,dt = \sum_{k=1}^{\infty} \frac{x^k}{k^2}$$

Elliptic integrals:

$$K(k) = \int_0^{\pi/2} \frac{d\theta}{\sqrt{1-k^2\sin^2\theta}} \quad \text{(complete, first kind)}$$

$$E(k) = \int_0^{\pi/2} \sqrt{1-k^2\sin^2\theta}\,d\theta \quad \text{(complete, second kind)}$$

Appendix E

Vector and Tensor Identities

E.1 Vector Algebra

For vectors **A**, **B**, **C**, **D**:

Dot and cross products:

$$\mathbf{A} \cdot \mathbf{B} = \mathbf{B} \cdot \mathbf{A}$$
$$\mathbf{A} \times \mathbf{B} = -\mathbf{B} \times \mathbf{A}$$
$$\mathbf{A} \cdot (\mathbf{B} \times \mathbf{C}) = \mathbf{B} \cdot (\mathbf{C} \times \mathbf{A}) = \mathbf{C} \cdot (\mathbf{A} \times \mathbf{B}) \quad \text{(scalar triple product)}$$

BAC-CAB rule:

$$\mathbf{A} \times (\mathbf{B} \times \mathbf{C}) = \mathbf{B}(\mathbf{A} \cdot \mathbf{C}) - \mathbf{C}(\mathbf{A} \cdot \mathbf{B})$$

Vector quadruple product:

$$(\mathbf{A} \times \mathbf{B}) \cdot (\mathbf{C} \times \mathbf{D}) = (\mathbf{A} \cdot \mathbf{C})(\mathbf{B} \cdot \mathbf{D}) - (\mathbf{A} \cdot \mathbf{D})(\mathbf{B} \cdot \mathbf{C})$$

Levi-Civita identities:

$$\begin{aligned}
\epsilon_{ijk}\epsilon_{lmn} &= \det \begin{pmatrix} \delta_{il} & \delta_{im} & \delta_{in} \\ \delta_{jl} & \delta_{jm} & \delta_{jn} \\ \delta_{kl} & \delta_{km} & \delta_{kn} \end{pmatrix} \\
\epsilon_{ijk}\epsilon_{lmk} &= \delta_{il}\delta_{jm} - \delta_{im}\delta_{jl} \\
\epsilon_{ijk}\epsilon_{ljk} &= 2\delta_{il} \\
\epsilon_{ijk}\epsilon_{ijk} &= 6
\end{aligned}$$

E.2 Vector Calculus Identities

For scalar fields ϕ, ψ and vector fields **A**, **B**:

Gradient:

$$\nabla(\phi\psi) = \phi\nabla\psi + \psi\nabla\phi$$
$$\nabla(\mathbf{A}\cdot\mathbf{B}) = \mathbf{A}\times(\nabla\times\mathbf{B}) + \mathbf{B}\times(\nabla\times\mathbf{A}) + (\mathbf{A}\cdot\nabla)\mathbf{B} + (\mathbf{B}\cdot\nabla)\mathbf{A}$$

Divergence:

$$\nabla\cdot(\phi\mathbf{A}) = \phi(\nabla\cdot\mathbf{A}) + \mathbf{A}\cdot\nabla\phi$$
$$\nabla\cdot(\mathbf{A}\times\mathbf{B}) = \mathbf{B}\cdot(\nabla\times\mathbf{A}) - \mathbf{A}\cdot(\nabla\times\mathbf{B})$$

Curl:

$$\nabla\times(\phi\mathbf{A}) = \phi(\nabla\times\mathbf{A}) + \nabla\phi\times\mathbf{A}$$
$$\nabla\times(\mathbf{A}\times\mathbf{B}) = \mathbf{A}(\nabla\cdot\mathbf{B}) - \mathbf{B}(\nabla\cdot\mathbf{A}) + (\mathbf{B}\cdot\nabla)\mathbf{A} - (\mathbf{A}\cdot\nabla)\mathbf{B}$$

Second derivatives:

$$\nabla\cdot(\nabla\phi) = \nabla^2\phi \quad \text{(Laplacian)}$$
$$\nabla\times(\nabla\phi) = \mathbf{0} \quad \text{(curl of gradient is zero)}$$
$$\nabla\cdot(\nabla\times\mathbf{A}) = 0 \quad \text{(divergence of curl is zero)}$$
$$\nabla\times(\nabla\times\mathbf{A}) = \nabla(\nabla\cdot\mathbf{A}) - \nabla^2\mathbf{A}$$

E.3 Integral Theorems

Divergence theorem (Gauss):

$$\int_V \nabla\cdot\mathbf{A}\,dV = \oint_S \mathbf{A}\cdot d\mathbf{S}$$

Curl theorem (Stokes):

$$\int_S (\nabla\times\mathbf{A})\cdot d\mathbf{S} = \oint_C \mathbf{A}\cdot d\mathbf{l}$$

Gradient theorem:

$$\int_V \nabla\phi\,dV = \oint_S \phi\,d\mathbf{S}$$

Green's identities:

$$\int_V (\phi\nabla^2\psi + \nabla\phi\cdot\nabla\psi)\,dV = \oint_S \phi\nabla\psi\cdot d\mathbf{S} \quad \text{(first)}$$
$$\int_V (\phi\nabla^2\psi - \psi\nabla^2\phi)\,dV = \oint_S (\phi\nabla\psi - \psi\nabla\phi)\cdot d\mathbf{S} \quad \text{(second)}$$

E.4 Curvilinear Coordinates

Cylindrical (r, ϕ, z): $x = r\cos\phi$, $y = r\sin\phi$, $z = z$

$$\begin{aligned}
\nabla f &= \frac{\partial f}{\partial r}\hat{\mathbf{r}} + \frac{1}{r}\frac{\partial f}{\partial \phi}\hat{\boldsymbol{\phi}} + \frac{\partial f}{\partial z}\hat{\mathbf{z}} \\
\nabla \cdot \mathbf{A} &= \frac{1}{r}\frac{\partial (rA_r)}{\partial r} + \frac{1}{r}\frac{\partial A_\phi}{\partial \phi} + \frac{\partial A_z}{\partial z} \\
\nabla^2 f &= \frac{1}{r}\frac{\partial}{\partial r}\left(r\frac{\partial f}{\partial r}\right) + \frac{1}{r^2}\frac{\partial^2 f}{\partial \phi^2} + \frac{\partial^2 f}{\partial z^2}
\end{aligned}$$

Spherical (r, θ, ϕ): $x = r\sin\theta\cos\phi$, $y = r\sin\theta\sin\phi$, $z = r\cos\theta$

$$\begin{aligned}
\nabla f &= \frac{\partial f}{\partial r}\hat{\mathbf{r}} + \frac{1}{r}\frac{\partial f}{\partial \theta}\hat{\boldsymbol{\theta}} + \frac{1}{r\sin\theta}\frac{\partial f}{\partial \phi}\hat{\boldsymbol{\phi}} \\
\nabla \cdot \mathbf{A} &= \frac{1}{r^2}\frac{\partial (r^2 A_r)}{\partial r} + \frac{1}{r\sin\theta}\frac{\partial (\sin\theta A_\theta)}{\partial \theta} + \frac{1}{r\sin\theta}\frac{\partial A_\phi}{\partial \phi} \\
\nabla^2 f &= \frac{1}{r^2}\frac{\partial}{\partial r}\left(r^2\frac{\partial f}{\partial r}\right) + \frac{1}{r^2\sin\theta}\frac{\partial}{\partial \theta}\left(\sin\theta\frac{\partial f}{\partial \theta}\right) + \frac{1}{r^2\sin^2\theta}\frac{\partial^2 f}{\partial \phi^2}
\end{aligned}$$

E.5 Differential Forms

Exterior derivative:

- On 0-form (function): $df = \partial_i f\, dx^i$
- On 1-form $\omega = \omega_i dx^i$: $d\omega = \partial_i \omega_j\, dx^i \wedge dx^j$
- On k-form: $d(d\omega) = 0$ (Poincaré lemma)

Wedge product properties:

- $\alpha \wedge \beta = (-1)^{pq}\beta \wedge \alpha$ for p-form α, q-form β
- $d(\alpha \wedge \beta) = d\alpha \wedge \beta + (-1)^p \alpha \wedge d\beta$

Hodge dual in $\mathbb{R}^3$:

$$*1 = dx \wedge dy \wedge dz, \quad *dx = dy \wedge dz, \quad *(dx \wedge dy) = dz$$

Generalized Stokes theorem:

$$\int_M d\omega = \int_{\partial M} \omega$$

E.6 Tensor Index Identities

Metric operations:

$$g^{\mu\rho}g_{\rho\nu} = \delta^\mu_\nu$$
$$V_\mu = g_{\mu\nu}V^\nu \quad \text{(lowering)}$$
$$V^\mu = g^{\mu\nu}V_\nu \quad \text{(raising)}$$

Covariant derivative:

$$\nabla_\mu V^\nu = \partial_\mu V^\nu + \Gamma^\nu_{\mu\rho}V^\rho$$
$$\nabla_\mu \omega_\nu = \partial_\mu \omega_\nu - \Gamma^\rho_{\mu\nu}\omega_\rho$$
$$\nabla_\rho g_{\mu\nu} = 0 \quad \text{(metric compatibility)}$$

Christoffel symbols:

$$\Gamma^\rho_{\mu\nu} = \frac{1}{2}g^{\rho\sigma}(\partial_\mu g_{\nu\sigma} + \partial_\nu g_{\mu\sigma} - \partial_\sigma g_{\mu\nu})$$

Riemann tensor symmetries:

$$R_{\rho\sigma\mu\nu} = -R_{\sigma\rho\mu\nu} = -R_{\rho\sigma\nu\mu} = R_{\mu\nu\rho\sigma}$$
$$R_{\rho\sigma\mu\nu} + R_{\rho\mu\nu\sigma} + R_{\rho\nu\sigma\mu} = 0 \quad \text{(first Bianchi)}$$

Contracted quantities:

$$R_{\mu\nu} = R^\rho{}_{\mu\rho\nu} \quad \text{(Ricci tensor)}$$
$$R = g^{\mu\nu}R_{\mu\nu} \quad \text{(Ricci scalar)}$$
$$G_{\mu\nu} = R_{\mu\nu} - \frac{1}{2}g_{\mu\nu}R \quad \text{(Einstein tensor)}$$
$$\nabla^\mu G_{\mu\nu} = 0 \quad \text{(contracted Bianchi)}$$

Index

End of "Mathematical Tools for Theoretical Physics"

www.ingramcontent.com/pod-product-compliance
Lightning Source LLC
Chambersburg PA
CBHW041747220926
51847CB00013B/131
* 9 7 8 2 4 8 8 6 1 6 1 2 6 *